WEAPONIZING MAPS

Also Available

**Five Billion Years of Global Change:
A History of the Land**
Denis Wood

**Making Maps, Second Edition:
A Visual Guide to Map Design for GIS**
John Krygier and Denis Wood

The Power of Maps
Denis Wood

Rethinking the Power of Maps
*Denis Wood
with John Fels and John Krygier*

WEAPONIZING MAPS

Indigenous Peoples and Counterinsurgency in the Americas

Joe Bryan and Denis Wood

THE GUILFORD PRESS
New York   London

© 2015 The Guilford Press
A Division of Guilford Publications, Inc.
370 Seventh Avenue, Suite 1200, New York, NY 10001
www.guilford.com

All rights reserved

No part of this book may be reproduced, translated, stored in
a retrieval system, or transmitted, in any form or by any means,
electronic, mechanical, photocopying, microfilming, recording,
or otherwise, without written permission from the publisher.

Printed in the United States of America

This book is printed on acid-free paper.

Last digit is print number: 9 8 7 6 5 4 3 2 1

Library of Congress Cataloging-in-Publication Data is available
from the publisher.

ISBN 978-1-4625-1991-0 (paper)
ISBN 978-1-4625-1992-7 (cloth)

Just as none of us is outside or beyond geography, none of us is completely free from the struggle over geography. That struggle is complex and interesting because it is not only about soldiers and cannons but also about ideas, about forms, about images and imaginings.
—Edward Said (1993)

Know the Earth . . . Show the Way . . . Understand the World.
—Motto of the U.S. National Geospatial-Intelligence Agency

Acknowledgments

Working on this book, we spent a lot of time in the library or with its online equivalent. Historical documents referred to here come from the Library of Congress, George Washington University's National Security Archive, the Chief George Manuel Library at the Center for World Indigenous Studies, the excellent "Sandino Rebellion" archive compiled and maintained by Michael J. Schroeder (*www.sandinorebellion.com*), and elsewhere.

But we did plenty of primary research too, initially as participants in the México Indígena controversy. Later we were invited to participate in the 2011 World Human Geography Conference that the American Geographical Society (AGS) hosted at Haskell University. Along the way we repeatedly engaged Dobson, Demarest, and Herlihy in public debate. We also visited the Sierra Juárez in February 2012, where we interviewed key personnel with direct knowledge of the México Indígena project, including residents of Tiltepec, Yagila, and Yagavila. With John Krygier, Denis Wood visited the AGS's Brooklyn offices in 2012.

We have benefited from the help and advice of many people in the course of writing this book. In Oaxaca, we would especially like to thank the community authorities of San Miguel Tiltepec, San Juan Yagila, and Santa Cruz Yagavila for their willingness to discuss the México Indígena project with us. We are also grateful for the guidance and assistance we received there from Salvador Aquino-Centeno, Melquiades Cruz, Oliver Froehling, Aldo González, Jaime Martinez Luna,

Wilfredo Mendoza, Santiago Peréz, Simón Sedillo, Gustavo Ramírez, Gregorio Urbano, and Holly Worthen, as well as a number of others who wish to remain anonymous.

For reading early drafts of pieces of the book, we want to thank the late Neil Smith, Michael Schroeder, the late Joseph Sonnenfeld, John Pickles, Michael Watts, Derek Gregory, Jeremy Crampton, Zoltán Grossman, and others.

Thomas Berger, Philippe Bourgeois, Mac Chapin, Peter Dana, Georg Grünberg, Charles Hale, Joseph Sonnenfeld, Stefano Varese, and others took time to respond to queries about dates and details. We are grateful for every contribution to the project.

For support and feedback, we want to thank Brenda Baletti, Zoltán Grossman, John Krygier, Catherine d'Ignazio, George McCleary, Sharlene Mollett, Donald Moore, Tad Mutersbaugh, Tim Oakes, Tracey Osborne, Linda Quiquivix, Alvaro Reyes, Joel Wainwright, and Emily Yeh.

Unless otherwise credited, Joe Bryan took the photographs.

Last, but not least, our thanks go to Karl, Leif, and Sonya; and, as always, to Ingrid.

Contents

	List of Figures	xi
	A Narrative Table of Contents	xii
1	In the Rincón of the Sierra Juárez	1
2	The Decline and Fall of the Once August American Geographical Society	16
3	"Red Mike" Edson's U.S. Marine Patrols Up Nicaragua's Río Coco in 1928–1929 and the Development of the *Small Wars Manual*	34
4	The Birth of Indigenous Mapping in Canada	54
5	Maps, Guns, and Indigenous Peoples	74
6	From Territory to Property: Indigenous Mapping after the Cold War	96
7	Counterinsurgency and the Rise of the "Warrior Scholars"	127
8	The AGS, the Bowman Expeditions, and the México Indígena Project	142

Coda. Kill the Insurgent and Save the Man: Indigenous Peoples and Human Terrain	163
A Note on Maps	177
Notes	181
Bibliography	237
Index	260
About the Authors	272

List of Figures

FIGURE 1.1.	In Tiltepec, February 7, 2012.	2
FIGURE 1.2.	México Indígena map of Yagila.	10
FIGURE 2.1.	Former headquarters of the AGS.	18
FIGURE 2.2.	The door of the AGS's offices.	31
FIGURE 3.1.	Edson's map of the Wanks Patrol.	42
FIGURE 3.2.	The Bragman's Bluff map Edson used.	46
FIGURE 4.1.	A map from the *Inuit Land Use and Occupancy Project*.	62
FIGURE 4.2.	A map from the Fort George Resource Use and Subsistence Economy Study.	65
FIGURE 4.3.	A map of traplines collected by Phoebe Nahanni's team of Dene researchers.	70
FIGURE 5.1.	Map of the Miskito, Sumu, and Rama populations from Bourgois and Grünberg's report.	85
FIGURE 5.2.	Yapti Tasba.	92
FIGURE 6.1.	Detail from *The Coexistence of Indigenous Peoples and Natural Resources*.	101
FIGURE 6.2.	Bernard Nietschmann's map of states and nations of Central America.	103

FIGURE 6.3. Multicommunal tenure blocs, North Atlantic Autonomous Region (RAAN), Nicaragua.   119

FIGURE 6.4. Map showing Awas Tingni's use and occupancy of land.   122

FIGURE 6.5. Village of San Marcos, *The Maya Atlas*.   125

FIGURE 8.1. Editorial cartoon: *Él que parte y reparte se queda con la mayor parte*.   150

FIGURE 8.2. Sign declaring that private property does not exist in Ixtlán de Juárez, Oaxaca.   152

FIGURE 8.3. Street art protesting biopiracy in Oaxaca.   154

A Narrative Table of Contents

Bearing an eagle staff and an American flag, a Color Guard of Native American Veterans opened the World Human Geography Conference at Haskell Indian Nations University in Lawrence, Kansas, on September 15, 2011. Standing in front of the audience assembled in the auditorium, the Color Guard implored all present to speak the truth freely and respectfully. It was an auspicious beginning for the event.

Though Haskell was the official host of the conference, the event had been organized by the Department of Geography at the University of Kansas, Haskell's cross-town sibling, and by the American Geographical Society (AGS). Passed off as an academic conference about research ethics, the event was wholly funded by the U.S. Army Research Office. Indeed most of the presentations over the course of the two-day conference were given by researchers working with funding from the U.S. Army or otherwise aligned with the military. Out in front of them—and they included representatives from the Foreign Military Studies Office, the Human Terrain Systems program, and the U.S. Army's Command and General Staff College—was a pair of geography professors from the University of Kansas: Jerome Dobson and Peter Herlihy.

In his capacity as president of the AGS, Dobson had secured funding in 2005 for the organization's first Bowman Expedition, aimed at mapping indigenous[1] lands in Mexico. Herlihy led the expedition, employing "participatory research mapping" techniques he had perfected during two decades of research in Central America. In spite of Herlihy's claims to be using maps to advance indigenous rights, he failed to disclose to a number of participating communities that the U.S. Army's Foreign Military Studies Office was bankrolling the entire expedition. Nor did he tell these communities that all the data collected were being transmitted to a military contractor, Radiance Technologies, which bills itself as "creating innovative solutions for the warfighter." In the end, the project concluded in storm of controversy, ignited when the Zapotec communities mapped by Herlihy found out about the Army's involvement. In a series of sharply worded declarations, communities both individually and collectively demanded the return of all data collected by Herlihy's expedition.

Among other things, the Haskell conference was intended to put an end to this controversy by distinguishing the AGS's Bowman Expedition from other, more controversial efforts by the U.S. military to enlist social scientists in counterinsurgency campaigns. At the same time, the AGS actively sought the implicit approval for its work that holding the event at Haskell might convey. To that end, the conference included presentations from a number of Native American educators as well as from Herlihy's collaborators on the latest Army-funded project, the Bowman Expedition to the "Borderlands Region" in the predominantly indigenous region of eastern Honduras known as La Mosquitia.

As the conference wore on, the choice of Haskell to host it became more and more disconcerting. Founded in 1884, Haskell originally opened as the U.S. Indian Industrial Training School, together with other institutions designed to assimilate Native American children into U.S. society through education. Many of the children who attended Haskell were refugees of the Indian wars launched from nearby Fort Leavenworth, located not an hour's drive away, hard on the banks of the Missouri River. In an 1892 speech before the U.S. Congress, Indian war veteran and boarding school advocate Captain Richard H. Pratt described the goal of the schools this way:

> A great general has said that the only good Indian is a dead one, and that high sanction of his destruction has been an enormous factor in promoting Indian massacres. In a sense, I agree with the sentiment, but only in this: that all the Indian there is in the race should be dead. Kill the Indian in him, and save the man.[2]

Pratt saw himself as an alternative to the military campaigns that culminated in the massacre of as many as 300 Lakota men, women, and children at Wounded Knee not two years prior, on December 29, 1890. Pratt's good intentions were not enough to save Indian children from the mass graves used at Wounded Knee. In a corner of the Haskell campus, a short walk from the auditorium where the World Human Geography Conference convened, lies a cemetery where many of the children who died at the school are buried. Most of them died within five years of Haskell's opening, their names and lives commemorated on the 103 Army-issue white marble headstones lining the cemetery. Countless more children died out of sight of school administrators in the swamp that lies beyond the cemetery.

In the shadow of that history, Native Americans themselves have slowly transformed the former boarding school into Haskell Indian Nations University, turning the grim reminders of the murderous past into the basis for a new way of life. That transformation was emphasized by the Color Guard's entrance, demanding the truth while reminding all of their service to the nation that had colonized them. It was all enough to make one wonder what the conference was *really* about.

Answering that question requires putting the AGS's World Human Geography Conference into a larger historical and geographical context, tracing an arc that connects *internal* colonialism with the *external* expansion of U.S. power through the Banana Wars in Central America, Cold War proxy battles, and, most recently, the so-called war on terror. In what follows, we trace the links among these seemingly disparate contexts in terms of the tactics and strategies of counterinsurgency. In all of them, the U.S. military has confronted a series of unconventional armed threats, both real and potential, posed by rebel organizations, criminals, and others not content to simply bow before the demands of U.S. security. Throughout, the U.S. military has been at pains to define the terrain of struggle, one that too often spills off the battlefield into the forests, fields, and cities where people make their everyday lives. Under such conditions, *all* of society becomes a potential battlefield.

Maps have long been an important means of knowing this terrain, showing the locations of towns, where people farm and obtain food, and the trails and waterways they use to move from place to place. To borrow Mao Zedong's aphorism, the insurgent must move in this everyday landscape "the way the fish swims in the sea," but this means being intimately acquainted with it. Counterinsurgency relies on the same approach to identify threats to security and to manipulate the vulnerability of life in settings where the battlefield is everywhere. In most cases, this involves detailed mapping. Indeed it is often only through maps that the U.S. military has been able to aggregate individual lives into populations, defining societies in a manner capable of identifying threats to them, both external and internal. At the same time, maps have come to play an indispensable part in indigenous peoples' own efforts to secure protection of their rights as distinct populations. In these efforts, maps showing indigenous peoples' knowledge of the territory, fashioned from their use and occupancy of a particular area, have become an indispensable means of countering state claims to authority. As is often the case, a quintessentially colonial instrument has become a weapon for liberation.

This book concerns itself with this tension between military application and political advocacy in the practice of indigenous mapping. By this latter term, we refer to the broad field of practices used to make maps of, for, and occasionally by indigenous peoples for a broad range of political purposes. As divergent as those purposes have been, and as they continue to be, they share in an undeniably colonial logic that locates indigenous peoples as historically antecedent to and outside the sociospatial order guaranteed by states through institutions such as citizenship and property rights. Indigenous mapping, then, constitutes the grounds for the recognition of indigenous peoples' basic human rights to territory, self-determination, and self-government.

But it also helps bring the conquest *home* in important ways. This happened in the halls of Haskell Indian Nations University during the AGS's conference, where indigenous rights activists, academic geographers, and the U.S. military were brought together. Here we explore these connections further, examining the conditions under which indigenous mapping has come into existence, the problems it has been used to identify, and the political and military interventions it has produced.

Our narrative weaves together a motley array of characters and institutions. We chronicle the decline and fall of the once august AGS; "Red Mike" Edson's Río Coco patrols in Nicaragua in 1928–1929 and his role in developing the U.S. Marine Corps' *Small Wars Manual*; the rise of indigenous mapping with indigenous rights movements in Canada and Latin America; the more recent rediscovery and revision of the *Small Wars Manual* in the wake of U.S. wars in Iraq and Afghanistan; the AGS's quest to rebuild itself through military-funded "Bowman Expeditions" aimed at compiling "geographical intelligence"; the alliance of the AGS, the University of Kansas Department of Geography, and the Foreign Military Studies Office in the México Indígena

("Indigenous Mexico") Project; the response of the Zapotec of the Sierra Juárez; and the World Human Geography Conference at Haskell Indian Nations University.

This configuration can be distilled into three strands. There's a military strand that traces the creation and modification of counterinsurgency tactics within the Marine Corps and Army from the Río Coco to Kandahar and beyond. There's an academic strand that follows the transformation of indigenous mapping from a method for collecting data on land use and occupancy to a vehicle for participatory research. And there's an indigenous strand that traces indigenous peoples' own experiences with the use of mapping as a tool for advocacy. This last strand is often invoked as justification for mapping, but it glosses over an indigenous discomfort with the very idea of mapping. For this reason, our account begins in the Rincón de Ixtlán of the Sierra Juárez, above the Mexican city of Oaxaca. Residents of the Rincón were the first to denounce the Bowman Expeditions' coupling of military interests and academic geography. Their words serve as a reminder of what's really at stake, shifting the focus from the defense of academic reputations on the Haskell stage to Zapotec lives. We follow this discussion with a mosaic of chapters that interweaves the three strands of our story.

Our approach is genealogical, tying together sites where indigenous mapping has emerged as method and political tool. Though the chapters are organized chronologically, the links between them are not causal. But to assert that the relationship between these sites is simply a historical one or, worse, the outcome of a singular strategy is to oversimplify their complexity and overwrite their specific contributions. The Indian wars of the 19th century are not the same as the proxy wars of the 1980s in Central America or the counterinsurgency campaigns in Iraq and Afghanistan in the 2000s. Instead, each serves as an important site for understanding the emergence of indigenous or tribal areas as particular kinds of space, defined as much by a collective way of life as by the particular approach to war fought there. Each setting provides a new set of challenges and conditions, addressed through the application of lessons learned from past wars through the innovations of new tactics and strategies.

Chapter 1. In the Rincón of the Sierra Juárez

The immediate origins of the AGS conference at Haskell lie in the Zapotec towns of the Rincón de Ixtlán of the Sierra Juárez (p. 1). In 2006,

Peter Herlihy led a U.S. Army–funded Bowman Expedition into the Rincón (p. 4) under the pretext of mapping the region's complex system of communal land tenure. Instead of mapping an isolated, mountainous "corner" (*rincón*) of Mexico, Herlihy found himself navigating the complexities of a Zapotec society knit together by communal land ownership and a healthy skepticism of outsiders. Herlihy took his offer to map traditional lands directly to the communities of Yagavila, Yagila, and Tiltepec (p. 6). Yagavila dropped out of the program almost immediately, but the México Indígena team succeeded in mapping Yagila and Tiltepec—*and* in transmitting the full results of its investigations to the Foreign Military Studies Office of the U.S. Army, which, in fact, had funded the entire expedition. With this becoming clearer and clearer, on January 14, 2009, the Union of Organizations of the Sierra Juárez of Oaxaca (UNOSJO) issued a proclamation denouncing México Indígena for engaging in geopiracy; on March 17, 2009, the community of Tiltepec issued a proclamation demanding an apology from Herlihy, the University of Kansas, the AGS, and the Foreign Military Studies Office (p. 7); and on July 24, 2011, the communities of Yagila, Yagavila, Tepanzacoalco, Zoogochi, and Teotlaxco joined in denouncing the México Indígena expedition (p. 13).

Chapter 2. The Decline and Fall of the Once August American Geographical Society

Why was the AGS sending an expedition into the Sierra Juárez in 2006? Founded in 1851 by 31 wealthy New Yorkers (p. 17), the AGS (as geographers call it) is still the oldest organization of U.S. geographers, though today it exists as a shadow of its former self. In its day, it mounted "expeditions" to "far-off places" (p. 18) and played an important role in preparations for the Paris Peace Conference, to which the society's director, Isaiah Bowman, accompanied Woodrow Wilson. In 1904, however, a number of academics founded the Association of American Geographers (p. 20), which has become, by a very great margin, the largest and most important association of professional geographers in North America, perhaps in the world (it has members from over 60 different countries). As the AAG (as geographers call it) has grown, the AGS has shriveled (p. 21). The AGS was forced to sell its resplendent headquarters building on Manhattan's Audubon Terrace (p. 30); it had to give its famous library to the University of Wisconsin–Milwaukee; and it teetered on the brink of *complete* irrelevance until it struck a deal to collect intelligence for the Army (p. 32).

Chapter 3. "Red Mike" Edson's U.S. Marine Patrols Up Nicaragua's Río Coco in 1928–1929 and the Development of the *Small Wars Manual*

Meanwhile, in Nicaragua, U.S. Marine Corps Captain Merritt A. "Red Mike" Edson was leading a series of patrols up the Río Coco during the U.S. occupation of the country between 1926 and 1931 (p. 34). By the later 1920s, the occupation had devolved into a "messy guerrilla conflict" with General Augusto Sandino and his adherents (p. 36). In Edson's effort to enlist the indigenous Miskitos to the U.S. cause, he stirred an anthropological element into the politically sensitive and highly personal diplomacy required by his mission (p. 40); and he and his immediate superior, Major Harold Utley, later taught this method at Quantico and elsewhere (p. 47). Utley incorporated the method into his *Small Wars Operations*, renamed the *Small Wars Manual* in 1940 (p. 49). For most of the Cold War, the *Small Wars Manual* languished in the Marine archives (p. 52), dusted off occasionally during the Vietnam War and again in the 1980s at the height of the Reagan Administration's support for proxy wars in Central America and Afghanistan, wars in which "tribes" and "ethnic groups" characteristically played enormous roles.

Chapter 4. The Birth of Indigenous Mapping in Canada

In 1967, Frank Arthur Calder and the Nisga'a Nation Tribal Council initiated an action that led the Canadian Supreme Court to rule for the existence of an *aboriginal title*, one dating to a Royal Proclamation of 1763 (p. 54). This decision prompted the still young liberal federal government of Pierre Elliot Trudeau, eager to recover from an initial "misstep" in Indian affairs, to attempt to extinguish these titles by negotiating treaties with those indigenes who had never signed them; and in 1974, it began supporting work capable of leading to such negotiations (p. 57). Maps showing patterns of indigenous land use and occupancy, traditional ties to the land, and cultural cohesion proved invaluable to framing the negotiations, so much so that by middle of the decade, the Canadian government itself was financing mapping projects (p. 60). Leery of cooptation, some Indian organizations in Canada broke from that mold, insisting on their status as nations properly subject to international—as opposed to domestic—law (p. 64). That approach was no less reliant on maps to demonstrate the status of Indian peoples as nations, but it also helped conjure a political vision of a "Fourth World" linking indigenous nations around the globe (p. 71).

Chapter 5. Maps, Guns, and Indigenous Peoples

Among the Fourth World allies that Indians in Canada found were the Miskito people of eastern Nicaragua, previously mapped by "Red Mike" Edson's Marines (p. 74). Miskito political mobilization in the 1970s hinged in part on demonstrating their historical rights to land and resources. Like First Nations in Canada, maps made by anthropologists and geographers of Miskito use and occupancy of land and resources proved to be an invaluable tool for political mobilization (p. 75). In particular, they made use of maps made by then cultural ecologist Bernard Nietschmann, transforming his data into evidence of their political claim to territory (p. 80). Reluctant at first, Nietschmann became one of the more visible advocates of the Miskito position during their armed struggle with the Sandinista government during the Contra War in the 1980s (p. 88). Following the Miskitos' lead, Nietschmann pioneered the use of maps as a means of bringing the Fourth World into reality by using them to represent indigenous national struggles for territory and self-determination. In spite of their anticolonial stance, both Nietschmann and the Miskito crossed paths with the Reagan Administration's policy of supporting proxy battles waged by "oppressed minorities" and "freedom fighters" against Communism (p. 87). The Reagan Administration controversially slotted the Miskito into that geopolitical vision, revisiting the terrain mapped by "Red Mike" Edson on his Coco Patrols in 1928–1929 and in the first version of the *Small Wars Manual* (p. 90). The events proved pivotal in weaponizing maps as both a tactic for indigenous mobilization and a new approach to small wars in terms of "counterinsurgency" (p. 93).

Chapter 6. From Territory to Property: Indigenous Mapping after the Cold War

Nietschmann's combination of mapping and advocacy provided a template for the diffusion of indigenous mapping in the 1990s (p. 96). But instead of charting the contours of a great wave of decolonization, as Nietschmann predicted, the decade culminated in the mainstreaming of indigenous mapping (p. 98). In the hands of conservationists and development experts, mapping indigenous communities became a key strategy for recognizing indigenous rights to property as opposed to territory, as citizens rather than nations (p. 100). This transformation was first driven by appeals to conservationists to see "the coexistence of indigenous peoples and natural ecosystems" concentrated in tropical forests such as those in Central America (p. 101). Among the geographers advocating this approach was Peter Herlihy, the man who would later lead the AGS's first Bowman Expedition to Mexico (p. 103). However, conservationists soon soured on the idea, insisting on the importance of science, and not politics, in guiding their efforts (p. 107). Instead, indigenous mapping was taken up by advocates of "ethnodevelopment" at the World Bank and similar institutions (p. 108). Recognizing community rights to property proved to be an effective way of absorbing indigenous challenges to development projects, extending efforts to transfer state lands to private ownership consistent with neoliberal economic reforms (p. 113). This mainstreaming of indigenous mapping helped roll out a kind of slow-motion counterrevolution that neutralized (or at least tried to neutralize) indigenous demands for territory and autonomy (p. 115). This effort was critical to new efforts to economically integrate Mexico and Central America through the Plan Puebla–Panamá and fashion new regional approaches to security through the Mérida Initiative. Mapping indigenous property rights also paved the way for the technique's return to prominence among U.S. military officials and security experts, who made it a key counterinsurgency tactic in the wars in Afghanistan and Iraq.

Chapter 7. Counterinsurgency and the Rise of the "Warrior Scholars"

The emphasis on recognition of indigenous rights to property, facilitated by mapping, coincided with approaches to Latin America security that emphasized formal recognition of property rights as necessary to the functioning of markets (p. 127). Peruvian economist Hernando de Soto popularized this approach, claiming that the Peruvian state could defeat the Shining Path, a Maoist insurgency, through formal recognition of de facto property rights (p. 130). De Soto's idea became tremendously influential at the World Bank, where it reinforced free-market policies. Military personnel also took note, seeing in property a means of ordering the complex terrain of urban warfare, and in forested areas a means of identifying security threats (p. 132). Geoffrey Demarest folded de Soto's emphasis on property into his own experience as a military attaché to Guatemala in the late 1980s and to Colombia in the 1990s. In particular, Demarest called for comprehensive mapping of areas without mapped property records as an effective counterinsurgency technique (p. 133). As Demarest argued, "to succeed in both counternarcotics as well as the suppression of lawlessness, an indispensable starting point is the knowledge of ownership and the value of land." Demarest presented his argument in 1998, reprising it again in 2003 with specific regard to Colombia, just as the U.S. Army was rediscovering the value of "counterinsurgency" as a military tactic in Afghanistan and Iraq. The Army incorporated this approach into its new *Counterinsurgency Field Manual*, compiled by General David Petraeus and published in 2007 (p. 134). Among other points, the *Manual* highlighted the importance of mapping the "human terrain" as a critical aspect of counterinsurgency, revising "Red Mike" Edson's vision of the battlefield in the face of an expanding "war on terror" (p. 136).

Chapter 8. The AGS, the Bowman Expeditions, and the México Indígena Project

Demarest's vision for a global cadaster registering property ownership might never have made it out of the military archives were it not for the AGS's singular approach to "saving" geography from its academic practitioners (p. 142). In 2001, geographer Jerome Dobson left a 26-year career at the Oak Ridge National Laboratories in Tennessee to take a position at the University of Kansas (p. 144). In 2002, Demarest submitted a proposal to the Defense Intelligence Agency for determining the feasibility of creating a digital database of property ownership in Colombia (p. 146). The proposal identified the AGS as an ideal academic partner for the project. That year, Dobson became the president of the AGS. One year later, in 2003, the University of Kansas signed a joint research and education agreement with the U.S. Army Command and General Staff College and the Foreign Military Studies Office, both housed at nearby Fort Leavenworth. By 2005, the AGS had secured a pair of contracts worth $281,213 from the Foreign Military Studies Office to launch its Bowman Expeditions program aimed at gathering "open source intelligence" on foreign countries (p. 147). Peter Herlihy led the first expedition to Mexico that same year, mapping indigenous communities in the Huasteca Potosina in central Mexico (p. 148). Following a second round of funding that brought the cash total for the project over $700,000, Herlihy expanded the project south to the Sierra Juárez of Oaxaca (p. 149). Under the terms of his contract, Herlihy's team submitted all data gathered by the expedition to a third party, Radiance Technologies, a military contractor known for generating intelligence databases (p. 158). In 2009, the project exploded in controversy following a series of public statements made by organizations and communities in the Sierra Juárez, claiming that Herlihy had never informed them of the military's role in the project (p. 159). Aware that the U.S. Army could easily share that information with the Mexican Army, the communities accused Herlihy and the AGS of geopiracy and demanded the return of the data collected (p. 160). The AGS, however, was undeterred, expanding its Bowman Expeditions program to the Antilles, Jordan, and Colombia with geographers from Louisiana State University, the University of Akron, Western Kentucky University, and elsewhere.

Coda. Kill the Insurgent, Save the Man: Indigenous Peoples and Human Terrain

The AGS's World Human Geography Conference in 2011 brought the three strands of our account—indigenous mapping, counterinsurgency, and academic geography—into stunning relief (p. 163). An elaborate charade to end the controversy over the México Indígena project, the conference's real purpose was to trumpet the virtues of applied human geography for producing intelligence for the massive geospatial intelligence complex now linking the military with security and intelligence agencies throughout the U.S. government (p. 165). Correspondingly, the military has funded further Bowman Expeditions, sending Herlihy and Dobson to Central America (p. 171). The new project, "CA [Central America] Indígena," expands the approach taken by the México Indígena project (p. 173). It also further outlines the strategic orientation of their approach, expanding the AGS's efforts to map the U.S. "Borderlands Region" that includes "all of Latin American countries bordering the Gulf of Mexico and the Caribbean Sea." Herlihy again claims to be acting in the best of interest of indigenous peoples, providing them with detailed, accurate maps that they can use to make themselves visible to state agencies, while saying nothing about the escalating U.S. military presence in Central America, much less addressing indigenous concerns with the militarization of their territories. These challenges raise pressing issues about the continued importance of mapping to indigenous peoples' struggles for territory and autonomy (p. 174).

1

In the Rincón of the Sierra Juárez

Picture this: a high green slope, steep, rising a couple of thousand feet from an unseen valley floor. Wisps of cloud drift past, torn from those wreathing the heights. Snagged now and then by a tree, by a protruding mass of rock, they're torn again: handkerchiefs of moisture. The green is mostly that of trees: pines, oaks, and more pines. There are epiphytes, ferns, and mossy beards. It's hard to see these from across the valley, but the patches cut for *milpas* are easy to make out, those slipping into fallow, those in maize, each a different shade of green.

We're standing, looking down and out, at the edge of a kind of formal yard or court (Figure 1.1). It's framed by a secondary school, by the Agencia Municipal de San Miguel Tiltepec used by the town's authorities, by the offices of the Comisariado de Bienes Comunes tasked with coordinating communal property rights, by an elementary school, and by a road. All these wrap around a basketball court where, at the moment, boys and girls are lined up at the free throw lines waiting their turns to shoot. Across the road three more elementary school buildings have already begun to climb the slope leading up to the ridge on this side of the valley, equally unseen another thousand feet higher. Kids are everywhere, tumbling, running, jumping, playing basketball. Their noise infects the wisps of cloud, the drifting mist, the cool and the damp with sunshine, with warmth.

The promising smell of cooking fires fills the air.

FIGURE 1.1. In Tiltepec, February 7, 2012.

The slope we're on is dotted with houses, their yards teeming with hens, with roosters, with turkeys. The gabble-gabble is unstopping. The crowing is continuous, here, there, now at our feet, then further off. A donkey brays, almost a shriek. Outhouses, sheds, other buildings crowd around the homes. It's a farming community and the men, the few who'd come in from the *ranchos* they maintain at their *milpas*, were long since at work, often as distant as a two (or more) hours walk. They're harvesting coffee this time of year, and the sunless damp isn't helping with the drying of the berries.

The little kids laugh and run. The older boys swagger, saunter, slouch. The older girls exhibit propriety.

Coffee below the town, maize above it, though neither much disturbs the sense that the valley walls are a seamless green. They grow sugarcane, too, much farther down; and of course squash and beans among the maize. Dooryard gardens are thick with flowers, greens, herbs. Though plenty of money is wired home from those working in the States, the town is otherwise self-sufficient. It eats what it grows and it cherishes what it eats. The corn is theirs. They know where it comes from, that it's clean, that it's free of herbicides, of pesticides. It's their own seed, too. It has been . . . for centuries.

These people—well, them and their parents, and *their* parents, and *their* parents—we could go on. For a long time. Because these people have lived here for a long time. A thousand? Two thousand? Three thousand years? Hard to be much more precise about this particular valley,[1] but the Zapotec have been growing corn in these mountains for 5,000 years.[2] Not the coffee or the cane. Those are crops that came with the Conquest—Old World crops—though the people raise them with equal care and concern for cleanliness, purity, quality, and taste. So it's their *own* coffee they give you in their homes. They sweeten it with *panela*, sugar that they've refined from cane that they've grown. They make a point of telling you they wouldn't think of using . . . white sugar.

We could get enthusiastic about the Zapotec contributions to the domestication of corn ("the most remarkable plant breeding accomplishment of all time"[3]); allude to Monte Alban resplendent on its hilltop,[4] and to Zapotec writing, one of the first in the Americas[5]; mention the commitment of those in the Sierra Juárez to communal property rights; to their insistence on the preeminence of the assembly of the whole people, the *asemblea*; and to their sense of community, of *comunalidad*[6]; and go on to sketch a culture node of global significance that is also a kind of paradise on earth. But that would be ridiculous, would open us to accusations of essentialism, romanticism, and plain old idiocy; and in the pervasive irony of these postmodern years, no one buys the idea of paradise anyway.

Besides, paradise is given, not made; and these people have made this durable place, made it and maintained it against every odd. The death and disease of the Conquest should have destroyed it, or then the Catholic conversions and the slavery, the turmoil of independence, the extractions of the dictatorship of Porfirio Díaz during the late 1800s and early 1900s, or the ongoing tolls of capitalism, of the North American Free Trade Agreement (NAFTA). At least the language should have been wiped out and the quaint custom of communal ownership of the land. Instead it's still all here, a testament to their efforts to preserve their collective way of life. The communities of the Rincón have worked tirelessly to defend themselves against the onslaughts of commercial logging, the genetic pollution of their corn crops, the conversion of their land into state-controlled national parks. These people, thoroughly aware of the world—fully *participating* in it—yet certain theirs is the only *really* good life, have persevered, to say it again, against every odd.

It's been a *long* struggle. It's no accident at all that San Miguel Tiltepec has been here as long as it has.

It's an Ongoing Struggle

In 2006, Peter Herlihy led an expedition into town. God knows he wasn't the first. The first to mount an expedition into the Sierra—at least on record—seem to have been the Aztecs, though they signally failed to subdue the Rincón to their tribute empire: the terrain was *forbidding*, they lost a battle here in Tiltepec, and the Rincón wasn't rich enough to encourage further effort.[7] After the Aztecs came Hernán Cortés. In 1521, he launched an expedition into the Sierra under Gonzalo de Sandoval in search of the gold Moctezuma had assured them was here. Sandoval, in turn, deputed a Captain Briones, whose "expedition came to an abrupt halt at the Zapotec town of San Miguel Tiltepec, where over a third of the Spanish troops were wounded in battle."[8] In 1523 and 1524 Cortés sent Rodrigo Rangel. The first year, Rangel was defeated by the rain and the terrain; the next year by the Zapotecs, who "defended themselves with fifteen-foot lances."[9] Next to die in Tiltepec was Luis de Barrios, he and six other Spaniards. Only in 1526 were the Spaniards finally able to subdue the Sierra, though "many incidents took place in the 1530s, the worst in 1531 in Zapotec Tiltepec," where another seven Spaniards bit the dust. And once the Spaniards *had* subdued the Sierra? They never got much out of it. During the 1530s the colonial landlords, the *encomenderos*, complained "that they could not get enough to eat and that they often had to buy food because their Indians would not give it to them."[10] So it went.

For the residents of Tiltepec, none of this abstract detail. They have their *own* chronicle of events, some recorded in illustrations and Zapotec script on a 13-foot-long cotton cloth, or *lienzo*, from 1591, narrating their battles with the Spanish and their alliances with other communities.[11]

Anyhow, it's been a *long* struggle.

In the late 19th century a whole new kind of conquistador rode up into the mountains. Safe to say that Frederick Starr was many things, but in the Sierra Juárez he was an anthropologist, an ethnographer.[12] He came to photograph people, measure them (along 14 different dimensions), and make plaster casts of their heads. He wasn't subtle about it:

> We then told the *presidente* of the work we had before us, and informed him that, because his town was so small, we should ask for only thirty-five

men for measurement, and that these must be ready, early in the morning, with no trouble to us. The *presidente* demurred; he doubted whether the people would come to be measured; we told him they would not come, of course, unless he sent for them. When morning came, although everything had been done for our comfort, there was no sign of subjects. That no time might be lost, we took the *presidente* and three or four other officials, who were waiting around the house; then, with firmness, we ordered that he should bring other subjects. The officials were gone for upwards of an hour, and when they returned, had some ten or twelve men with them. "Ah," said I, "you have brought these, then, for measurement?" "On the contrary, sir," said the *presidente*, "this is a committee of the principal men of the town who have come to tell you that the people do not wish to be measured." "Ah," said I, "so you are a committee, are you, come to tell me you do not wish to be measured?" "Yes." Waiting a moment, I turned to the officials and asked, "And which one particularly does not wish to be measured of this committee?" Immediately, a most conservative-looking individual was pointed out. Addressing him, I said, "And so you do not wish to be measured?" "No, sir," said he, "I will not be measured." "Very good," said I. "What is your name?" He told us. I marked it down upon my blank, and wrote out the description of his person. Then, seizing my measuring rod, I said to him quite sharply, "Well, well! Take off your hat and sandals. We must lose no time!" And before he really realized what we were doing, I had taken his measurements. . . . [13]

It was worse with the plaster castings—much worse—but anyhow, a hundred years later, when Herlihy showed up to measure their *land*—and in between there'd been other ethnographers, Mexican development experts, representatives of timber interests, mining concessions,[14] folk concerned with carbon credits and biodiversity conservation—Herlihy was seen in the Sierra as but the most recent in a long, long line of initiatives against *serrano* communities. Though it didn't start that way and it need never have come to that.

But what did he *want*, this geographer from the University of Kansas? We mean, it's clear enough what the Aztecs were after (gold, bird feathers), Cortés (gold, slave labor), Starr (physical anthropological data), and the rest (timber, mining concessions for gold and silver), but this geographer, it wasn't clear what he was after. What he said was, *he wanted to give the communities maps*.

Now, okay . . . if we're not going to have paradise, certainly we can't have the tooth fairy either. And even the tooth fairy gets to keep the tooth. What was the tooth this fairy flew away with? And, in fact, *that's pretty clear*. Having read the reports and the papers and the flying accusations (and we'll be combing through those in detail), and, more importantly, having heard in Tiltepec and neighboring Yagila and

Yagavila—where Herlihy also led expeditions—from *agentes, presidentes,* and *comisariados* past and present; from the young men, the *jovenes,* that Herlihy trained (the ones who did the actual mapping), and from others he worked with; from the involved government authorities, local biological and social scientists, and still others; and having seen the maps he gave the communities; it's plain that what he flew away with was most of the data and gave it to the U.S. Army.[15]

What Herlihy left the communities was the *topographic* data (at 1:50,000) that he'd acquired, at minimal cost, from the Mexican National Institute of Statistics and Geography (INEGI), along with the *plan* information—that is, borders—from Mexico's National Agrarian Registry (RAN), *plus* the toponymic data—place names—the residents had given him themselves. In other words, Herlihy didn't give Tiltepec or Yagila—the two *serrano* towns in which he was allowed to complete his work—anything they didn't already have . . . *except for the map artifact itself.* In Tiltepec this map is hanging on the wall behind the desk in the office of the *comisariado* (there's a second copy standing on end in the adjacent storeroom and a third we didn't see). In Yagila the Agente Suplente—we'd meet the Agente later—didn't seem to know they even *had* a map, though a former *agente* located it for us, rolled up, in a storage room. But then, why should he know they had a map? It's comparatively useless.[16]

In fact, we recognized this map—the framed one—hanging on the wall in Tiltepec. It was an English estate map of the 16th and 17th centuries—a kind of portrait of the land—the sort of thing wealthy English landlords commissioned itinerant mapmakers to make for them: "microcosmic symbols of landed wealth," as J. B. Harley put it.[17] Except . . . this wasn't paid for by a wealthy landlord. It was paid for by the U.S. Army, which, god knows, is no sort of tooth fairy at all.

Not that it says so on the map. *Anywhere.* There's a huge credit block—it occupies about a fifteenth of the map's surface—and it's got the names and logos of a couple of dozen institutions and individuals on it. None of them alludes, even *distantly* alludes, to the Army. There're a bunch of PhDs (Dr. Herlihy, Dr. Smith, Dr. Dobson, Dr. Robledo, Dr. Brady, each with his university affiliation), and a raft of students (mostly from the Autonomous University of San Luis Potosí and the University of Kansas); an acknowledgment of the *presidente* of the Comisariado de Bienes Comunes of Tiltepec and the local investigators (the Tiltepec *jovenes* Herlihy had trained); a local labor coordinator; and four logos: those of the AGS, the University of Kansas, the Autonomous University of San Luis Potosí, and SEMARNAT, Mexico's Ministry of the Environment and Natural Resources (the Secretaría de Medio Ambiente y Recursos Naturales). All that . . . but no room for the logos of the U.S. Army's Foreign Military Studies Office ("an open source research organization of the U.S. Army") that commissioned the work, or of Radiance Technologies ("Creating Innovative Solutions for the Warfighter") that administered the project.

Bizarre.

Especially given how unbelievably proud of the work everyone is at home.

Among some in Tiltepec today, this is all sort of a yawn. "We got one GPS [global positioning system] unit, two *jovenes* trained to use it, and three copies of a map," says the present *agente*, and that's pretty much how he wants to leave it. Others are less sanguine. They're still heated about the comparative worthlessness of the map they got and the paucity of equipment (*one* GPS unit?), to say nothing of the fact that Herlihy did not tell them that he was transmitting all maps and the data he collected to the U.S. Army.[18] And we're talking about a *lot* more data than he included on the maps he gave Tiltepec and Yagila. We're talking about individual farmers' parcel data. (Now who in the Army could want that? And what for?) Anyhow, here's how Tiltepec put it back in 2009, how the *entire community* of Tiltepec put it:

> To the General Public:
> To the Media:
>
> The citizens of the community of San Miguel Tiltepec, through our Municipal Authority and Commissioner of Communal Lands, wish to make our opinion publicly known with respect to the research project known as México Indígena. Begun in 2006 and completed in July of 2008, the project produced a map containing toponyms and other cultural and geographic information provided by residents of our community.
>
> The researchers and their students (Derek Smith, John Kelly, Aída Ramos, and others) led by Peter Herlihy, appeared before our General Assembly in the community, informing us that the only goal of their research was to study the impacts of PROCEDE on indigenous communities. They never informed us that the data they collected in our community would be turned over to the Foreign Military Studies Office (FMSO) of the United States Army. Nor did they inform us that this institution was one of the sources of funding for the project. For this reason, we believe that the researchers deceived our General Assembly in order to take information from us that served their own purposes.

The Community did not request the research. Instead it was the researchers who convinced the community to approve the project. Accordingly, the research was never based in the needs of the community. Instead the México Indígena project researchers designed the project's methods in order to gather the kind of information that they were interested in.

Without our knowledge, stories have been circulated in the press and on the internet alleging that our community was satisfied with the results of the research. These stories have been circulated by the México Indígena project researchers (Peter Herlihy) and the president of the American Geographical Society, Jerome Dobson.

For the reasons indicated, we would like to declare to the general public our complete disagreement with the research done in our community. We were never properly informed of the project's true goals, the uses of the information gathered, and its sources of funding.

We demand that those responsible for the México Indígena Project, the American Geographical Society, the Foreign Military Studies Office of the US Army, the Autonomous University of San Luis Potosí, and the University of Kansas, as well as all other involved institutions who participated without our knowledge comply with the following:

- Cease all use of the information gathered in our community
- Return all information gathered in our community to us
- Immediately destroy all information in your possession regarding our community and provide us with proof of that destruction
- Immediately remove from the internet all information published about the research done in our community
- Publicly apologize to the community for your violation of our rights as indigenous peoples and for violating the American Geographical Society's code of ethics that you profess to respect.

Lastly, we wish to alert all indigenous peoples and communities in Mexico and the rest of the world that they should not be caught unawares by the researchers and the Bowman Expeditions or by any other researcher interested only in pursuing their interests or those of the groups they represent. Instead, it should be the communities and peoples themselves who decide what research interests them and who should do it.

San Miguel Tiltepec, Ixtlán de Juárez, Oax., March 17, 2009

RESPECTFULLY YOURS
Rogelio Hernández, Agente de Policía Municipal San Miguel Tiltepec
Bernardino Montaño Mendoza, Presidente del Comisariado de Bienes
    Comunales San Miguel Tiltepec

At first reading there's something amusing about the tiny town of Tiltepec tilting against the combined might of the U.S. Army, its Foreign

Military Studies Office, Radiance Technologies, the AGS, and faculty and students of Mexican and U.S. universities. But then, you think, given its track record, given its deep history of resisting—for centuries!—the encroachments of Aztecs, the Catholic Church, Spanish enslavement, private property, capitalism, NAFTA, and transgenic maize, that Tiltepec will be able to tolerate the insults of these latest attentions too.[19] For isn't that what they amount to, *insults*? What could this amalgamation of institutions possibly have to offer this truly perdurable community?

Free Computers

Perdurable? Undoubtedly. But wealthy? By no means. In economic terms Tiltepec is among the poorest communities in the poorest state in Mexico. Which is not to say they go hungry in Tiltepec. They're subsistence farmers, after all, who sell much of their coffee for cash. But neither they nor their often wealthier neighbors—Yagila, Yagavila, Zoogochi—have a lot to spare for computer hardware, for geographic information systems (GIS) software, for GPS units. And yet . . . the idea that the Sierra could benefit from its own GIS competence had been growing in the Rincón for some time: The region *is* among the most biodiverse in the nation and has other resources—and, in the summer of 2006, a local biologist, Gustavo Ramírez, at once a consulting biologist to Tiltepec *and* working on his master's degree in Morelia, heard Peter Herlihy give a talk there about his ongoing work in the Huasteca Potosina of San Luis Potosí. There Herlihy had already initiated México Indígena, the prototype for the U.S. Army-funded Bowman Expeditions named for the former AGS president and advisor to U.S. Presidents Wilson and Roosevelt, Isaiah Bowman.[20] To Ramírez it sounded as though Herlihy were offering GIS equipment, training, and maps in exchange for access to local communities, so he alerted Aldo González's Union of Organizations of the Sierra Juarez of Oaxaca (UNOSJO) about the opportunity. González followed through, inviting Herlihy to make a presentation to the organization.[21] Herlihy *immediately* accepted González's offer and, under UNOSJO auspices, Herlihy and his team made presentations that summer in Guelatao, Tiltepec, Yagila, Yagavila, and Zoogochi. And in exchange for access to the communities—and in the end, three of them agreed to participate: Tiltepec, Yagila, and Yagavila—Herlihy promised them GPS equipment and training, together with a map of each community that would be the projects' ostensible justification (Figure 1.2).[22]

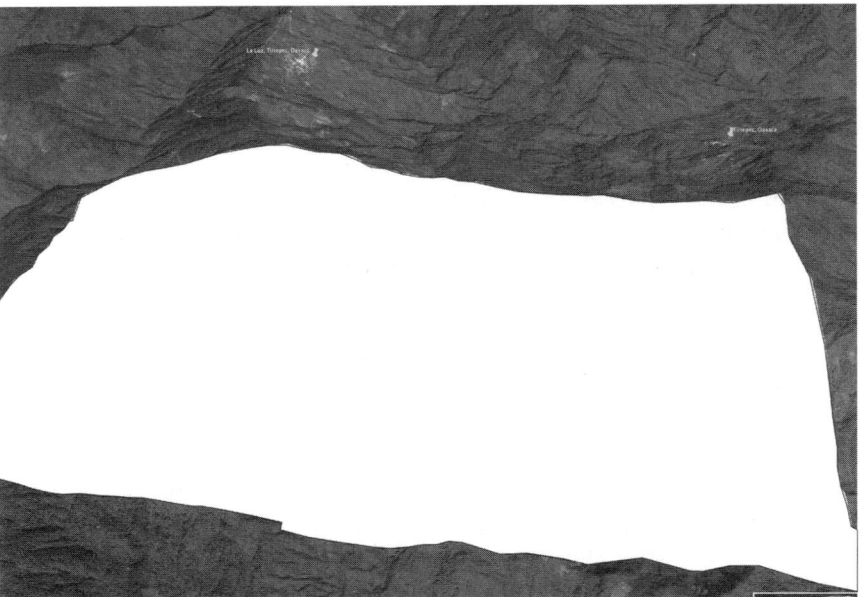

FIGURE 1.2. The México Indígena map of Yagila, hidden behind a coat of white to respect the wishes of Yagila. Here, the México Indígena team had superimposed its map of Yagila over Google Earth imagery of the relevant portion of the Sierra Juárez. We were still able to access the original map on August 17, 2013.

There was a collision of interests here. The communities were interested in free gifts of computers, GIS software, and handheld GPS units, and training. Aldo González wanted both to feather his political nest but also to begin building the GIS infrastructure UNOSJO needed to help with numerous resource use and allocation issues: proposed mines, timber concessions, carbon credits. Herlihy needed indigenous communities to map because that's his niche as a geographer and because he and the Bowman Expeditions were funded by the U.S. Army's Foreign Military Studies Office. The Foreign Military Studies Office had two interests. On the one hand, it wanted to test the possibility of using U.S. social scientists to gather open-source intelligence the Army didn't otherwise know how to gather, while, on the other hand, it was attempting to build a worldwide cadaster that would allow it to monitor property transfers, especially in *indigenous areas*, in *centers of "organized outlawry,"* and in *the large urban slums* where the Foreign Military Studies Office believed the military threats of the future would originate.[23]

None of these interests coincided. From the perspective of the Foreign Military Studies Office the whole thing—the Bowman Expeditions, México Indígena—was just a prototype, a proof-of-concept test of one approach to building a global, "place-based GIS," where *place-based* is code for *parcel-level property data*, what Herlihy calls "producing the 'digital human terrains' of indigenous Mexico."[24] From the perspective of the AGS, the exercise was its sole reason for continuing to exist, indeed a high road back to the role in American geography it had once played but long-since forsaken. For the AGS, México Indígena was the prototype, the proof-of-concept test for what *it* envisioned as a flotilla of Army-funded Bowman Expeditions girdling the globe. For Herlihy, the Army's support was the largest single source of funding he'd ever had for a project, and it was a chance to cement his status as a founding figure in the field of "participatory research mapping."[25] For Ramírez and González—soon to face off in a contest for political office—bringing México Indígena to the Sierra Juárez was a way of raising their political profiles, especially if it resulted in a regional GIS competence, which both saw, quite independent of any political ambitions, as extremely valuable for Sierra communities confronting an array of resource use and allocation decisions.[26] This consideration was paramount for Tiltepec, Yagila, and Yagavila, though each saw things from its own perspective. All three, however, expected to get out of their participation nothing less than a map and some sort of GIS, the latter being the tool they'd need for navigating the negotiations they foresaw over the use and allocation of their land. At the very least they expected some sort of hardware, and training in its use, the only sorts of things they could imagine México Indígena offering them. Certainly they didn't expect to learn anything about their land. They already *knew* their land. Indeed, those who *didn't* know the land at that point included, uhh, let's see . . . Peter Herlihy? The AGS? The U.S. Army?

Now, everyone involved recalls what happened—*knows* what happened—only from his or her own perspective. And *we*—Joe Bryan and Denis Wood—can order and recount these recollections only from *our* perspective. And every perspective is interested. Though we've struggled to track down everything relevant and to have ordered it all as best we can, there is still . . . *no truth*, no single story that tells it all, that tells it all as, you know, *it really was*. But, that said, here's what Gregorio Urbano, of Yagavila, recalls:

> They introduced themselves and made a proposal. They did this at a meeting and suggested that it would be nice if every community had their own

> map and that every place that is known could be written in a book. At the community assembly they commissioned me and another to join in learning the use of GPS. Then afterward we went to a plot of land. Nobody wanted to loan us their plot so we went to mine and then they examined it point by point. We were there all day with them and we asked what is the importance of them doing this? Would it actually benefit the community or not? But they never gave me the answers I was looking for. So then I went to consult with various people in the community. I told them that I thought this was crooked, corrupt, that it was not going well. There was another meeting where they introduced themselves again at which Aldo [González] was also present, and Peter [Herlihy] again said he was Aldo's friend. Then they had an argument during lunch. Aldo asked Peter why he told the community that he came with his consent. Herlihy answered, "We saw each other that one time," but Aldo responded, "You didn't tell me what you had come for." They came back to the community one more time but the people were conscious. So everyone told them to go to hell. We said, "When we decide to, we are going to make our own maps however they be, but they're going to be ours and belong to us." But some of the other communities were convinced and I think that, to this day, the mappers did not leave here in good terms.[27]

Now the thing about Aldo González is interesting—it's all interesting, but the thing about González is especially interesting—because if you leave him out of the story, as Herlihy is wont to do, a mystery arises: how Herlihy gained access to Tiltepec, to Yagila, to Yagavila. We mean, it's not like it was in the days when Starr could just march in and demand subjects to measure and cast. And even Starr, even if he didn't bother to show them all the time (as in the instance quoted), even he had credentials, had a letters from Fomento and the Governor of Oaxaca.[28] You don't just drop in on a bunch of Zapotec communities and start measuring farm parcels.

The way Herlihy tells the story, *he* involved González and UNOSJO in *his* project, not the other way around. In his own words: "UNOSJO has done community development projects in support of the indigenous communities of the Sierra Juárez. That is why in 2006 we initially involved them in our project, and we held a training workshop for potential local investigators at their Guelatao office. . . . "[29] Now listen to Rigoberto Perez, Yagila *agente* at the time, tell it from his perspective:

> They first supposedly arrived with UNOSJO. UNOSJO is a non-profit based in Oaxaca. This organization's main objective is to look out for the indigenous people in the Sierra and to keep an eye on their development. The mappers supposedly arrived with the consent of Aldo González. He has always been on our side and we recognize his work as being very valuable.[30]

Okay, now given that we're never going to know what *really* happened, not only did we hear this same story from a whole raft of people who were there at the time, but something like it *had to have happened*. In the Huasteca Potosina, Herlihy was working with faculty and staff at the Autonomous University of San Luis Potosí: *They* had the contacts, *they* had the access. But except for UNOSJO, which worked and continues to work with and for Sierra communities, Herlihy had no contacts in the Sierra at all. Without the meeting with Ramírez and the invitation from González, Herlihy's presence in the Sierra becomes . . . *magical*, as though all he had to do was wave a wand and, hey, *presto*! He'd be there collecting parcel data in Yagila.

But with UNOSJO and González *in* the story from the beginning, *and as the trusted props for the indigenous communities that they remain*, we can dispense with UNOSJO as the snake in the Garden of Eden story that Herlihy tells, in which he plays Adam and Tiltepec plays Eve.[31] UNOSJO wasn't some sort of interloper, an outside agitator. It had actually introduced Herlihy to these communities and was embarrassed by the revelation of the Army's role. Besides, the communities didn't need to know about the Army's involvement to smell a rat. Urbano had caught on right off, and he was enthusiastically seconded by his entire community. And Zoogochi hadn't even let Herlihy's crew get started. What knowledge of the U.S. Army's involvement gave them was a skeleton around which they could drape the flesh of their anxieties. But the anxieties were there from the beginning. They grew from the simple question we all ask when anyone offers to give us something for nothing: What do you want in return?

Which is the simple question Herlihy has yet to answer, which Herlihy has *still* yet to answer. In this vacuum the Zapotec communities of the Rincón have found this to say:

> We, the undersigned, Municipal Authorities and Commissioners of Communal Lands from the communities of San Juan Tepanzacoalco, Santa María Zoogochi, Santa Cruz Yagavila, Santiago Teotlaxco, and San Juan Yagila, meeting in the community of San Juan Yagila on July 24, 2011 in the town hall meeting room, after having reflected upon what happened in the communities of San Juan Yagila and San Miguel Tiltepec in 2006 when the México Indígena project was carried out, which formed part of the global project called the Bowman Expeditions, promoted by the American Geographical Society and the Foreign Military Studies Office belonging to the US Army, state the following:
>
> - We do not agree with the manner in which the geographical studies were carried out in the communities of San Juan Yagila and San Miguel

Tiltepec by the México Indígena project team between the years of 2006 and 2008, because they did not inform these communities as to the origins of the resources they used to carry out this research, specifically concealing the participation of the US Army, and in this way violating the right to free, prior, and informed consent which for us as indigenous communities is recognized in the United Nations Declaration on the Rights of Indigenous Peoples; as well, we support both of the communities in any problems they may have later on as a result of these investigations.

- We declare the creation of a single, organized front amongst the communities in our region, known as the Rincón de Ixtlán.
- We will seek to acquire the necessary information regarding the pros and cons of governmental and non-governmental projects and programs which are offered to our communities, with the aim that before deciding to accept them or not, the principle of free, prior, and informed consent is applied.
- We demand that our communities are paid, in an unconditional and compensatory manner, sufficient economic resources for the conservation of the forests that exist in them, as it is proven that it is the indigenous communities who have conserved the forests and jungles of Mexico, as well, that this be done with public resources, so as to not fall into the hands of transnational businesses who are only interested in profiting from our lands and to wash themselves of their blame for the climate crisis which they have caused on the planet.

Sincerely,
Municipal Authority of San Juan Yagila, Rodrigo Perez
Municipal Authority of Santa María Zoogochi, Julián Santiago Cervantes
Municipal Authority of Santa Cruz Yagavila, Sadot Gómez Santiago
Commissioner of Communal Lands of San Juan Yagila, Wilfrido Ramos Hernández
Commissioner of Communal Lands of Santiago Teotlaxco, Pascasio Gerónimo Hernández
Commissioner of Communal Lands of San Juan Tepanzacoalco
Commissioner of Communal Lands of Santa María Zoogochi

What *Do* You Want in Return?

So why doesn't Herlihy answer this simple question?

Because he can't.

Why? Because, as he writes in one of his status reports to Radiance Technologies ("Creating Innovative Solutions for the Warfighter"), "For simplicity's sake, we now call the two interrelated AGS/FMSO projects by the single name *México Indígena*," where, to recall, AGS is the American Geographical Society and FMSO is the U.S. Army's Foreign

Military Studies Office. This is to say, quite simply, that México Indígena is a military project, *not* an academic project that just happens to be funded by the military.

What would have happened had Herlihy requested access to a community in the Rincón of the Sierra Juárez and told the assembled members he wanted to come in and map out individual farmers' parcels along with their names and other characteristics and hand it all over to the U.S. Army?

The Zapotec are a courteous people. They would, very courteously, have told him to, as Urbano put it, *go to hell*.

So, whatever Herlihy says, this is not what he did. What he did, as Urbano suggests, was to say that nations have maps, states have maps, cities like Oaxaca de Juárez have maps, and you should have maps too. He may also have mentioned his personal interest in the effect of the Mexican government's Program of Certification of Ejido Rights and Urban Lots (PROCEDE) program on *ejidal* land tenure, but this wouldn't have signified in the Rincón, where there are no collectively owned agricultural lands or *ejidos* recognized by Mexican officals, where property, for time out of mind, has been communal. So he probably didn't stress that slant in the Sierra.

And, then, of course, he was offering the GPS units and the training of the *jovenes*, the local investigators.

But of course all this did was raise more insistently the question concerning what he *really* wanted.

And that he couldn't tell them. And caught up in the story he's been spinning ever since, he hasn't been able to tell anyone. Not straight. Not like he does in the status reports to Radiance Technologies.

The AGS is in a similar bind. If it just admits it no longer does much more than act as a conduit and administrator of Army money, that it's sort of an academic version of Radiance Technologies ("Geographic Knowledge for the Warfighter"?), it will lose every shred of academic respectability, and along with it, its sole utility to the Army. Which is precisely to cloak in respectability the intelligence—and oh, yes, it's all open source!— that it gathers through its Bowman Expeditions.

How did the AGS, America's oldest geographical society, end up in this situation? Since there's no understanding of any of this without understanding the situation of the AGS, we turn in the following chapter to the decline and fall of the once august American Geographical Society.

2

The Decline and Fall of the Once August American Geographical Society

Here, some names: Strabo, Ptolemy, Marco Polo, Columbus, Magellan, Da Gama, Humboldt, Dias, Cabot, Davis; Hudson, La Salle, Mercator, Cartier, Balboa, Baffin, Coronado, Drake, Ritter, De Soto; Champlain, Vespucius, Livingstone.

Who are these people? A few—Ptolemy, Mercator, Humboldt (Alexander von), Davis (William Morris), and Ritter (Karl)—could be called scholars, but the rest were what a generous spirit might be tempted to call explorers, though theirs was a notably rapacious brand of exploration. Columbus explored the Bahamas, Greater and Lesser Antilles, and the Caribbean coasts of Central America and Venezuela, and claimed them all for the Spanish crown. Balboa explored the Isthmus of Panama and its Pacific littoral, and claimed the Pacific Ocean and adjoining lands for the Spanish crown. Cartier explored the St. Lawrence and claimed what is now Canada for France. La Salle explored the Great Lakes, Mississippi, and the Gulf of Mexico, and claimed the entire Mississippi watershed for France. Cabot explored part of the coast of North America, and claimed it all for the King of England. Drake was a straight-out pirate. He was also a slaver, as was Columbus. Columbus even cut off the hands of the Taínos who failed to deliver their quota of gold to him, and decimated the native population of the island of Hispaniola.[1] They were all men, and with the possible exception of Ptolemy (who *was* a Roman citizen), they were all born in Europe. Most were bloody, violent men. Livingstone may have been a Protestant missionary, but given his motto, "Christianity, Commerce, and Civilization," it's fair to say he was pretty much cut from the same cloth.

They're also the names carved deeply into the white Indiana limestone of the frieze surmounting the massive Ionic columns that march around the façade of the headquarters built for the American Geographical Society in 1911 by Archer Milton Huntington (Figure 2.1).[2] Archer Milton Huntington was the stepson of Collis P. Huntington, who not only built the Central and the Southern Pacific Railroads, but, with Tom Scott, created the modern corporate lobby.[3] Huntington's is a name that could easily have gone up there on the frieze.

It should come, then, as no surprise to learn that during the Society's first 20 years, more papers were delivered on the extension of a railroad to the Pacific "and other aspects of the westward movement than on any other comparable topic."[4] Nor was the Society's interest limited to the expansion of U.S. territory. It was equally ardent about the expansion of U.S. trade; and men associated with the "opening" of Japan, Siam, and Paraguay all spoke before the Society. Its first lobbying effort had been directed toward the opening of Paraguay to American agricultural and industrial products, and the Society successfully memorialized the Secretary of the Navy for a survey of the Río de la Plata in Argentina. Toward the end of the 19th century the Society displayed particular interest in the route of a trans-isthmian canal, with the Society's president, Charles Daly, arguing that the Society was the most appropriate body for studying the comparative merits of the Panamanian and Nicaraguan canal routes because the Society had "greater facilities than any other body in the country . . . for disseminating throughout the world the information elicited by such an investigation"—though he failed to mention that he was vice-president, legal counsel, and a principal investor in the Nicaragua Canal Company, of which the Society's vice-president, Francis Aquila Stout, was president.[5]

But then the advancement of geographical science was only one of the Society's two stated purposes, the other being the promotion of the business interests of that "great maritime and commercial city," New York.[6] An unstated but constant third purpose was "the service of philanthropic and religious causes," and nothing promoted all three purposes like an expedition: "Exploring expeditions were supported in the expectation that they would yield not only new scientific data but facts of practical use to the merchant or missionary."[7] Because the Society's

FIGURE 2.1. Former headquarters of the AGS on New York's Audubon Terrace, west side of Broadway, between 155th and 156th Streets, in 1920.

fortunes were less than stable during its early years, it heard about more expeditions in its lecture hall than it actually supported, but it nonetheless managed to sponsor expeditions to the Open Polar Sea by Kane, Hayes, Hall, and Schwatka, Hall returning with a family of Eskimos who not only appeared on the platform with Hall at Society meetings . . . but in Barnum's Museum.

This too should occasion no surprise for it may be doubted whether there were ever men, anywhere, who were surer of the superiority of their race and civilization (and gender[8]) than the members of the great explorer societies. The Société de Géographie of Paris, the Royal Geographical Society of London, the Imperial Geographical Society of St. Petersburg, and the National Geographic Society of Washington were, like the AGS, all brought into being as gentlemen's clubs for philanthropists, prominent jurists, notable clergymen, and accomplished academics with shared interests in exploration and travel.[9] Like the rest of these the AGS of New York shared a great willingness to shoulder the "white man's burden," or at least as much of it as the shoulders of New York merchant philanthropists could bear—which, with the waning of the 19th century and their mind-boggling accumulations of vast wealth, were capable of bearing more and more.

Certainly by 1911, when Archer Milton Huntington erected the AGS's grand limestone headquarters, the days of the Society's penury were well behind it. Yet despite its new address and its increasingly impressive library of books and maps, the Society stood in an awkward position vis-à-vis its potential publics. Before the cornerstone had been laid for the building on Audubon Terrace, the National Geographical Society had run off–and we mean *run off*–with the *popular* public, whereas the fledging Association of American Geographers was sneaking away with the *scholarly* one.

Founded in 1888, the National Geographic Society began immediate publication of the *National Geographic Magazine*.[10] Initially a scholarly journal, the magazine was transformed into the popular monthly we know today well before the century was out; and since you could only receive the magazine as a benefit of membership, the rapidly growing popularity of the increasingly well-illustrated magazine soon drove the National Geographic Society's membership into six figures. With the typical cabal of wealthy patrons, explorers, and the like at its helm—to say nothing of Alexander Graham Bell, who was the new Society's second president—the National Geographic Society sponsored the usual expeditions; and when the Society-sponsored Robert Peary reached the North Pole and the Society itself was accepted as the arbiter of competing claims, the National Geographic Society achieved the popular preeminence that has only grown and solidified over the years.[11]

At the same time, William Morris Davis, on the faculty at Harvard from 1879 to 1911, was arguing that the popular orientation of geographical societies threatened the discipline's claim to be a science. The public lectures, flamboyant support of expeditions, and endless bestowing of gold medals that were the stock in trade of the AGS, the National Geographic Society, the Geographic Society of Philadelphia, and even the Appalachian Mountain Club, contrasted sharply with Davis's sense of geography as an academic discipline. The pomp and circumstance that defined the societies "reinforced the sense both outside and inside the academic community that geography was either a utilitarian tool for surveyors or a curious lecture pastime of travelers,"[12] rather than a serious academic subject. "Geography," Davis argued, "will find a place in our colleges and universities very soon after it is shown to be a subject as worthy of such a place as are the subjects whose position is already assured,"[13] and nothing, he felt—he's speaking in 1903—could so advance the cause of geography as a decent professional organization:

"No greater assistance to the development of mature scientific geography lies within our reach than the establishment of a geographic society which shall take rank with the Geological Society of America, for example, as a society of experts, in which membership shall be open only to those whose interests are primarily geographical and whose capacity has been proved by published original work in a distinctly geographical field."[14] In case any should have missed the point, he went on:

> While we must not overlook the excellent work that our geographical societies have done, neither must we overlook the fact that in making no sufficient attempt to require geographical expertness as a condition for membership, there is a very important work that the societies have left undone. They have truly enough cultivated a general interest in subjects of a more or less geographical nature, but they have failed to develop geography as a mature science. Indeed it may be cogently maintained that the absence of any standard of geographical knowledge as a condition for society membership has worked as seriously against the development of mature scientific geography as has the general abandonment of geographical teaching in the secondary schools. Large membership seems to be essential to the maintenance of good libraries in handsome society buildings, and it is certainly helpful in the collection of funds with which journals may be published and with which exploring expeditions may be equipped and sent out.[15]

But, continued Davis, "large numbers of untrained persons are not found necessary to the maintenance of vigorous societies in which other sciences are productively cultivated, and it is therefore reasonable to believe that large numbers would not be essential to the formation of a geographical society of high standing."[16]

With this paper—which 20 years later Albert Perry Brigham recalled as "a paper of historic significance"[17]—Davis proposed the establishment of what he called the American Geographers Union,[18] and in 1904, Davis, Brigham, and some 60 others founded the Association of American Geographers. "*Sneaking* away with the scholars," we said, and in fact, by 1924, 20 years after its founding, the Association still hadn't attracted as many as 150 members, whereas the AGS had almost 4,000 and the National Geographic Society, over a million.[19] For Davis this was less a problem than a badge of honor: "The organization of this small society, even now with a membership list of less than 150 names, may be regarded as one of the greatest steps of geographical progress yet made in America," for "it is believed to be the only geographical society in the world which limits its membership, as astronomical and geological societies usually limit theirs, to proficient and productive workers in its own field."[20]

Larger and far wealthier, the AGS patronized the scholars' Association from the get-go, hosting the first of their joint meetings at Audubon Terrace in 1914 (where it treated the Association members to lunches), and paying for the publication of the Association's *Annals* from 1914 to 1922.[21] These years marked the apogee of the AGS's prestige. After moving into its new building on Audubon Terrace in 1911, the AGS sponsored the Transcontinental Excursion of 1912 (see below). In 1915, it appointed Isaiah Bowman director, hiring him away from his tenured position at Yale.[22] In 1916, the AGS launched *Geographical Review*, a scholarly journal intended to compete with the Association's *Annals*. But 1917 revealed the true measure of AGS's influence when, with the U.S. entry into World War I, President Woodrow Wilson appointed Bowman and a select group of AGS members to "The Inquiry." Building on Bowman's offer to "put the Society's facilities at the disposal of the government,"[23] the Inquiry transformed AGS headquarters on Audubon Terrace into a center for the gathering and analysis of information to shape U.S. negotiating positions at the Paris Peace Conference. In 1918 Bowman himself accompanied Wilson to Paris as an advisor, armed with the detailed maps on which the AGS envisioned a geographical fix to the twin problems at hand.[24] The first involved the territorial dissolution of the Austro-Hungarian and Ottoman Empires into racially and linguistically defined nations. The second task was to extend this approach to the rest of Europe, fashioning a post-imperial order comprised of nation-states. Then, immediately after the peace, the AGS launched its *Research Series*—those publications for which its reputation will endure—and Bowman made clear his cartographic vision of the world by committing the Society's wealth and prestige to making a *Map of Hispanic America on the Scale of 1:1,000,000* ("the millionth map"). But by 1929, when the Society acquired its Fliers' and Explorers' Globe, its membership was already peaking (at something less than 6,000), and from then on the AGS slumped into a gradual decline, watching its membership dwindle as the Association of American Geographers waxed in importance, giving away the library and map collections that were its heart and finally selling off its Audubon Terrace building. Today, the AGS has substantially fewer members than the number of *abstracts* the Association of American Geographers received for presentations at its 2012 meetings in New York (1,500 abstracts).

What happened?

In the first place, the society was *never* what it seemed. Part of its genius was simply its name, the *American* Geographical Society, when in fact it should have called itself the Manhattan Explorers Club, where *Manhattan* would have accurately characterized its reach, *club* would have made plain its roots in fashionable male sociability, and *explorers* its commitment from day one to exploration with all its hardships, thrills . . . and potentials for new markets. That is, all along the AGS was actually just another vehicle for advancing the wealth and prestige of its members. In 1854, in the Society's inaugural "Annual Address," Matthew Maury, "father of modern navigation and ocean science," catalogued most of the larger geographical problems and developments that were of special interest to the Society in its formative years[25]:

> These, broadly, were: Arctic exploration in the American sector; the establishment of better connections between the eastern United States and the Pacific through a transcontinental railroad and one or more Isthmian canals, the exploration of South America, and the extension of the telegraph, which, like the railroad was then a new and exciting invention.[26]

Expanding at length on some 10 expeditions, Maury established a model for years to come. Explorers were popular people—they still are—and the Society was often forced to rent large halls when they spoke. Although this raised the Society's profile, especially in New York, it did little to increase membership or to establish the Society as a geographical institution on any sort of firm footing. In fact, as a geographical institution there was hardly any footing at all, with Maury's topics making much more sense for a chamber of commerce than a gathering of the geographically inclined. But the Society wasn't a chamber of commerce either. In the end it lacked identity.

Instead, the Society built its prestige through what were essentially honorary appointments. This is how, for example, Maury, Alexander Bache, and John Frémont became vice-presidents in the 1850s. Maury did give the annual address we've just quoted from, but none of them ever attended a council meeting, and even the Society's official historian admits that Frémont was "window dressing," pure and simple.[27] There were always slews of honorary members too—elected by the council as no more than an honor—and over the years these have included Livingstone, John Murray, Humboldt, Ritter, August Petermann, A. E. Nordenskiöld, Dom Pedro III (Emperor of Brazil), Ismaïl Pasha (Khedive of Egypt), the Grand Duke Constantine of Russia, and so on and on. The Society also loved giving medals—often the recipient would grace the Society with a talk (though as often the recipient would receive his or her medal at a U.S. embassy)—and it had a number of medals to bestow: the Cullum Geographical Medal, the Charles P. Daly Medal, the David Livingstone Centenary Medal, the Samuel Finlay Breese Morse Medal, and the George Davidson Medal. With each oversized gold medal, the Society burnished its claims to prestige. Though the Society did award medals to geographers such as William Morris Davis, Vidal de la Blache, Ellen Churchill Semple, Carl Sauer, Sir Halford Mackinder, and the Society's own Isaiah Bowman, more often it hung them around the necks of Peary, Nansen, Stefanson, Rasmussen, Scott, Shackleton, Amundsen, Byrd, Ellsworth, Sven Hedin, and Theodore Roosevelt. Again, high-flying company, but shy of substance.

The building on Audubon Terrace is another case in point. The land it stood on was offered to the Society by its then president, Archer Milton Huntington, on behalf of his mother, Arabella Duval Huntington, on the condition that the Society construct a building on the site "in harmony with the general scheme of the buildings of the Hispanic Society" that Huntington had already constructed there. Huntington paid for the building and hired his cousin, Charles P. Huntington, to design it.[28] Later on he replaced the original oak doors "with splendid bronze doors designed by his wife, Anna Hyatt Huntington, one of America's foremost sculptors."[29] There's a sense in which the whole of Audubon Terrace, where the AGS, in the shadow of a large equestrian statue of El Cid sculpted by—you guessed it—Anna Hyatt Huntington, in company with the Hispanic Society, the Museum of the American Indian, and the American Numismatic Society, all similarly endowed by Archer Milton Huntington in buildings designed by Charles P. Huntington, was nothing but a giant Huntington vanity project—which would make the AGS Huntington's vanity geography society. To put it another way, the impressive building was less the sign of a healthy, robust organization than of a charity case in the hands of an exceedingly wealthy benefactor with a penchant for collecting.

The Transcontinental Excursion of 1912 was a similar story. The idea was originally Davis's. In 1908, he'd led a group of American and European university students on an essentially physiographic tour of Italy and France, an experience he proposed, at the 1909 meetings of the Association of American Geographers, to replicate in the United States. Finding no takers among his scholarly colleagues, he turned to the AGS. Archer Milton Huntington agreed to foot the bill on the condition that it take place under the banner of the AGS in celebration of its 60th anniversary and new building. With Huntington's money, Davis chartered a train of two standard Pullman cars, two Pullman observation cars, a dining car, and a baggage car. As though to confirm Huntington's status as the son of a railroad baron, the entourage was usually accompanied by a car carrying railroad officials who looked after the party's comfort. The train became a moving social event. "Governors of states, mayors of cities, university presidents, chambers of commerce, scientists, and ordinary citizens everywhere lavished hospitality"; at the Muir Woods the tour was met by John Muir himself, Luther Burbank, and David Starr Jordan (former Sierra Club director and leading eugenicist); there were side trips in automobiles and on steamers; and in the end a *Memorial Volume of the Transcontinental Excursion of 1912*.[30] The distinguished European geographers were the real draws for the excursion (their expenses paid for by the Society), and with their American counterparts (traveling on their own dime), they were quite the spectacle at Niagara Falls, Yellowstone, Crater Lake, the Grand Canyon, the Petrified Forest, the Meteor Crater, the Great Smokies, Monticello, and Washington, DC. In their elegant, mobile quarters, great friendships were struck up that were to stand Bowman and the others in good stead when they came to redraw the boundaries of Europe during their work for the Inquiry; and "the excursion won repute for the Society at home and abroad."[31] No doubt. But an excursion is a brief pleasure trip and in their Pullmans with the dining car, the receptions and the banquets, the champagne and the toasts, that's what this was: eight lavish weeks of "geography," AGS-style.

Did the Society ever do *anything* serious?

Certainly the Society's work with Woodrow Wilson's Inquiry Commission had serious consequences, but whether it was geography has to be questioned. There's no question that *it drew on* geography, but wasn't it Davis's point that geography is a science? The current president of the AGS, Jerry Dobson, clearly seems to think so: "Geography is a dimensional science and humanity, based on spatial logic in which location, flows, and spatial associations are primary evidence of earth processes, physical and cultural."[32] That is, geography *studies* things. Now, there seems to be little doubt that Bowman, with the AGS at his back, all but drew the boundaries that the Americans adopted in the treaty negotiations following World War I. Among other things he was in charge of putting together the notorious Black Book that Wilson relied on during the treaty negotiations. This is Bowman's claim to fame. The politics of that project were entirely suspect. But how was this science? How was it geography? *In any way?*

It's not that Bowman drew boundaries. That's what geographers do.[33] Geographers draw boundaries around every conceivable phenomenon. In 1919 they were drawing them around linguistic and racial predominance.[34] What transformed their work from science to political policy mongering was turning, say, a linguistic boundary into a political one, especially when the political boundary the geographers recommended would invariably depart at this or that critical juncture to take into account racial factors, political concerns, or "natural associations."[35] At that point the geographers turned into politicians, into policy hacks. That's fine. But it's not without consequences. Bowman's idea that race could be used to bound discrete territories figured in the nationalist conflicts that shaped the remainder of the 20th century, from the rise of the Nazi Party in Germany through to the Balkan Wars of the 1990s.[36] It was also precisely the kind of geography that mitigated against the field's aspirations to be taken as the serious university discipline Davis sought to make it. Instead of shoring up their ranks as a science, geographers were steadily killing their discipline, lured on by the desire for fortune and fame that the AGS stoked among its Manhattan boosters.

The AGS's desire to be useful, to put its amassed knowledge to work for the city's commercial magnates, for the national planners of transcontinental railroads, for the builders of international canals, led it straight into war. Indeed, as the era of exploration drew to its close, the AGS *staked its reputation* on war. After deprecating the Society's involvement in the American Civil and Spanish Wars, J. K. Wright observes that, "Of the two World Wars the record is different. In both, the Society rendered important services to the Government in connection with the military effort," and as we've seen, "As soon as the United States entered the First World War Mr. Bowman put the Society's facilities at the disposal of the government." That quote continues: "At the start a suite of rooms on the third floor was turned over to the organizers [of the Inquiry] and kept under lock and key in an atmosphere of secrecy that became far more familiar in the Second World War. Eventually the Inquiry expanded to take over practically the entire building and its resources."[37] We're not sure, but whatever this is, it doesn't sound too much like geography.

Here, then, with its Inquiry work, where the Society managed something serious, it turned out not to be terribly geographic, at least not in the scientific way for which Davis had hoped. And that's more or less the problem with the *Map of Hispanic America on the Scale of 1:1,000,000*, the famous "millionth map." In conception this was genuinely geographic, but in its merely topographic implementation it remained resolutely imperial. Nothing less had been envisioned than "the review and classification of all available scientific data of a geographical nature that pertains to Hispanic America": topography, climate, population, *everything* that affected human distribution, activity, and economic welfare. It was to include soil, vegetation, and land classification maps at various scales to look back into history and forward into the future. Bowman claimed that this conception had a definite date, late 1912 or early 1913, when he was preparing, with AGS support, for his 1913 expedition to the Andes. When he turned to the Society's map collection, Bowman "came to see the disordered state of even the best collections in the field of Latin American cartography," and he determined to do something about it after his return. But then he was appointed director of the AGS and the war soon followed, and it wasn't until the war's close "that I was able to get down to business upon the problem. The rest of it came along as a matter of course."[38] Well, that's one thread.

Here's another: "While Dr. Bowman was at the Paris Peace Conference a long-standing boundary dispute between Guatemala and Honduras came to a head, and the republics nearly went to war. The two governments appealed to the Secretary of State of the United States to act as mediator. Mr. Lansing, thereupon, asked the Society to suggest a form of settlement,"[39] and Bowman proposed an economic survey of the territory in dispute. The Society carried this out and it led through "entirely peaceful" negotiations to a 1933 treaty. The Special Tribunal used the Society's map throughout and ultimately adopted it as the official map of the boundary. Therefore, Wright concludes, "had it not been for the Inquiry, Mr. Lansing would not have turned to Dr. Bowman when the dispute became acute. The economic survey, by bringing sharply before the Society the need of better maps of Hispanic America, was a factor leading up to the decision to make the Millionth Map."

Okay, but here's a third. Archer Milton Huntington's first love was, without any question, Spain and all things Iberian. He traced this passion to a childhood reading of George Borrow's gypsy romance, *The Zincali*, but he pursued it almost single-mindedly from then on, learning Spanish, Arabic, translating *El Cid*, collecting books, paintings, coins, and founding, funding, and housing the Hispanic Society on Audubon Terrace. It was his extraordinary collection of Visigothic coins, the largest ever assembled, that sparked his interest in the American Numismatic

Society—the AGS and Hispanic Society neighbor on Audubon Terrace—and it was his interest in the Iberian colonial legacy that fired his commitment to support the mapping of Hispanic America.[40] And support it he did, with a start-up grant of $25,000 in 1920. Support then fell to James B. Ford—Commodore of the Larchmont Yacht Club and an officer of the United States Rubber Company (one of the original 12 stocks in the Dow Jones Industrial Average)— whose contributions of $192,000 supported the project until Ford's death in 1929. At that point Huntington stepped in with another grant of $200,000. Between the two of them, Ford and Huntington defrayed three-quarters of the map's cost. The Rockefeller Foundation picked up most of the balance, enhancing their role as "Missionaries of Science" through campaigns to improve agricultural efficiency and eradicate disease in Latin America.[41]

The 107 sheets of the map took 25 years to complete at a cost of almost half a million dollars. It was a wholly new compilation, but, with the exception of an expedition to the sources of the Amazon in 1928, it derived from existing surveys. Part of the enormous labor lay simply in the assembly of the source materials. It's to take nothing from the remarkable mapmaking, however, to point out that the project fell several light-years short of its ambitions. Those had been indubitably geographic. The end product, laudable though it might be, was simply topographic mapping. This distinction was rooted deeply in geography's 19th-century German background (Humboldt, Ritter, Ratzel, Penck), but nowhere more explicitly than in the work of Max Eckert—work, like so much of Davis's, directed toward the establishment of geography *as an academic science*. Essential to Eckert's program was the division of maps into two overarching categories: general-purpose (or reference) maps and special-purpose (or thematic) maps. This division gave prestige to a practice of small-scale, often statistical mapmaking that was easy to justify as a subject in a university curriculum. Supporting this was the division of labor between mere *technicians*—concerned with "practical and handicraft cartography" and responsible for the reference (the topographic, the base maps, the *Map of Hispanic America*)—and *scholars*, who were supposed to create the thematic (the special-purpose, the applied) maps that constituted the unbroached body of the original Hispanic mapping proposal.[42]

In his history of the Society, Wright expends pages on the backgrounds of the technicians who worked on the map: They had been artists, craftsmen, a former army officer, five former naval officers, a lithographer, a surveyor. He goes into raptures over their abilities:

"Map drafting of the quality needed for making the fair drawings for the Millionth Map requires a patient temperament, a steady hand, and good eyes."[43] He details the steps from compilation to printing plates (up to as many as 12 plates per sheet). He's ecstatic about their ability to letter: "The most time-consuming operation and the one that called for the greatest skill in draftsmanship was the lettering of the names, of which there are at least 200,000 on the 107 sheets. . . . Perhaps ninety-eight or ninety-nine per cent of the 200,000 names were lettered by hand with a pen—something that amazes uninitiated persons. The lettering is so perfect as to be almost indistinguishable from print."[44] What is perfectly distinguishable in Wright's history is the gulf between these meritorious technicians and . . . the *geographers* who supervised them. The AGS's production of the *Map of Hispanic America* was practically an illustration of Eckert's division of labor. On one side sat the technician "who has no time to care about scientific problems . . . because he is already totally occupied with his manual, but nevertheless scientifically guided work."[45] On the other side strode the guide, the geographer, the scientist, wholly consumed with problems of analysis. *Except* . . . that the geographic, the scientific part never got done: "As originally planned, each sheet was to be accompanied by a handbook on the geography of the area covered, but only one handbook [out of 107 intended], a volume accompanying the La Paz sheet which Ogilvie wrote while at the Society. This was published in 1922 under the title *Geography of the Central Andes*."[46]

Okay, so again, terrific stuff, just not geography, just not geography *by the Society's own disciplinary lights*. The Society knew what geography looked like. *It published it!* It looked like Ogilvie's *Geography of the Central Andes* (1922), it looked like Bowman's *Desert Trails of Atacama* (1924), it looked like Wright's *Geographical Lore of the Time of the Crusades* (1925), it looked like G. M. McBride's *Chile: Land and Society* (1936), it looked like Owen Lattimore's *Inner Asian Frontiers of China* (1940), it looked like R. H. Brown's *Mirror for Americans: Likeness of the Eastern Seaboard, 1810* (1943).[47] These were all exemplary volumes on geography, serious, still valuable. Some have become classics. Lattimore's *Inner Asian Frontiers* has been reprinted more than once. Wright's *Lore* is still in print.

That is, it wasn't that the Society's members *couldn't* tell geography from topographic mapping, from pleasure trips, from policy mongering. The problem was that the Society's identity remained so tied up with the New York society from which it drew its pay and prestige that geography was usually playing second fiddle. Any scholarly ambitions the Society might have had perpetually played second fiddle to the honorary officers (Frémont had just run for president of the United States), to the honorary members (the Emperor of Brazil! the Khedive of Egypt!), to the medal pinnings (on the chests of Theodore Roosevelt and Albert I, Prince of Monaco), to the excursions, to the dinners, to the national service, to the topographic mapping, to . . . the library! Sure the library was important—core!—but in fact it was also just another collection to which rich New Yorkers could give things. Francis Stout— "of old New York stock and large inherited means"[48]— liked to give the library things like a rare Ortelius atlas of 1573, a 1633 Mercator, a 1671 Montanus (with its early view of New Amsterdam), and the costly facsimile *Monuments de la géographie*, which Edme François Jomard had assembled. Charles Daly, the New York City judge after whom the medal was named, not only endowed the library with things like a 1508 folio by Ptolemy but threw in his whole 700-volume geography collection.[49] The collections for which the library is renowned? Almost every one of them is a monument to a wealthy New Yorker.

And when that wealth found other places to bestow itself—flashier places, places where people could *see* your name on a wing or a gallery wall: the Metropolitan Museum of Art, the Museum of Modern Art, the New York Public Library—then there was no other source of income to prop up the increasingly unfashionable institution.[50] Among other problems, Audubon Terrace found itself in less and less "respectable" surroundings: Washington Heights was flooded with Latin American migrants after World War II, first Puerto Ricans and Cubans, then Dominicans and Mexicans. By the time the 1980s rolled around, the neighborhood boasted New York City's largest concentration of immigrants; and drugs, coke, then crack, more and more came to people's minds when the neighborhood was mentioned. Even as early as the mid-1960s, when Harvey Flad worked at the AGS, there was "a 'chop shop' for stolen cars driven over from New Jersey" next to his apartment not two blocks from the AGS, and Flad writes that "Audubon Terrace had begun to seem somewhat 'out of place.'"[51]

The Society still had some life in it, however. David Lowenthal was a research associate from 1956 to 1972 doing interesting work in what would become, thanks largely to his initiative, environmental perception; to say nothing of writing his truly seminal "Geography, Experience, and Imagination: Towards a Geographical Epistemology."[52] For many of those years William Warntz was a research associate too

(1956–1966), publishing *Toward a Geography of Price* (1959) and *Macrogeography and Income Fronts* (1965) before moving on to Harvard to become a professor of theoretical geography and regional planning.[53] At the end of his tenure he even had Michael Woldenberg working with him,[54] and Woldenberg even published his "Energy Flow and Spatial Order: Mixed Hexagonal Hierarchies of Central Places" in the *Geographical Review*.[55] It was all serious geography, of the very highest order, and, with the surprising late blooming of J. K. Wright's reputation following the 1966 publication of *Human Nature in Geography: Fourteen Papers, 1925–1965*, the AGS almost seemed . . . substantial.[56]

It was an illusion. The glitter was like the iridescent surface of a soap bubble and inside it was hollow. Even before Lowenthal left in 1972, it had been decided to shut the thing down. It was in the early 1970s that University of Wisconsin–Milwaukee geographer Barbara Borowiecki approached Bill Roselle, director of the University's library, "with the news that the AGS was looking for a home for the Library and Map Collection and raised the question, 'Why not UWM?'"[57] You can imagine the eyes rolling in the boardroom back on Audubon Terrace at the thought of giving away their prized collection—amassed by the best of New York society—to a public university at the far end of the Great Lakes' Rust Belt. But a 1974 expansion of the UWM library provided "30,000 square feet of modern climate-controlled library space, able to be transformed into a stand-alone, secured unit, and—of no small importance—also able to support the great concentrated weight of a large map collection."[58] Apparently this "bird-in-the-hand advantage" outweighed competing but vague or promissory offers from more prestigious institutions. In 1976 the AGS agreed to give the collections to Milwaukee, and they were shipped there in 1978. Two years later the AGS sold the Audubon Terrace building to the recently chartered Boricua College (a private, nonprofit geared to the needs of Puerto Ricans and other Latinos),[59] and "moved its staff, by then much reduced in size, to more modest quarters, first at 15 West 39th Street in 1981, then at 156 Fifth Avenue in 1984, and then at 120 Wall Street in 1997."[60] Since then it's moved what staff remains to 32 Court Street (Suite 201), in Brooklyn where three or four rooms, with nothing much in them, squat behind a wooden door at the end of a linoleum hallway. The plaque with the association's name on it is bent where someone's tried to jimmy it from the door (Figure 2.2).[61]

In any case it's not much staff. The downsizing caused the restructuring of "the publication of the Society's journals, which moved from

FIGURE 2.2. The door of the AGS's offices at 32 Court Street (Suite 201), Brooklyn, New York. Photo courtesy of John Krygier; reprinted with permission.

in-house production in New York to a leaner and geographically dispersed operation.... By 1996 that reorganization process was complete, with the appointment of the first *Geographical Review* editor based outside the Society's New York City headquarters."[62] Well, they're not really headquarters anymore either, really just a mail drop and receptionist, but none of it matters. The significance of the *Review* has faded to the point that it's hardly read by academics. They certainly don't publish in it, not if they're interested in promotion and tenure: The *Review* has an impact factor of 0.6 and ranks 50th, out of 65, on the Institute of Scientific Insititutes's (ISI) citation ranking of journals in geography.

The fact is that for the past 40 years the AGS hasn't really been a scholarly organization. When you get down to it, it's a subscription list today, a contracted editorial and production services provider, a mail drop, a website, and a handful of officers scattered around the country. Indeed the signature image of the AGS is no longer its splendid, columned building but the "Fliers' and Explorers' Globe," which is easier to tote around and store. This is a globe that John Huston Finley, an associate editor of the *New York Times* in the 1920s and 1930s, had used for collecting autographs from the great explorers and fliers to which his position gave him ready access. It's like an autograph book, only it's an autograph globe. On several occasions, "he carried the globe down to the wharf in order to meet the home-coming explorer and get his signature and his route while the news was still fluid."[63] From 1925 to 1934, Finley was president of the AGS and in 1929 he gave the Society his globe. As with the signed ball of an exuberant baseball fan, the autographs gave the globe a numinous luster, but it's worth noting that the globe gets only a half-page photo and a third of a page of text in Wright's 488-page history of the Society's first 100 years because today it's the *sixth* link on the homepage of the AGS's website. There you can download a 14-page PowerPoint with extensive coverage of recent globe signing events, including photos of the March 2004 New York Globe Signing Event, the May 2004 Atlanta Globe Signing Event, and information about the February 2008 University of Delaware Globe Signing Event (when Captain Lawson W. Brigham "pen[ned] his name on the globe"). Toward the front there's a slide that points out that "the AGS has developed successful corporate sponsorships in support of the Globe," and at the end there's another that reminds viewers "interested in corporate sponsorships of the Globe, or in a Globe display event for your corporation or museum event," to get in touch with the AGS. It's like the current leadership of the AGS is channeling that old Audubon Terrace spirit, but with none of its resources.[64] Instead of chiseling names in stone, they're selling opportunities to witness the signing of a globe. Given this level of desperation, it can come as little surprise to learn how eagerly the AGS embraced an invitation from Fort Leavenworth to do a little something for the Army. The tragedy is that, other than the globe, the only resource the AGS had left was to offer itself up as a conduit from the Army to geographers in U.S. universities. On the other hand, the AGS has a long history here.

A Little Something for the Army

We'll be dealing with that invitation and the AGS creation of the Bowman Expeditions in Chapter 8, but in what we've already seen of Herlihy's work in the Rincón of the Sierra Juárez, which we'll also detail further in Chapter 8, we have a pretty clear picture what the little something was that the Army hoped to get out of its association with the AGS. Probably not coincidentally, it's awfully close to what Bowman himself had hoped to get out of his Hispanic American "millionth map" project. Remember that? Bowman wanted nothing less than all available scientific data of a geographical nature: topography, climate, population, *everything* that affected human distribution, activity, and economic welfare; soil, vegetation, and land classification maps at various scales; and history and projections for the future. This is exactly what the Army wants, except that the Army wants it particularly for those *indigenous areas*, those *centers of "organized outlawry,"* and those *large urban slums* that the Foreign Military Studies Office believes are going to be the incubators for the military threats of the future. This is to say, the Army wants it for those areas where the Foreign Military Studies Office thinks insurgencies are going to arise, that is, where the "small wars" of the future are going to be fought—what the Army these days prefers to call "Military Operations Other Than War." Which is to say that the little something the Army's looking for from the AGS is the intelligence it's going to need to combat insurgencies, the intelligence it's going to need to engage in counterinsurgency warfare.

To be straightforward, the Army wants to turn the AGS into an intelligence agency; perhaps not into a *secret* intelligence agency—it's "all" open source, after all—but an intelligence agency nonetheless.

The Army's idea about the kinds of intelligence it needs to fight counterinsurgency wars has a long history, but as *doctrine*—that is, as the Army's authoritative body of statements on the conduct of operations using a common vocabulary, as the Army's official word on action—these ideas about intelligence root themselves in the late 1920s, in Marine Capt. Merritt A. "Red Mike" Edson's pursuit of Augusto Sandino through the Caribbean "jungles" of Nicaragua. It's to these origins that we turn next.

3

"Red Mike" Edson's U.S. Marine Patrols Up Nicaragua's Río Coco in 1928–1929 and the Development of the *Small Wars Manual*

Thinking back six or seven years, U.S. Marine Major Merritt A. "Red Mike" Edson recalled a night along Nicaragua's Río Coco in July of 1928:

> It had rained steadily since our arrival at Bocay, culminating in an especially heavy downpour during the afternoon and night of the twenty-fifth. The night was as dark and black as any I have ever seen. In spite of posting extra sentries over the boat moorings on the beach and around the buildings occupied by the Indians, our largest bateau and a smaller pitpan broke loose and were swept away in the rapidly rising river and several of the local Indians slipped into the bush between the sentries. It was impossible to hold the sentries accountable for this. The river rose so quickly and the current increased to such an extent that, if there had been no special post over the boats, our entire flotilla would have carried away. There was less excuse for the loss of the boatmen, but under the conditions which prevailed, a cordon of sentries would have been required to have stopped them.[1]

Edson was only a captain when he held command along the Río Coco—and a newly minted captain at that—but he seemed undaunted by his demanding assignment and the ever-worsening conditions. Certainly these included an unwillingness on the part of the Miskito Indians—in whose territory he was operating—to manage the boats only they had the skills to handle, but the rain and the rapidly rising river were an even deeper concern:

> The river rose steadily, and by the evening of the twenty-eighth had passed the flood mark reached three days before. Our rate of march was slowed to only a half mile an hour. Late that afternoon, we made camp on a small, flat space on the north bank of the river, just large enough for our bivouac, some six feet above the water. In accordance with customary procedure, the advance guard continued on up the river for at least a mile above the camp site before returning to bivouac, and shore patrols were pushed out into the brush on both banks to look for any trails which might exist. The boat ferrying the south bank patrol across the river "capsized [here Edson begins quoting from his own report], resulting in the loss of two (2) Browning automatic rifles and one (1) Springfield with belts and ammunition. Unable to retrieve due to deep and swift water. Lost 104 lbs. corned beef (No. 1 tins) and other rations." Outside of a thorough wetting, there was no injury to personnel and all boat equipment was salvaged.
>
> By nine o'clock that night, the water began to overflow the bivouac area. Between then and daybreak, we were forced to move three more times, each move higher up the bank and onto steeper and more unsuitable terrain. In a little over twelve hours, the river had risen about twenty feet. It was the highest flood in over thirty years, the water actually covering some of the permanent buildings on the south bank of the river at Waspuc. Not only was the water so high that, to travel at all, we would be forced to cut our way through the trees and brush, but the raging torrent was so swift and so full of every sort of debris that any movement of the patrol was too dangerous to be considered. It was necessary, however, to find some place suitable for a camp.[2]

They managed, of course, to find such a place and a few months later they also managed to successfully complete their mission. As Edson recalled it, this had been "to proceed up the Coco River, to drive the bandits from the river valley, destroying such groups as we might encounter, [and] to take . . . Poteca."[3]

Bandits? Edson and his marines were enduring these hardships for some *bandits?*

Bandits

When did the Marines become a constabulary force? And if they weren't, what were Marines doing chasing bandits around Nicaragua? In fact, what were they doing in Nicaragua at all? It's interesting: None of these questions appears to have occurred to Edson as he was writing up his Nicaraguan experiences, no, not in all the 22,000 words of his recollections.[4] The closest he came to reflecting on the purpose of what was going to become the template for U.S. counterinsurgency actions was this: "*USS Denver* was steaming north from Colon to Cape Gracias á Dios at the northeastern tip of Nicaragua. Our Christmas holiday in the Canal Zone had been disrupted by orders to proceed immediately to investigate reports of banditry against an American citizen residing in Cape Gracias."[5] Really? Marines disrupt a Christmas holiday to investigate a report of *banditry* against *an* American citizen? Evidently! As Edson continues, "Everyone in the Special Service Squadron was interested in the attempts of the Fifth Marines to combat Sandino's bandit operations in Nueva Segovia."

Sandino's bandit operations? Of course, Edson knew well—and later even grudgingly admitted[6]—that, in fact, Sandino was *General* Augusto C. Sandino, and his bandits were actually *soldiers* in a revolutionary army. Edson knew that the bandit operations were an *insurrection*, a revolution against the government of Nicaragua, the Standard Fruit Company, and the Yankee Marines sent there to prop it up.[7] The problem, despite the Marines' active presence for 15 years in a country smaller than New York state, was not only that the Corps had signally failed in every one of the general missions assigned to it; was not only that every day Sandino's revolution gained sympathy around the world (in 1928, e.g., a division of the Chinese Kuomintang Army renamed itself the Sandino Brigade[8]); but that six months earlier Sandino's forces had launched a dramatic assault on a Marine garrison at Ocotal, and although rebuffed, the attack had drawn intense scrutiny at home to an occupation already opposed by powerful Congressional forces and increasingly unpopular among U.S. allies and the American public.

That is, the Marines badly needed a way to demonize Sandino's army while, at the same time, minimizing the threat his revolution posed to Nicaraguan stability and so to American interests. As David Whisnant puts it:

> The problem of how to fight Sandino, it turned out, was inseparable from what to call him, and *that* problem was as much cultural as political. After toying briefly with the designation "guerrillas" in early 1928, the marines decided (as General Lejeune told the Senate Foreign Relations Committee, "for lack of a better term") to fall back on the culturally loaded (and revealing) designation that was already common in the popular press: "bandit."[9]

Sandino was not amused: "Tell your people," Sandino said to an American reporter, "that there may be bandits in Nicaragua, but they are not necessarily Nicaraguans."[10]

Not only did the pejorative sting, but in the complicated landscape of Nicaragua, it had real utility. The term had long since been in widespread use to describe the liberal and especially conservative gangs that wrecked havoc in the Nicaraguan countryside—particularly in Nueva Segovia—during the 18 months between the Espino Negro Accords of early 1927 and the "free and fair," U.S.-supervised elections of late 1928.[11] Conservatives mobilized more than 20 armed gangs *in the Segovias alone*, the gang members referred to by almost everyone as *bandits*: "a group of bandits attacked the [liberal] civic guard that we have here," "around Consuelo there are various bandits," "a large group of armed bandits," "those who were with Anastacio raped me, the same with my little niece. Julián Sevilla along with Anastacio gave orders to have me shot. The bandits were Ismael Gómez . . . ," and so on.[12] And, although most of these may have been sponsored by conservative interests, liberals had gangs too. Indeed, this simple distinction is inadequate, for each party had factions. Conservative followers of U.S.-backed General Emiliano Chamorro sent bandits against their conservative "allies" as happily as they did against their liberal enemies. As Nicaraguans who had accepted the Espino Negro Accords, and so the continued presence of U.S. Marines, Sandino—until then a leading liberal general, but one who rejected the treaty as a violation of Nicaraguan sovereignty—opposed them all, conservatives and liberals alike, as Nicaraguan traitors. *They* might have been bandits; *his* were soldiers in a nationalist, anti-imperialist crusade, the Army Defending the National Sovereignty of Nicaragua.[13]

What the label of *bandit* did was elide these shifting, cross-cutting realities, was obscure a social geography not only complicated and very deeply rooted, but responsible for the presence of the Marines in the first place. While it's true that the U.S. Navy had patrolled the Middle American, and especially Caribbean, coasts throughout the 19th century—landing the Marines whenever it felt like it (e.g., in the Dominican Republic, in 1800; Puerto Rico, 1824; Mexico, 1847; Nicaragua, 1852, 1853, 1854; Panama, 1856; Colombia, 1860; Mexico, 1870; Panama, 1885; Nicaragua, 1894; Colombia, 1895; Cuba, Puerto Rico, Nicaragua, 1898; and so on[14])—U.S. attention had sharpened with the growth of interest in a trans-isthmian canal (especially after the California Gold Rush of 1848–1849); with the Spanish–American War of 1898; and with the Banana Wars that followed in that war's wake, say 1898–1934.[15]

Many things motivated this attention—and they varied over time—but in Nicaragua the driving force was the trans-isthmian canal. As we might infer from the deep involvement of the AGS in the Nicaragua Canal Company,[16] the initial concern was *protecting* a Nicaraguan route—a bill supporting a Nicaraguan canal was introduced in the U.S. Senate as late as 1902—but once it had decided to back the Panamanian route, U.S. government interest shifted to protecting the Panamanian canal and *preventing* the construction of one in Nicaragua. Ensuring all this required a U.S. hegemony in the region, one the more easily maintained if "law and order" prevailed. In Nicaragua the most serious threat to "law and order" was the unresolved differences between the liberals and conservatives, differences reaching back to the 16th-century rivalry between liberal León (founded by low-ranking Spanish foot soldiers) and conservative Granada (founded by upper-class Spaniards).[17] It was to intervene in this ongoing local war—essentially as State Department cops—that the Marines landed in 1910 and then occupied the country from 1912 to 1933, though U.S. geostrategic interests remained focused on the Panama Canal that had finally opened in 1914. Nicaraguan politics, much less local banditry, could never be more than a pretext for landing troops in the service of maintaining a geopolitical climate that supported the canal.

Eastern Nicaragua

Almost none of this history unfolded in the theater where Edson found himself, for Nicaragua not only had a convoluted history but a complicated geography. Most of this history unfolded along the Pacific, in the western lowlands with their lakes and big cities (León, Managua, Granada). Separating these from the Caribbean lowlands were cooler, wetter central highlands (Matagalpa, Nueva Segovia). These highlands rise to constitute a nearly impassable barrier to movement between the east and the west, except in the very south along the Río San Juan. The San Juan marks the country's southern border with Costa Rica, drains Lake Nicaragua, and was most of the route of the proposed canal. North of the San Juan and east of the central highlands stretched a broad coastal plain largely covered in pine savanna. Large rivers tumbled out of the rainforest-covered slopes of the central highlands, cutting across the savanna to empty into lagoons rimmed by mangroves and the Caribbean Sea beyond. In comparison to the agricultural fields and coffee plantations of the west, the region has historically been politically and economically peripheral to the rest of Nicaragua. Its inhabitants speak an array of indigenous languages and Creole English, the latter signaling its strong historical ties to British rule in the Caribbean.

On this Mosquito Shore, as the region was often known, the British practiced the form of indirect rule they would later perfect in their African colonies and India. In 1631, they crowned a Miskito king and came to rule the region as the superintendency (1749–1786). With the demise of Spain's American empire, the British strengthened their rule, rechristening the Mosquito Shore as a protectorate (1837–1860). Finally, they promoted it to a quasi-independent Mosquito Reserve (1860-1894).[18] These efforts at governing shaped the understanding of the region's ethnic identities. Originally the term *Miskito* had been more geographical than ethnic, referring broadly to the amalgam of escaped slaves and indigenous peoples living along the coast; but under British rule it gave shape to distinct ethnic groups, separating Afro-Caribbean Creoles from indigenous Miskito, Sumu, and Rama populations.

These subdivisions were boosted in 1847 when missionaries from the Moravian Church arrived in Bluefields. Given their work among slaves on Jamaica, British administrators saw the Moravians as unusually qualified for educating the English-speaking Miskito king. Alternately known as the *Unitas Fratrum* or more simply "the brethren," the Moravians proved themselves suited to the task, and by 1900 had succeeded in extending their reach over most of the Mosquito Shore, in no small part due to their willingness to preach in the Miskito language.[19] This further helped to distinguish the region from Nicaragua's Catholic, Spanish-speaking, and mestizo west, endowing the cartographic boundary drawn by the British with a social and cultural reality.

Not that this stopped the Nicaraguan elite—who knew the region largely from maps—from claiming it as their own. Aided by U.S. diplomatic efforts to push the British out of the Caribbean, Nicaraguan troops led the so-called "reincorporation" of the Caribbean lowlands in 1894, asserting sovereign control over an area they saw as rich in natural resources. Nicaraguan officials' vision for colonization of the region relied on increasing foreign investment in the hope that it would bring economic order to the frontier.[20] They looked to the United States to supply that investment. This was a pragmatic decision. Speculators from the United States were already combing the former Mosquito Reserve in search of gold and mahogany.[21] The presence of Americans "on the spot" so shaped U.S. officials' sense of the region's significance that they played a central role in negotiating Britain's withdrawal. Nicaraguan reincorporation of the region pushed this development, and by 1920 both Standard and United Fruit Companies had established presences there, with concessions to grow bananas and log on vast areas of land. Both companies used the exclusive right to log and grow bananas in the concession areas to build enclaves in which they exercised near total economic control over everything from labor to housing, save owning the land outright. The biggest of these enclave economies was in the far northeast, around the logging camp of Puerto Cabezas, on a bluff overlooking the Caribbean. Capital from Standard Fruit expanded the mill's capacity and fueled the camp's growth. To feed the mill, Standard began construction of a railroad line inland and a pier for loading logs onto ships bound for New Orleans.

But as the camp grew into a town and the concession into an enclave, the very lawlessness of the frontier, initially conducive to U.S. investment, became a liability. In Puerto Cabezas, Standard relied heavily on segregation to discipline a workforce composed of West Indian migrants and indigenous Miskitos. The former organized themselves in alliance with Marcus Garvey's United Negro Improvement Association (UNIA), itself a product of labor dynamics in banana concessions in Costa Rica and Honduras.[22] Nicaraguan officials conflated this alliance with fears of Communist "bolshevism" spread by migrant laborers, even sending an official to investigate unrest in 1925.[23] And then there were the Chinese merchants and German missionaries, to say nothing of the continued influx of speculators from the United States and Canada. Sandino himself arrived in Puerto Cabezas in 1926 on his way back from a stint in the oil fields of Mexico.

In 1928, all this was in play in the pine-studded savannah and tropical forests stretching north and inland from Puerto Cabezas to the Río Coco: liberals and conservatives, rumors of bandit gangs, Sandino's insurrection, Moravians, Miskito Indians, U.S. geopolitical ambitions and private commercial interests with, just over the horizon, the 1929 crash and the onset of the Great Depression.

Indians

It was into this roiling complexity that Edson launched his patrols up the Río Coco. In the closing months of 1927, steaming north toward Cape Gracias á Dios, Edson recalled that

> a Christian Brothers' map hung on the bulkhead in the Exec's office and daily we plotted all reported movements of both Marines and bandits and

tried to foretell what would happen next. Two things on this map impressed me: the Coco River, its source only a few miles from the Gulf of Fonseca [on the Pacific coast], flowing eastward through all of Segovia and emptying into the Caribbean at Cape Gracias; and Santa Cruz, in the heart of Sandino's territory, labeled as the limit for boat transportation on that river. It seemed to me that here was a feasible supply route for the outlaw forces or, in case our Marines made things too hot for them, an excellent way for them to escape from the country. We had no information about the Coco River or the country through which it flowed. Why not send a reconnaissance patrol upstream from Cape Gracias?[24]

In the end, Edson was to lead three patrols up the Coco, or *Wanks* as the river was also called: first, the Wanks Reconnaissance, an initial probe of three weeks; then a more sustained effort, known as the Wanks Patrol (Figure 3.1), to garrison the lower river; and finally, a still more serious expedition, known as the Coco Patrol, into the heart of Sandino territory.

It's true that his commanding officer, Major Harold Utley, had given Edson explicit orders "to establish cordial relations with the inhabitants," but the fact is that Edson warmed to the Miskito from the beginning.[25] This is evident not only in the respect he pays their boat-handling abilities—that made mockeries of his men's attempts[26]—but even more in his adoption of their ways. One of these was sleeping in leaf lean-tos, whose construction he described in a letter to his five-year-old son, Austin:

> You are probably asking if these Indians live in tents, aren't you? They do not use tents, but lean-tos when stopping for only a few days. These lean-tos are made like this. Four bamboo poles are cut and tied together at the top [picture of triangular structure]. Then on the side towards the wind where the rain will come, they put up a roof or a wall of leaves something like this [illustration in letter]. The floor is the sand, and their beds are made of big green banana leaves laid on the sand. Then they put down a blanket made from the bark of a tree, and that is their sleeping plan. It is not a bad bed either, for your Daddy has slept several nights just like that.[27]

The use of these lean-tos undoubtedly allowed Edson to lighten the loads his patrols had to carry—"Shelter halves were left behind. Experience had taught us that leaf lean-tos could be built in about ten or fifteen minutes and that they were actually more waterproof and more comfortable than shelter tents"[28]—but something else was at play here as well, something that showed up when he first shot the rapids and delighted in "the tenseness of the boatmen as they crouch in the

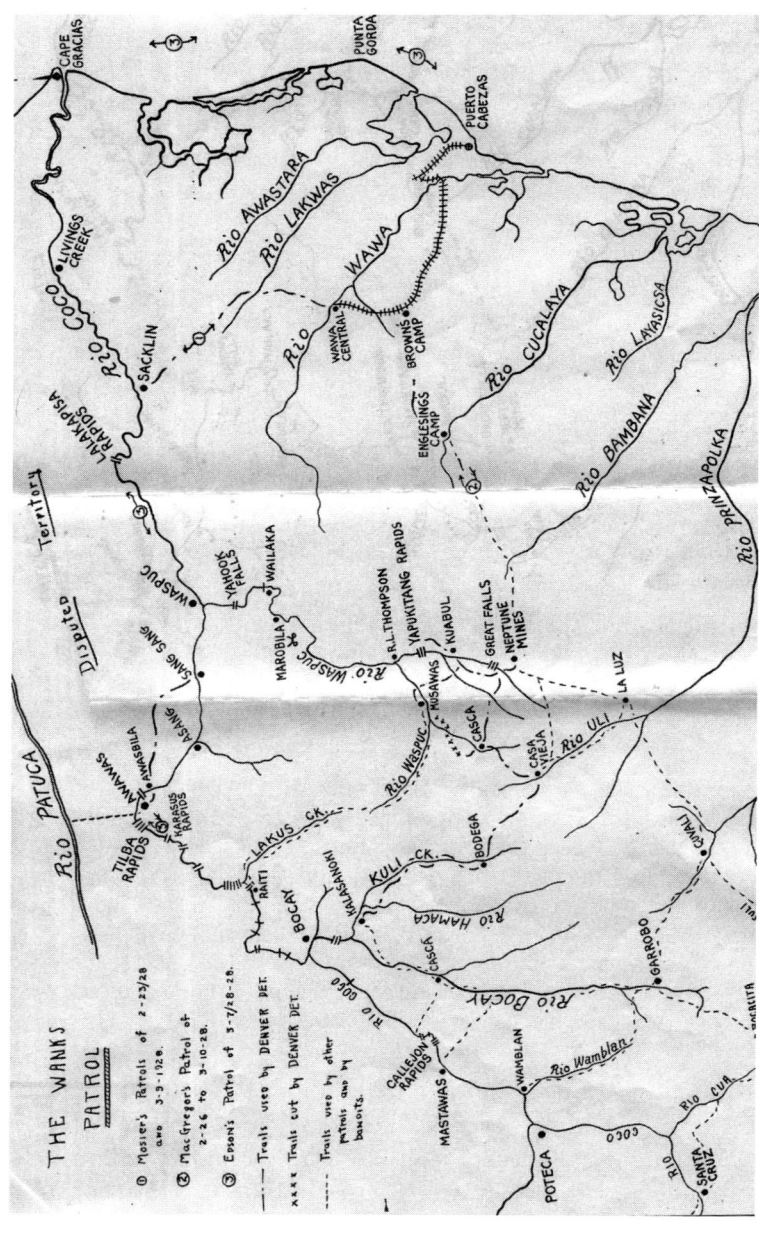

FIGURE 3.1. Edson's own map of the Wanks Patrol. Source: U.S. National Archives (downloaded from *www.sandinorebellion.com/EastCoast/ATL-1927/Edson-WanksPatrol.jpg*).

bow to fend the boat away from rocks, crags and hidden debris; the shouts and cries of the Indians as they wager their skill, quickness of eye and deftness with pole and paddle against the river"[29]; something that showed up in his pride in speaking Miskito (however poorly he spoke it); and something that showed up with his satisfaction in undoubted friendships.

Whatever this something was, Edson was never unaware of the military advantages it brought—"During all of my activities along the river I found these Indians to be entirely loyal and dependable to me" he wrote—and he was aware that his intuitive interest always had a military motivation: "By learning enough native words to make my wants known to them; by showing an interest in them and their mode of living; and by always treating them fairly, I believe that I succeeded in that part of my mission 'to establish cordial relations with the inhabitants.' "[30] It's important to observe that this cordiality, never based on a commonality of interest, waxed and waned with subtle but continuous transformations in the immediate situation. For example, while Edson launched the Coco Patrol with 43 Marines and 46 Miskito boatmen, these Miskito engaged themselves for a trip up the Coco, not for an engagement with Sandino. In the first place they had no beef with Sandino, who had treated the Miskito well during his time on the Coco the year before; but perhaps more importantly, they had a healthy respect for his abilities, something they lacked for those of the Americans.[31] So when it became apparent that Edson was actually leading them into Sandino territory, the Miskito balked, and the desertions began that Edson described in our opening quotation, continuing desertions, which Edson also believed involved sabotage with the loss of boats, supplies, and weapons: "They were certainly very much frightened. They expected an ambush momentarily. They doubted our ability to protect ourselves and they thought that we would be surrounded, annihilated and all hands massacred."[32] Yet once it became apparent that the Americans could handle themselves,[33] these fears dissipated and most of the Miskito accompanied Edson all the way to what had been rumored to be Sandino's headquarters at Poteca: "There were no causalities among the Indian boatmen. They were under fire twice and conducted themselves exceptionally well."[34]

The fact is that without the Miskito pitpans and the Miskito boatmen, the Marines could scarcely have advanced up the Coco, much less harassed Sandino's troops to the extent that the Marines ended up in control of the entire Coco valley. As David Brooks put it:

> When Edson left the Eastern Area a little over a year later, in March 1929, the Marines firmly controlled the Río Coco and had established forward bases along its banks deep in the interior of the country, including one at the site of the rebel chief's former headquarters. They accomplished this, in large part, by winning the cooperation of the local Indians. How Edson's leadership helped secure the Río Coco is more than a story of innovative military tactics. It is an example of how an enterprising Marine Officer, working with limited manpower in a geographically forbidding area, pioneered the development of a form of warfare which combined political sensitivity, personal diplomacy and even a kind of militarized anthropology in order to achieve Washington's counterinsurgency goals.[35]

Perhaps more important than the fact that it was geographically forbidding was the fact that it was, as Edson pointed out, geographically *unknown*. We've emphasized the contributions of Miskito technology—boats and shelter—and Miskito labor—especially their handling of the boats—but no less significant was Miskito knowledge of Coco valley geography. This was especially critical with regard to trails—without knowing them, Edson was confined to moving up and down waterways (as his opponents slipped across from watershed to watershed)—but Edson desperately needed to know the names of things, where villages stood, commercial operations, and, given the way the river capsized his boats and drowned supplies, the possibilities for living off the land.

Little of this information appeared on the maps Edson had available to him. Once, when armies more or less lived off the land, maps of this kind were common. The Duke of Alba was among the first to commission them for his 1573 march from Milan to Brussels: The maps are not elegant but they show everything an army on the march needed to know.[36] In France this sort of knowledge was the province of military lodging masters, of *maréchaux des logis*. From 1590 on, Jacques Fougeu drew hundreds of maps covering France and parts of its neighbors. Essentially sketch maps—"scrappy-looking little maps," David Buisseret calls them—they recorded, village by village, the numbers of hearths officers could count on finding along any line of march.[37] Over time military foraging has decreased—"By and large the story of logistics is concerned with the gradual emancipation of armies from the need to depend on local supplies"[38]—but it never disappeared; and here in Nicaragua again and again it proved necessary.[39]

Yet foraging was but a single example of the way in which insurgency, and so counterinsurgency, claimed the entire landscape as a battlefield and turned all of society into combatants; and this meant that maps at least as penetrating as those of Fougeu would once again be required. As Edson moved up river, he would have liked to have known the feelings toward Sandino of the people he would encounter, whether they were liberal or conservative (which colored how he read the information they gave him), what kinds of Indians they were (Edson also operated in Sumu territory), what their numbers were, how they felt toward the Marines, the conditions of trails and where they went, and a million other things beyond where the river was wide and calm enough for an amphibious airplane landing. That is, Edson would have liked to have known precisely the sorts of things Bowman wanted his Hispanic American "millionth map" project to provide, the sorts of things Herlihy's maps stripped from the Sierra Juárez and transmitted to the Foreign Military Studies Office, the sorts of things Letitia Long, Director of the National Geospatial-Intelligence Agency, calls . . . *human geography*.[40]

Instead, Edson didn't even have a decent map, much less the kind of detailed, large-scale maps he needed. Preparing for the Wanks Reconnaissance, he noted that

> neither the Ham or Christian Brothers maps were suitable for our use. The Bragman's Bluff Lumber Company had in its possession a map of northeastern Nicaragua compiled by the Moravian Missions, blueprints of which were made and used on this patrol. So far as the Coco River valley from Cape Gracias to Awawas was concerned, this map was found to be extremely accurate although serious discrepancies were discovered when later operations took us away from the lower river.[41]

Both the Christian Brothers' and Ham maps were small-scale maps of Nicaragua, but the Bragman's Bluff maps, which were used in the management of timber and banana concessions (Bragman's was a subsidiary of Standard Fruit Company), were of a sufficiently large scale to be useful.[42] To put this into perspective, a map of Nicaragua on a typical atlas page is in the vicinity of 1:3,000,000. The large "Official Map of Nicaragua," published for the government in six sheets, showed Nicaragua at a scale of 1:370,000. Bragman's maps could range all the way up to 1:10,000. A detail of the map Edson refers to is illustrated in Figure 3.2 (at this size the whole map would just be a gray blob). The Coco runs along the top from right to left. Note the rapids marked on the Coco and Amak Rivers, the "old village site" in the upper left, and Kuli Creek Trail at left.[43]

Undoubtedly some of these details were on the map as the Marines received it, but Edson and others were continually adding to it: "Am

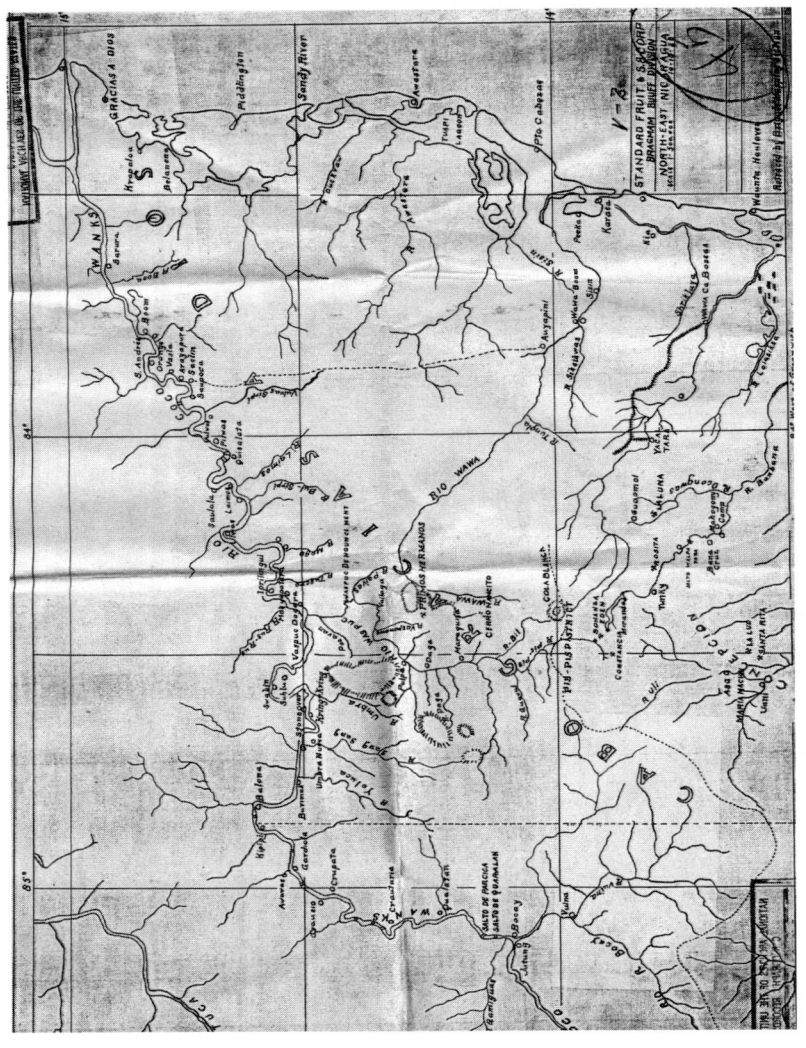

FIGURE 3.2. The Bragman's Bluff map Edson used. Source: U.S. National Archives (downloaded from *www.sandinorebellion.com/EastCoast/ATL-1927/StandardFruitCoMap-Stitch.jpg*).

sending herewith rough sketch showing trails this area as explained to or found by me," reads a typical field message to Utley. In order to achieve the flexibility in the field that Sandino's troops had, Edson needed every scrap of intelligence he could gather—"It was considered imperative that information be gained of existing trails and of conditions," he writes[44]—and where he lacked it, to take advantage of that of the Indians: "If Sumu boy you have knows trail to MUSAWAS you and two squads move there by trail," he orders a squad leader.[45] Without 19-year-old Arthur Kittle, of mixed U.S. and Miskito parentage and fluent in Miskito, Spanish, and English—"reared along the river and in the woods and without doubt the best man on the trail that I saw in Nicaragua"[46]—Edson would have been lost.[47] Of course, Edson took full advantage of all captured intelligence: "From the bandit correspondence, it was evident that Sandino was entirely ignorant of our occupation of Bocay and of the advance of our patrol. . . . He believed there was no danger of attack from the east and that his down river flank was absolutely secure."[48]

Despite the debilitating physical hardships—the incessant rain, wet clothes, ulcerating feet, lack of supplies (all lost to the river)—but thanks to his ever-sharpening knowledge of the terrain, of his opponent, and to Utley's complete support, Edson was able to establish and garrison a camp at Poteca in the very heart of Sandino country, a camp that gave the Marines control of all the navigable Coco and effective control of northeast Nicaragua. Edson led numerous patrols from this camp—which he commanded for seven months—to Santa Cruz, to the headwaters of the Cua River, even joint patrols with Marine units based in western Nicaragua.[49] But by then the "free and fair" elections of November 1928 were well behind them, and with this success came an increase in domestic opposition to the continuing occupation that finally mandated a reduction in force: Edson was ordered home.

The *Small Wars Manual*

It didn't take long for Sandino to rally, and by 1931 he had a base at Bocay and was raiding up and down the Coco all the way to Cape Gracias, but by then Edson had become an instructor in a course for new lieutenants at the U.S. Marine Corps' Basic School in Philadelphia, and was in correspondence with Utley, who had assumed responsibility for the small wars curriculum at the Marine Corps Base at Quantico, Virginia. In 1930, Utley visited Edson in Philadelphia, where he was devoting 130 hours of his course to tactics using a syllabus based on his experiences along the Coco (while bemoaning the fact that he was allowed only four hours for "Bush Warfare"). Utley had been reduced to "the time honored custom of 'bulling' on the subject" since the only "text"—Samuel Harrington's "The Strategy and Tactics of Small Wars" of 1921–1922—had fallen out of use.[50] Utley's efforts gave historical and geographical shape and form to the term *small wars*, which he defined as "those operations in which a trained regular force is opposed by an irregular and comparatively untrained enemy."[51] Carrying on, he wrote that "all of our campaigns against the Indians, the Boxer Rebellion, the Philippine Insurrection, the Punitive Expedition into Mexico in 1916, and the numerous campaigns and expeditions of the Marine Corps—except when serving with the Army in the Mexican and World Wars—fall under this category."

His net cast wide, Utley began a sustained effort to codify the lessons Marines had learned from these small wars. After hearing Edson deliver an impressive lecture on the Coco Patrol to staff and students at the Basic School, Utley asked him to repeat the lecture at Quantico. Edson's firsthand experience helped tie together Utley's readings of C. E. Callwell's 1896, *Small Wars: Their Principles and Practice*,[52] William Heneker's 1907 *Bush Warfare*,[53] E. H. Ellis's "Bush Brigades,"[54] Harrington's "Strategy and Tactics," and letters solicited about their experiences from other Marines. From those works, Utley cobbled together a document that he shared with Edson and some others. Edson responded with several long letters, and Utley refined all of this into "The Tactics and Techniques of Small Wars," which he published in the *Marine Corps Gazette* in 1931 and 1933.[55]

At the same time this was going on the Marines were in the middle of an identity crisis. Unless it could articulate an unambiguous role for itself, the Corps faced extinction. This pressure had been one of the reasons the Corps had been willing to act, as in Nicaragua, as State Department cops; but that position also chafed, and for political as well as strategic reasons, Marine Corps leadership increasingly argued that its real mission was *landing operations* (such as those that would occupy it in World War II). In 1933 the Marine Corps' Basic School suspended classes so that staff could devote their entire energy to the preparation of the Corps' first landing operations manual, which it published, seven months later, as *Tentative Manual for Landing Operations, 1934*. The school's commandant—and small wars advocate—General James C. Breckinridge, saw this as an opportunity to create the first small wars manual as well, and in 1935 the Marines published *Small Wars Operations*, all 900 pages of it.[56]

While leaning heavily on its predecessors, there was also plenty that was new. Its 32 chapters showed the influences of Callwell and Harrington, but also Utley and Edson, and most chapters included appendices with case studies and examples.[57] These were drawn from Callwell and the U.S. Army Indian campaigns, but also from the French in Indochina, the Napoleonic Peninsular War, and various other Marine operations.[58] As to the degree of detail, one appendix includes menus for field patrols. A 1938 revision, which cut 28 pages of text, set the stage for a major overhaul in 1939. Although four men had been appointed to the board assigned this project, Edson, by now Lieutenant Colonel Edson, did most of the work, since he had both the least demanding schedule and by far the most small wars experience. (It didn't hurt that his Coco Patrol reports had come out in 1936 and 1937.) The result of his work was the *Small Wars Manual* of 1940.[59]

At less than half the length of *Small Wars Operations*—its 32 chapters reduced to 15—this streamlined *Small Wars Manual* shed much of the detail, most of the examples, and all the appendices.[60] "Evidence indicates that most of the 1940 version, if not all of it, was a product of Edson's personal efforts," Edson lifting "a number of passages verbatim from his previous writings, both published and private."[61] Not to put too fine a point on it, the *Small Wars Manual* is the Coco Patrol writ large, with additional sections covering aspects Edson hadn't dealt with in Nicaragua: the supervision of elections, the establishment of a native constabulary, and force withdrawal. Otherwise the book often reads like a reprisal of his Río Coco reports: "The shelter-half can be dispensed with if materials are available in the field for the construction of lean-to shelters" is soon followed by "When necessary materials are available lean-tos can be constructed almost as quickly as shelter tents can be erected. They are roomier than the shelter tent and afford better protection during heavy rains. The lean-to consists of two forked uprights, a cross pole, and a rough framework which is thatched with large leaves, such as manaca, banana, palm, etc."[62] Sound familiar?

The book is larded with Edson's notions of the importance—of the *necessity*—of political sensitivity, personal diplomacy, and his brand of militarized anthropology; and while this advice is most prominent in opening pages devoted to psychology and intelligence, it shows up in *every* section. And so in the "Procurement of Animals" section of the "Mounted Detachments" chapter we find him advising: "It is well to utilize native dealers in procuring animals. The average cost per head thus will possibly be higher than it would be otherwise but this is offset by a saving in time and energy. Moreover, the native dealer will know where the desired types are to be found."

Yet Edson's concerns are most telling when he generalizes most broadly: "In major warfare, hatred of the enemy is developed among troops to arouse courage. In small wars, tolerance, sympathy, and kindness should be the keynote of our relationship with the mass of the population." It's in statements like these that the distance he's moved from Callwell, from Heneker, from Ellis and Harrington, is easiest to see, but this too is apparent in the smallest details. This is Heneker: "With regard to the native troops and their rations, they can in most cases subsist on the country"; whereas this is Edson: "If the patrol is living off the country, equitable treatment given to the natives attached to the patrol will usually be more than repaid by their foraging ability and by assistance in preparing palatable dishes out of the foodstuffs which are indigenous to the locality." Perhaps needless to say, but Heneker's "white troops" were not expected to eat foodstuffs indigenous to the locality.

Although a detailed analysis of the *Manual* would be worth carrying out, here we content ourselves with a few further quotations, almost at random:

> Small wars involve a wide range of activities including diplomacy, contacts with the civil population, and warfare of the most difficult kind.
>
> Cordial relationship between our forces and the civilian population is best maintained by engendering the spirit of good will.
>
> If sufficient information of the probable theater of operations has not been furnished, maps, monographs, and other current data concerning the country must be obtained, including information on the following: past and present political situation; economic situation; classes and distribution of the population; psychological nature of the inhabitants; military geography, both general and physical; and the military situation.
>
> Social customs in countries in which small wars operations usually occur differ in many respects from those in the United States. Violation of these customs, and thoughtless disrespect to local inhabitants, tend to create animosity and distrust which makes our presence unwelcome and the task of restoring law and order more difficult.
>
> Many companies have accurate and detailed maps or surveys on file which may be obtained and reproduced to supplement the small-scale maps available to the force.

> The application of the principles of psychology in small wars is quite different from their normal application in major warfare or even in troop leadership. The aim is not to develop a belligerent spirit in our men but rather one of caution and steadiness. Instead of employing force, one strives to accomplish the purpose by diplomacy. A Force Commander who gains his objective in a small war without firing a shot has attained far greater success than one who resorted to the use of arms.

> [In trail cutting] A fairly accurate map, an airplane mosaic or a previous air reconnaissance over the route would be of inestimable value, but probably none of them will be available. Native cutters should be employed, if possible, and they can usually be relied upon to select the best and shortest route.

> The motive in small wars is not material destruction. It is usually a project dealing with the social, economic, and political development of the people. It is of primary importance that the fullest benefit be derived from the psychological aspects of the situation. That implies a serious study of the people, their racial, political, religious, and mental development.

> A knowledge of the character of the people and a command of their language are great assets. Political methods and motives which govern the actions of foreign people and their political parties, incomprehensible at best to the average North American, are practically beyond the understanding of persons who do not speak their language.

And, of course, "At least two pairs of woolen socks; four pairs are recommended, if the patrol is to operate for 2 weeks or longer," and "For the upper rivers, the most suitable boats are those obtained locally from the natives."

Again, whatever else was at work, in the front of Edson's thinking were the military objectives. His weren't mere sentiments but practices that had proven their worth on the Coco again and again, proven their worth, ultimately, by permitting Edson and his men to banish Sandino from eastern Nicaragua. The proof was in the pudding. Some of Edson's success was due to his sheer tactical skill, but tactics can't be practiced in a vacuum. They require intelligence, appropriate tools, and adequate supplies, all things Edson largely depended on the environment for; and which the environment provided, thanks largely to Edson's ability to read the complicated political landscape, his diplomatic handling of the region's inhabitants (non-Nicaraguan managers and employees of commercial operations, Indians, other Nicaraguans), and his anthropological curiosity about the worlds in which he was operating. It was all this that he poured into *Small Wars*.

"The manual was a tour de force," writes Jon Hoffman, "for what amounted to a one-man-board. Regrettably, it was also irrelevant even before it went to the printing press. The Marine Corps was gearing up for a major conventional war and would soon be shunting the *Small Wars Manual* out of its mount-out boxes and into museums." [63] What Marines were reading instead was the *Tentative Manual for Landing Operations, 1934* or its immediate successor, 1938's *Fleet Training Publication 167 (Landing Operations Doctrine)*, and for joint operations, 1941's *Field Manual 31-5 (Landing Operation on Hostile Shores)*. Personal diplomacy, anthropological sensitivity . . . all that was vaporized in the atomic clouds that rose over Hiroshima and Nagasaki.

Small wars, however, were not vaporized in those two clouds. Indeed, they never went away at all. As Max Boot points out, "what is often forgotten is that along with the clash of big armies the 1939–1945 conflict saw plenty of guerilla operations by forces as disparate as the Yugoslav partisans and the French *maquis*—not to mention America's own Office of Strategic Services (OSS), forerunner of the Central Intelligence Agency (CIA)."[64] Following the war, the CIA assumed the roles Marines had formerly played in what these days are called "Military Operations Other Than War," and the Marines Corps continued to use the playbook it had developed in the Pacific. Indeed, the Marines, the Army, and the military as a whole continued to pursue conventional strategies even when they found themselves in what were manifestly small wars, with disastrous consequences: Vietnam, Lebanon, Afghanistan . . . though it's not that they abandoned the small wars idea altogether. More than a few of these small wars involved recruiting ethnic minorities as proxies and using their grievances to fuel insurgencies against national societies in order to advance U.S. objectives by proxy, as in the case of the CIA's "Secret War" in Laos.[65]

Edson's book, though difficult to access—it had been classified as Restricted[66]—had not been forgotten, and as military failure followed military failure it began to acquire a retrospective luster. In 1987, fully 47 years after its publication, the Marines reprinted it; and Edson's doctrine acquired widening influence, even before the 2003 invasion of Iraq and the grisly catastrophe of the Second Gulf War. Increasingly pervading the thinking at the U.S. Army Command and General Staff College at Fort Leavenworth—where the Foreign Military Studies Office is also located and not far from the University of Kansas, where AGS president Jerry Dobson and México Indígena's Peter Herlihy teach—the *Small Wars Manual*'s doctrine spread in many directions, ultimately spawning both the Army's Human Terrain System and the AGS's Bowman Expeditions, as well as taking on new form in 2006 as the *U.S. Army–Marine Corps Counterinsurgency Field Manual*.

We consider how all this comes together in Chapters 7 and 8, but before we can make much sense out of the Army's notion that it could use the AGS's Bowman Expeditions and México Indígena projects to create a prototype of *precisely the type of intelligence Edson lamented the lack of* when he embarked on his Río Coco patrols, we need to understand the origins of indigenous mapping and how it can be used to generate intelligence. These are the subjects of the next three chapters.

4

The Birth of Indigenous Mapping in Canada

If you go back to our Narrative Table of Contents, you'll find that our description of this chapter begins:

> In 1967, Frank Arthur Calder and the Nisga'a Nation Tribal Council initiated an action that led the Canadian Supreme Court to rule for the existence of an *aboriginal title*, one dating to a Royal Proclamation of 1763. This decision prompted the still young liberal federal government of Pierre Elliot Trudeau, eager to recover from an initial "misstep" in Indian affairs, to attempt to extinguish these titles by negotiating treaties with those indigenes who had never signed them; and in 1974, it began supporting work capable of leading to such negotiations.

And certainly that's *one* way to tell this part of our story.

The Nisga'a Land Question

It has the advantage of brevity, where brevity is held to be a virtue; but brevity always exacts a toll—in this case, toll after toll—and as usual it's the indigenous who pay. Let's begin with, "In 1967, Frank Arthur Calder and the Nisga'a Nation Tribal Council initiated an action." This was true enough—it was September 27, 1967 in fact—but put this way, it can't help but imply that something *started* in 1967 that had actually started a long, long time before.[1] To ignore this history is to create—well, to *reproduce*—a silence, for the ceaseless efforts of the Nisga'a to recover their land pretty much went unheard for a solid hundred years, which is to say, to ignore this history is to reproduce a silence *created by people who refused to listen.*

Britain, and so Canada, traced its claims to the land of what is now British Columbia back to Cabot's 1497 claim for the British Crown of all of North America, as well as to an undoubtedly illusory visit by Drake in 1579—you'll recall seeing both their names inscribed on the frieze of the AGS's Audubon Terrace headquarters—and if Nisga'a concerns weren't aroused then, or even by the 1793 appearance of George Vancouver in Observatory Inlet, they certainly were by the Hudson's Bay Company's construction of Fort Simpson in 1831 at the mouth of the Nass River; after all, *people of the Nass* is all "Nisga'a" means.[2] With the establishment of the colony of British Columbia in 1858 Nisga'a concern turned to alarm—an alarm that intensified with British Columbia's unilateral denial of aboriginal title in 1870 and with the passage in 1876 of Canada's long-detested Indian Act, which unraveled traditional governance systems and imposed constraints on aboriginal lives. The subsequent restriction of aboriginal fishing rights in 1877 and the construction of a cannery on the Nass propelled the Nisga'a in 1881 to send a protest delegation to the provincial capital in Victoria,[3] and in 1885 three Nisga'a chiefs traveled to Ottawa for a chat with Prime Minister John A. Macdonald over what was already being called the Nisga'a land question. Gaining nothing there, chiefs Arthur Gurney, John Wesley, and Charles Barton took a steamer two years later to Victoria, where British Columbia Premier William Smithe barred the Nisga'a and Tsimshian delegation from entering the legislature, letting them know that, "When the white man first came among you, you were little better than wild beasts of the field."[4]

What can you do?

In 1890 the Nisga'a established the Nisga'a Land Committee,[5] and nine years after that they joined with other northern coastal nations— the Nisga'a were hardly the only ones suffering these indignities—to form the Indian Rights Association. In 1913 the Nisga'a Land Committee unsuccessfully petitioned the British Privy Council for redress, though in 1924 they were finally allotted 76 square kilometers of reserve land. This needs to be put into perspective: They were allotted 76 of the roughly 25,000 square kilometers they'd heretofore regarded as theirs.[6] It only got worse: In 1927 Canada made it "illegal for any person to accept payment from an Aboriginal person for the pursuit of land claims," and that pretty much put a stopper on serious efforts at litigation.[7]

Even to have said "Frank Arthur Calder" understates this history, for *Job* Calder was one of those who addressed a Royal Commission of Inquiry sent to the Nass to equivocate with the Nisga'a in 1887. And in 1890 it was Job's son, *Arthur* Calder who helped found the Nisga's Land Committee.[8] Then the story goes that having

> lost his infant son while travelling on the Nass River, an elderly woman had a vision and announced that Arthur Calder's sister-in-law was pregnant with the reincarnated child. This baby was given to Arthur Calder and his wife to be raised as their own, as Frank Calder. The chief took his son at age four to his discussions with other chiefs about land claims, grooming him for the future. Frank Calder recalled, "Arthur Calder picked me up and stood me on the table and said to the gathering, 'And this is the boy I am going to send to the white area, and I'm going to make him speak like a white man, I'm going to make him walk like a white man, I'm going to make him eat like a white man . . . and he is the one that's going to bring this case to the highest court in the land.'"[9]

This was the Frank Calder, then, who in 1955 helped transform the Nisga'a Land Committee into the Nisga'a Tribal Council; who was elected its first president; and who, Parliament, in 1951, having rescinded the provision of the Indian Act that prohibited land claims, hired Thomas Berger in 1967 to represent the Nisga'a before the Supreme Court of British Columbia.

Berger and the Nisga'a lost there, they lost on appeal, and in February 1973 they lost before the Supreme Court of Canada. But there, in *Calder v. the Attorney-General of British Columbia*,[10] they did manage to get six Supreme Court justices to affirm the existence of an aboriginal title, one dating to a Royal Proclamation of 1763.[11] In doing so the Nisga'a not only set the stage for the signing of a treaty granting them 2,000 square kilometers outright, participation in the use of thousands more, and limited self-government (albeit 27 years later), but set in motion a transformation in Canadian policy that, to date, has resulted in the settlement of 15 comprehensive land claims, and as we type this, finds literally hundreds of other claims, if mostly specific, under negotiation.[12]

Calder Effects

Berger has said that the court's decision was handed down at "a politically auspicious moment." In 1969 Pierre Trudeau's liberals had embarrassed themselves with the publication of a white paper—*Statement of the Government of Canada on Indian Policy*[13]—that was intended to bring aboriginal status to an end by a thorough-going assimilation that included the denial of land claims. Not only was the paper universally repudiated by aboriginal organizations, but the decision in *Calder* was to deprive it of any constitutional foundation. Only a couple of months before the Court handed down its decision, Trudeau's liberals had been returned to power, but as a minority government:

> To remain in office, the Liberals would have to depend on the goodwill of the opposition parties, and so the question of aboriginal title was catapulted into the political arena. Both the Conservatives and New Democrats insisted that the federal government must recognize its obligations to settle aboriginal claims. The all-party Standing Committee on Indian and Northern Affairs passed a motion that approved the principle that a settlement of aboriginal claims should be made in regions where treaties had not already extinguished aboriginal title, and on August 8, 1973, Minister of Indian Affairs Jean Chrétien announced that the federal government intended to settle the claims.[14]

In 1974 the government began offering financial support for work leading to such negotiations, and in 1976 the Nisga'a Tribal Council opened negotiations with the federal government.

The Nisga'a may have been unusual in having so well-documented a record of advocacy for their territorial claims—and so unusually little need to document them further—but they were anything but unusual in struggling for control of their traditional territories. For the Dene this went back to the 1899–1900 signing of Treaty 8 (driven by the Yukon Gold Rush) when Chief Drygeese demanded a written promise that Canada was not preempting Dene occupation and use of the Mackenzie River Valley. Then, even after having signed Treaty 11 in 1921–1922 (driven by the discovery of oil), the Dene found themselves forced to send a letter reiterating their position in 1928. And so it would go through the subsequent discoveries of uranium, of yet more oil, of yet more gold. In the early 1970s the Dene found themselves threatened yet again, this time by a pipeline that would carry oil down the Mackenzie from the Prudhoe field.[15] At the same time, the James Bay Cree, on the eastern side of James Bay in northern Québec, found themselves threatened by a hydroelectric project.[16] Farther north the Inuit too had begun feeling anxious "about the rising number of nonrenewable resource development projects taking place in their hunting, fishing, and trapping territories and the resulting environmental damage caused by these projects' often poorly regulated activities."[17] In 1970 the Inuit formed a regional organization, the Committee on Original Peoples Entitlement, and then in 1971 a national Inuit organization, Inuit Tapirisat of Canada (now the Inuit Tapirit Kanatami).[18]

Prior to *Calder* and to Chrétien's announcement of the state's determination to settle aboriginal claims, neither the Dene, the Cree, nor the Inuit would have had much chance of successfully contesting these developments, though both the Dene and Cree were in court attempting to do so, and the Inuit in the early stages of negotiations. But with *Calder* and Chrétien behind them, the Dene and Cree were able to stop both proposed developments in their tracks, and the Inuit were ultimately able to achieve the self-governing territory of Nunavut. The novel ability of titular uncertainty around aboriginal claims to imperil highly capitalized developments suddenly made the settlement of aboriginal title claims an absolute imperative wholly independent of any ethical or legal considerations.[19]

If both legal and development concerns weighed on the Canadian government, it had, at the same time, an active interest in strengthening its sovereignty claims in the far north, especially over the islands and navigable waterways of the Canadian Arctic Archipelago. The concern dated to the close of the 19th century with the heavy presence of U.S. and Scottish whalers; to the Arctic explorations of the Norwegian Otto Sverdrup and the Americans Frederick Cook and Robert Peary (all capable of planting a flag); and to the habit of the Greenlandic Inuit—Danes in the eyes of the Canadian government—to hunt across Smith Sound to Ellesmere Island (a potential use claim). Canadian concerns sharpened in the 1920s with the discovery of oil in the western Arctic and reached a kind of fever pitch in the 1950s, thanks to the Soviets' Cold-War posturing and the growing presence of U.S. military forces on the DEW (Distant Early Warning) Line established to provide early warning of Soviet invasion.[20] In 1953 Canada went so far to solidify its claims as to forcibly plant a number of Inuit families from farther south on Ellesmere and Cornwallis Islands in the high Arctic.[21]

Sixty years later the Canadian government would apologize for this, but the point is that by the early 1970s the Canadian government had a host of powerful motives for its sudden eagerness to settle title claims it had ignored—or actively resisted—for a hundred years. Yet the Dene, Cree, and Inuit lacked the Nisga'a's long record of claims-making, and so each was obliged to document its claim with what have since become known as traditional use and occupancy studies. The Cree and Inuit received federal support for their studies in 1973, the Dene in 1974. Their studies took different forms, and each in its own way became a model for the tsunami of claims studies that were to follow. But without question the most influential was the landmark publication in 1976 of the three-volume *Inuit Land Use and Occupancy Project.*[22]

The Inuit Land Use and Occupancy Project

As we said, the Inuit Tapirisat of Canada was formed in 1971. Its members immediately entered into discussions about land claims with independent legal experts, with federal government officials, and with social scientists. Among the latter was a young anthropologist, Milton Freeman. During 1972 Freeman hammered out a proposal for "a comprehensive and verifiable record of Inuit land use and occupancy in the Northwest Territories of Canada."[23] In early 1973 the Inuit Tapirisat transmitted this proposal to the Department of Indian Affairs and Northern Development, where, given *Calder*, Chrétien approved it without delay. The funds started flowing in September of that year and soon there were 150 researchers in the field (120 of them Inuit), hard at work on the nine regional land use reports at the project's heart.[24]

It was a lot to do. In the 34 communities that Freeman studied, he was determined that every Inuit resident who wished to participate would be given the opportunity to contribute his or her own record of land use, and 1,600 chose to do so (the response ranging from 80 to 95%, depending on the community). The essential goal was determining the area used by the Inuit—land, sea, and sea-ice—though the nature of the use was a consideration too. The results, published in 209 maps and numerous tables in the report's first and third volumes (*Land Use and Occupancy* and *Land Use Atlas*), were untainted by even a single reference to previously published work, and this instantly distinguished the project from aboriginal land use studies such as the 1968 Federal Field Committee for Development Planning in Alaska report that had formed the basis for the Alaska Native Claims Settlement Act.[25] As Peter Usher, responsible for the western Arctic, put it:

> We were no longer mapping the "territories" of Aboriginal people based on the cumulative observations of others of where they were (as one would for mapping the ranges of wildlife species), but instead, mapping the Aboriginal peoples' own recollections of their own activities. The second innovation was to record peoples' own perceptions of the history and significance of their traditional lands. This was done through mapping geographical knowledge and oral history as exemplified by place names and ecological

knowledge, all of which were used as supplementary indicators of use and occupancy.[26]

Usher and the other researchers called these "map biographies," and in them, as Hugh Brody would put it, "hunters, trappers, fishermen, and berry pickers mapped out all the land they had ever used in their lifetimes, encircling hunting areas species by species, marking gathering locations and camping sites—everything their life on the land had entailed that could be marked on a map."[27] The method drew on an evolving tradition of applied anthropology, especially participant observation; and on a precedent history of the use of sketch maps in ethnographic research in anthropology and geography that dated to Franz Boas.[28] Later, Julian Steward used maps to illustrate his claim that cultures co-evolved with their environments, an approach that he called "cultural ecology."[29] Steward's maps of "culture areas," charting the contours of indigenous societies, laid the intellectual groundwork for the *Handbook of South American Indians* he edited between 1940 and 1947. Joseph Sonnenfeld later expanded this approach to focus on subsistence practices in his work among Alaska's Barrow Eskimo in 1956.[30] The tradition had been enriched during the 1960s by the mental maps movement in geography and planning[31]; and in anthropology by endeavors such as Evon Vogt's Harvard Chiapas Project, with its commitment to mapping and aerial photography, and Harold Conklin's work in the Philippines, which would result in the publication of *The Ethnographic Atlas of Ifugao*.[32]

But Usher's map biographies—pioneered in his 1971 dissertation[33]—were unlike anything that had existed before, and they inaugurated a new trajectory in the history of mapmaking. Hugh Brody, who had been part of the Inuit Land Use and Occupancy Project team, described collecting a map biography in a subsequent study with the Beaver Indians in northeast British Columbia:

> Joseph had his own agenda and his own explanations to give. He stood by the table, looked at the map, and located himself by identifying the streams and trails that he used. Periodically he returned to the map as a subject in its own right, intrigued by the pattern of contours, symbols, and colors and perhaps also by his recognition of the work that had brought us to his home. . . . As Joseph Patsah told his story, he searched the map until he found a particular bend in a river. . . . He sought the exact place where, in September or October, it is easy to catch fat rainbow trout. He traced the length of a trail that each year he and others used to travel from a spring beaver-hunting camp to the trading post at Hudson's Hope. He satisfied

himself that we understood the exact distance between the Reserve and the best of his winter cabins.... In the course of talking... Joseph had shown his hunting, trapping, and fishing areas on the map; had marked, with colored felt pens, all the places he had lived during a long life.[34]

It was in this and other equally intensive ways that the maps in the first and third volumes of the *Inuit Land Use and Occupancy Project* were made (Figure 4.1), and today variations of this process are in widespread use around the world.

In light of cartography's self-construction as a value-free transcription of the environment, doubt about the scientificity of these map biographies was almost reflexive. "Anticipation of possible challenges to the Indians' maps is defensive and may seem unnecessary," Brody would write in 1981, "But to refuse to anticipate criticism amounts to a more general rejection of social-scientific concerns," particularly the claim that "research done as part of a political process can actually be conducive to the most reliable results." From our perspective, this is key. "The [Beaver] Indians of British Columbia made maps, explained their system, gave detailed information about their economy, and took us into the bush with them," Brody wrote,

> because they believe that knowledge of their system will result in an understanding of their needs, and that this will in turn help establish and protect their interests.... The Indians' maps, like their explanations of them, are clear representations of their use of the land. The clarity comes from a wish to have others see and understand. There may be oversimplifications—lines and circles on 1:250,000 topographic sheets can scarcely do justice to the intricacies of which they are a distant overview. But they represent a reality and have an integrity that social science can rarely achieve.[35]

The maps' accuracy was attested to by appealing to hunting peoples' well-established preoccupation with the accuracy of information about the habits and location of their quarry; and by internal consistencies across numerous dimensions among maps produced independently by large numbers of individuals. Even more convincing was the fit of separate communities' aggregated maps, both with each other and with the terrain, for the individual map biographies, as Freeman recalled, were aggregated

> to create a composite map for each contemporary (and a number of now-abandoned) community's land use history. These composite maps—indicating the full areal extent of past and then-present hunting, fishing, and trapping practiced by community households over three blocks of social

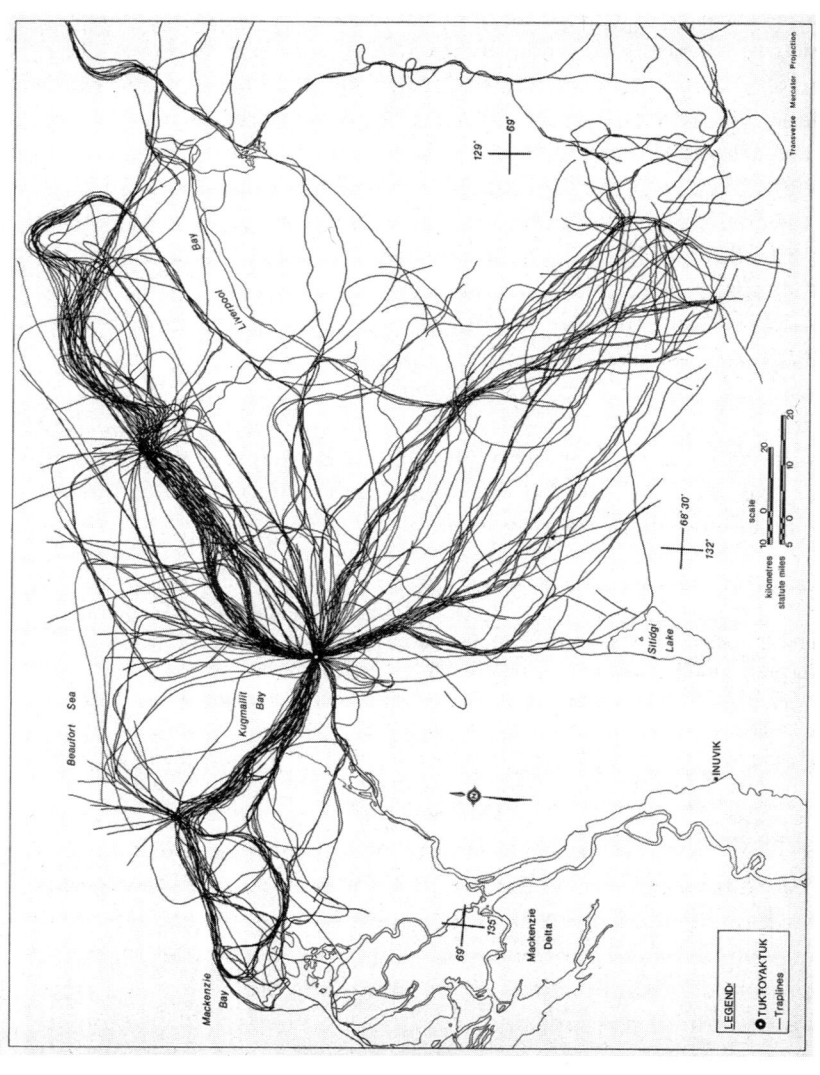

FIGURE 4.1. An example of a map from the *Inuit Land Use and Occupancy Project*. Source: Milton Freeman, ed., *Inuit Land Use and Occupancy Project Report: Vol. One: Land Use and Occupancy*, Supply and Services, Ottawa, ON, Canada, 1976a, p. 231 (Map 53).

time (as distinct from calendar time)—were scrutinized in open community verification meetings where any errors in representing use-areas were identified, discussed, and corrected.[36]

The maps *were* scientific, and if not in the vein of geodesy, geography, and psychology that professional cartographers fantasized about, then in that of ethnography, and indeed the practice has been called a kind of ethnocartography.[37]

But there was more to the *Inuit Land Use and Occupancy Project*, much more. In addition to the first volume's land use reports (with their 39 place-name maps), there was Hugh Brody's "attempt to show some part of what occupancy really means" in his "Land Occupancy: Inuit Perceptions," 50-some pages, along with 17 illustrative maps, of extracts from transcriptions of the commentaries Inuit had offered while making their map biographies. Lamenting the loss of richness resulting from the translation into English, Brody highlighted Inuit attention to the importance of ancestral land use, to intergenerational conflicts, to the linkage between Inuit identity and land occupancy, and to the variety of dangers in the Inuit's current situation. The second volume, *Supporting Studies*, does contain some previously published work in the 50 pages of its "Cultural Considerations" section, though often revised, edited, or excerpted; but the volume opens with 100 pages of background and technical considerations, continues with 65 pages on the prehistoric and historic evidence for Inuit use and occupancy, and concludes with a 45-page photo essay, "Inuit and the Land." In addition to its 145 land use maps, the third volume, *Land Use Atlas*, contains a valuable methodological introduction and eight regional summary maps as a sort of conclusion.

Acknowledging the scope of the project and "recognizing, *inter alia*, the important symbolic as well as instrumental value the report would provide to Inuit society," future prime minister Jean Chrétien increased the funding to ensure that "the project report would appear in an appropriately useful (e.g., full-color maps to increase readability) and respectful manner (e.g., engaging the services of a professional book designer and using high-quality paper and an external printing house)."[38] These enhanced production values showed off the high quality of the work and contributed substantially to the wide influence exerted by the *Inuit Land Use and Occupancy Project*. But it is the land use maps, representing "the distillation of an immense quantity of highly detailed and precisely mapped information about the use of the Arctic land within each Inuit respondent's own adult lifetime" that ultimately made the difference; that, and maybe even more importantly the fact that "the information that has gone into these maps has not come from a committee of cartographers, ecologists, anthropologists, and geographers; rather it is the compilation of thousands of real and personal land uses, the partnerships of individual Inuit people with the land's resources."[39]

These maps went on to play a key role in the negotiations that enabled the Inuit to assert an aboriginal title to the 2,000,000 square kilometers of Canada now known as Nunavut. In settling the claims the Inuit surrendered their aboriginal title for financial compensation; for exclusive ownership rights over a large part of Nunavut; and for decision-making power in the management and royalties from the resource exploitation of all of Nunavut.[40] Because the *Inuit Land Use and Occupancy Project* maps were too small-scale for the negotiations (and lacked any indication of *intensity* of use[41]), and subsequently published maps were at once too rich with information and too large-scale,[42] in 1985 the Tungavik Federation of Nunavut began the Nunavut Atlas Project, publishing the *Nunavut Atlas* in 1992.[43]

This substantial volume is, in its way, as significant as *Inuit Land Use and Occupancy Project*, capturing, as it does, archeological sites, campsites, subsistence and commercial fishing sites, outpost camps, major travel routes, intensity of Inuit land use, a host of wildlife information, and the Nunavut Settlement Boundaries in six fold-out maps of Owned Lands, 27 Community maps, and 118 Land Use and Wildlife Maps (heavily annotated). As in the *Inuit Land Use and Occupancy Project*, field workers interviewed hunters and elders in their homes, asking each to describe their land use directly on the maps, which were then, in consultation with hamlet councils and hunters' and trappers' associations, aggregated into the published maps. The result is an extraordinary portrait of Inuit land use in Nunavut, and it provided the basis for the detailed negotiations that transformed the agreement-in-principle of 1991 into the new territory of Nunavut in 1999.

The James Bay Cree and Dene Studies

At the same time that the Inuit were involved with their land use and occupancy project, the Cree were carrying out the Fort George Resource Use and Subsistence Economy Study (beginning in 1973) (illustrated in Figure 4.2), and the Dene were undertaking their own study of land use and occupancy (beginning in 1974). We noted that the three studies took different forms. On the one hand, this was because, unlike the

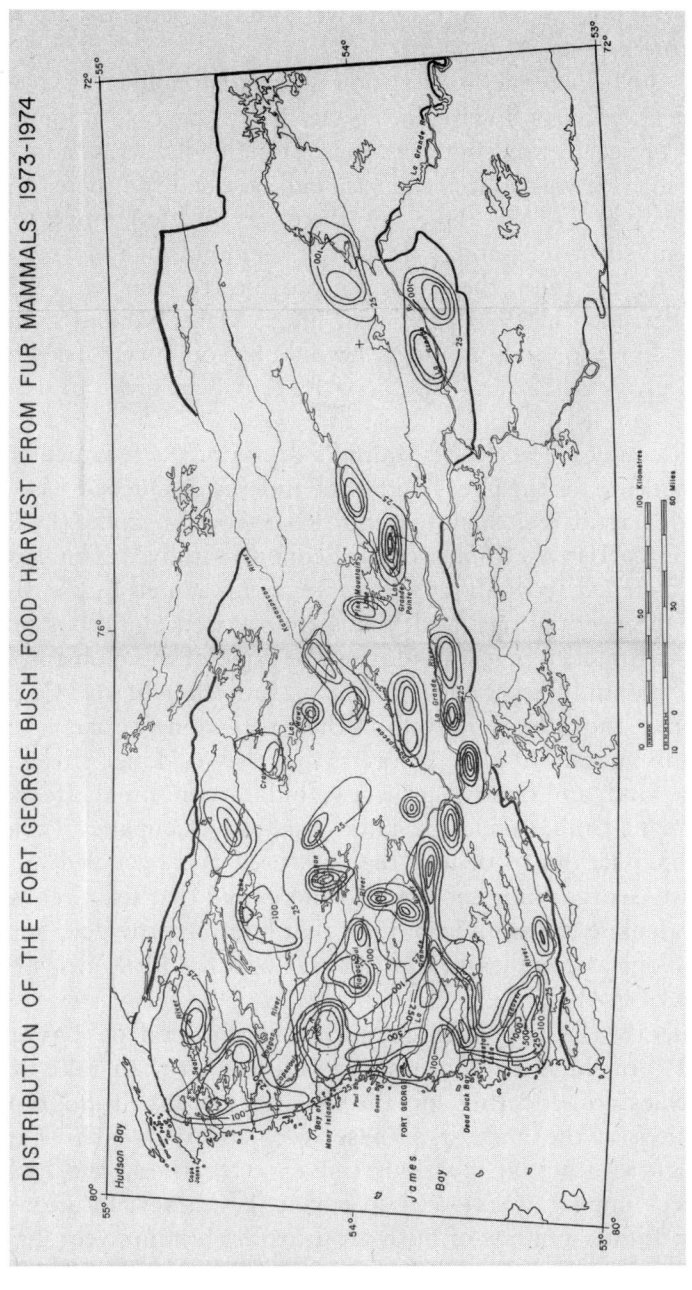

FIGURE 4.2. An example of a map from the Fort George Resource Use and Subsistence Economy Study. Source: Martin Weinstein, *What the Land Provides: An Examination of the Fort George Subsistence Economy and the Possible Consequences on it of the James Bay Hydroelectric Project*, Grand Council of the Crees of Québec, Montréal, QC, Canada, 1976.

Inuit, the Cree and the Dene were in court over their claims; but it was also because the Inuit, Cree, and Dene were unique peoples with very different histories.

In 1970, Québec Premier Robert Bourassa had announced plans for damming five large rivers flowing into James Bay as part of an immense hydroelectric project decked out with generators, spillways, roads, and transmission lines. It was conceived as though no one lived there.[44] The Cree and Inuit who *did* live there saw their way of life coming to an abrupt end—no surprise!—and, led by Billy Diamond for the Cree and Charlie Watt for the Inuit, they took the province to court in 1972 to stop the project. Late in 1973 Provincial Judge Albert Malouf granted an injunction doing so, and, although it would be soon overturned by a higher court, given *Calder* and Chrétien, the province agreed to negotiate.[45]

Within a couple of weeks of Malouf's decision, the technical support team, working for the Crees under the Indians of Québec Association, directed Martin Weinstein to begin what would become the Fort George Resource Use and Subsistence Economy Study.[46] The design of the study changed throughout its course, as mandated by the shifting needs of the negotiations, and Weinstein released numerous progress reports and interim maps to address the topic of the moment. But throughout, the study responded to the presumption of the Québec government that the Cree were well on their way along an irreversible transition from hunters to wage laborers, that they no longer lived primarily off the land, and that the project would have minimal effects on the environment. Unlike the Inuit Land Use and Occupancy Project, therefore, which set out to delimit the *area* used and occupied by the Inuit, the Fort George study set out to demonstrate that the Cree were hunters, to calculate the amount of food that hunting provided, and to detail the ecological complexities of the area and the likely disastrous consequences of its flooding.[47]

Consequently the Fort George study concentrated on harvests, surveying 92% of the men 19 years and older about their take of 65 different species and locating, on the usual 1:250,000 topographic sheets, the sites of the harvests. These were coded into three geographic categories—actual locations, relevant traplines, and hydroelectric project impact areas—and so coded the harvest locations of nearly half a million pounds of bush food for the hunting year 1973–1974. The Fort George report, *What the Land Provides*, is filled with maps and tables spelling out the catch per species, the food weight per animal, the food weight per species, the food weight by animal group per season, changes in bush food harvests from 1971–1972 to 1973–1974, and so on; along with 22 maps that plotted hunting lands, traplines, and the distribution of harvests by species (beaver, snow goose, ptarmigan, porcupine, and so on), these summed up in a map of the distribution of the total Fort George bush food harvest for 1973–1974.[48]

The study played a crucial role in the negotiations that led to the signing, in 1975, of the James Bay and Northern Québec Agreement. Since, given the alternatives, the Cree felt that the agreement would improve opportunities for maintaining their culture, society, and economy, they signed it even though they felt it was neither fair nor just. Indeed its implementation has been fraught with difficulties, as indicated by the need for a second treaty three years later—which incorporated the Naskapi along with the Cree and Inuit—and 20 additional accords expanding provisions and affecting details. To pick just one, in 2002 the Crees and Québec signed Complementary Agreement No. 13, "a nation-to-nation Agreement which strengthens the political, economic and social relations between Québec and the Crees, and which is characterized by cooperation, partnership and mutual respect, while remaining based on the respective commitments of the parties under the James Bay and Northern Québec Agreement," and so on.[49] In 1973 Québec, much less Canada, would hardly have acceded to a phrase like "nation-to-nation" for a people whose existence it wholly ignored when it acknowledged it at all.

While all this was going on in Québec, a whole other story was unfolding in the Mackenzie District of the Northwest Territories. In 1968 oil had been discovered in Alaska's Prudhoe Bay setting off the search for ways to move the oil south. In 1970 the Canadian government issued a proposal for a natural gas and/or oil pipeline corridor to run south from the Arctic. Arctic Gas, a consortium of 27 Canadian, U.S. (Exxon, Gulf, Shell, and so on), and other companies, outlined a route from Prudhoe Bay across the northern Yukon to the Mackenzie Delta and so south along the Mackenzie into Alberta. Foothills Pipe Lines sketched a shorter route from the delta to Alberta. Again, these were conceived as though no one lived there, or no one whose life wouldn't be destroyed by one of the largest construction projects in Canadian history.

Goaded by objections, protests, and lawsuits from aboriginal and environmental groups, in 1974 the federal government appointed by-then *Justice* Thomas Berger—the one who'd led the Nisga'a's legal case against British Columbia—to head a commission to evaluate the economic, environmental, and social impacts of the pipelines on the North.

Known as the Mackenzie Valley Pipeline Inquiry, the C$3.5 million project amassed better than 40,000 pages of testimony in some 283 volumes. The Inquiry produced a report, *Northern Frontier, Northern Homeland: The Report of the Mackenzie Valley Pipeline Inquiry*, nearly 500 pages in two volumes, and the best-selling book ever published by the Canadian government; and with the encouragement and support of the Indian Brotherhood of the Northwest Territories, Mel Watkins drew on these inquiry materials to produce a Dene version, *Dene Nation: The Colony Within*.[50]

Berger ran the Inquiry like none had been run before. In *Northern Frontier, Northern Homeland*—whose title reflects his acknowledgment that what was *frontier* to some Canadians was *home* to others— he writes:

> At the formal hearings of the Inquiry in Yellowknife, I heard the evidence of 300 experts on northern conditions, northern environment and northern peoples. But, sitting in a hearing room in Yellowknife, it is easy to forget the real extent of the North. The Mackenzie Valley and the Western Arctic is a vast land where people of four races live, speaking seven different languages. To hear what they had to say, I took the Inquiry to 35 communities—from Sachs Harbor to Fort Smith, from Old Crow to Fort Franklin—to every city and town, village and settlement in the Mackenzie Valley and the Western Arctic. I listened to the evidence of almost one thousand northerners.[51]

Some of these communities were pretty small. On one occasion Patrick Scott recalls: "When the meeting finally wrapped up in Trout Lake it was too dark to fly out of the unlit grass airstrip. So someone took Judge Berger to their home to sleep for the night while most of us moved the benches up against the wall and rolled out our sleeping bags on the floor to spend the night crammed together in the tiny community hall."[52] Scott adds that, "It was not unusual to have makeshift sleeping accommodations. Sometimes we found ourselves spread around the floor of a school gym or church. At other times it would be in the local transient center, usually with four bunk beds to a room."[53]

As was the case with the Cree, one of the most pernicious assumptions made by the pipeline advocates was that the Dene hunting economy was dead and that what the Dene needed were jobs.[54] Berger concluded, however, that the jobs the pipeline offered would be few and short-lived, and far more pertinently that, "income in kind from hunting, fishing and trapping is a far more important element in the northern economy than we had thought."[55] Helping him to this understanding were maps the Dene presented to the Inquiry. Scott recalls that

> during each of the community meetings a fieldworker from the Dene Nation would present a map to the Inquiry that illustrated all of the traplines of people from that particular community. The maps were a maze of black lines that looked like veins linking the entire traditional territory of the community. Each trapper would have been interviewed, sketching out on the map where they travelled on their traplines.
>
> At times a trapper would go up to the map and run their finger along the lines to show Judge Berger the many routes he would travel, setting snares and traps from one year to the next.[56]

Unlike the aggregate maps of hunting areas made for the Inuit or the aggregate maps of harvest locations made for the Cree, the Dene made these trapline maps themselves (Figure 4.3). Phoebe Nahanni recalled the year and a half's negotiations required to secure funding for the research; negotiations drawn out because although "we knew that anthropologists and white researchers had previously attempted to integrate a study of our land-based activities into their theses," the "community leaders and community people expressed their dislike of the invasion of their privacy by outsiders who didn't speak their language." Nahanni went on:

> We knew from our past experiences that government research by white researchers had never improved our lives. Usually white researchers spy on us, the things we do, how we do them, when we do them, and so on. After all these things are written in their jargon, they go away and neither they nor their reports are ever seen again. We have observed this and the Brotherhood resolved to try its best to see that, in future, research involves the Dene from beginning to end.[57]

Nahanni thus anticipated, by better than 30 years, Gregorio Urbano's resolution that when the Zapotec of Yagavila "decide to, we are going to make our own maps however they be, but they're going to be ours and belong to us." The Dene weren't interested in simply translating their lives into a cartographic language of land use and occupancy. Instead they were interested in harnessing mapping's potential for their own purposes, to create a space in which they could maintain a collective way of life.[58]

Given the time constraints, Nahanni and her team of two dozen decided to sample a third of the trappers from five regions 30 years of age or over, using interviewing and mapping techniques as rigorous as those used by Freeman's and Weinstein's teams.[59] They interviewed 26 younger people as well. In addition to making the maps, Nahanni testified at the Inquiry. Scott recalls that

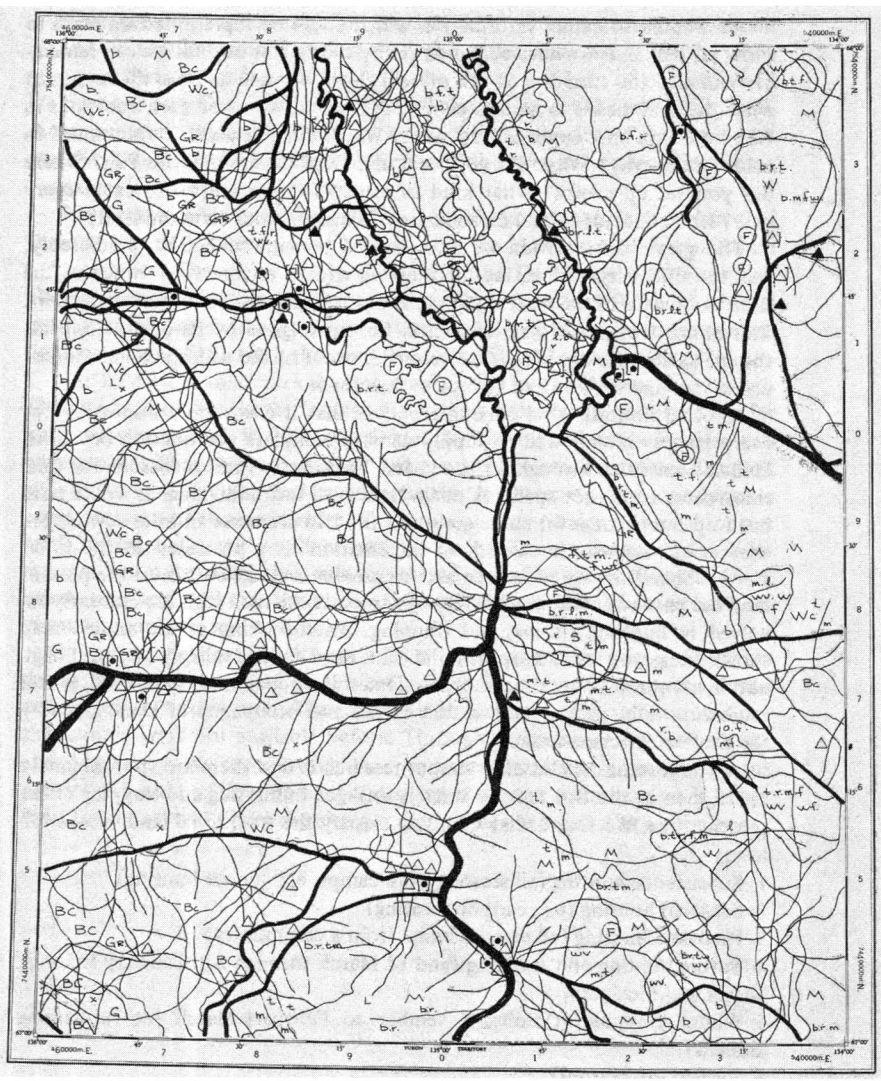

FIGURE 4.3. An example of a map of traplines collected by Phoebe Nahanni's team of Dene researchers. Source: Phoebe Nahanni, "The Mapping Project." In Mel Watkins, ed., *Dene Nation: The Colony Within*, Toronto, ON, Canada, and Buffalo, NY, University of Toronto Press, 1977, pp. 21–27; the map is on p. 25.

in the Dehcho (at the time of the Inquiry, it was called the Mackenzie Liard District), a young soft spoken woman from Fort Simpson named Phoebe Nahanni had worked with the trappers developing the maps. In an earlier visit to the community, over a three day period she was able to talk to seven trappers who were in Trout Lake out of 14 trappers to plot their traplines on the map. Her role at the Inquiry was to review the information on the map for the judge and to "interview" the trappers so they could tell their story.[60]

Nahanni might have been soft-spoken, but the conclusions she—and the Dene—drew from the maps were daunting: "The maps clearly show what the Dene have been saying all along before your legal institutions—that we have been here for hundreds and thousands of years; this is our land, and our life. This is the most graphic demonstration of the truth that we Dene own 450,000 square miles of land."[61]

Given the militancy of Nahanni's remarks, it comes as no surprise to learn that she was instrumental in the very formation of the Indian Brotherhood of the Northwest Territories and was an original signatory to the Dene Declaration. This was a statement of rights passed at the Second Joint General Assembly of the Indian Brotherhood and the Métis Association of the Northwest Territories in July 1975. The Declaration signaled the Dene's awareness that what was at stake in their struggle against the Mackenzie Valley pipeline was not simply a question of land, but a "struggle for the recognition of the Dene Nation by the Government and people of Canada and the peoples and governments of the world."[62] The Dene had come to see themselves as a part of the Fourth World, that world of aboriginal peoples brought to attention by the 1974 publication of George Manuel's *The Fourth World: An Indian Reality*. Manuel championed his ideas in Canada as the head of the National Indian Brotherhood, an alliance that incorporated regional organizations like the Indian Brotherhood of the Northwest Territories with whom the Dene were allied. Months after the Dene issued their declaration, the National Indian Brotherhood expanded their vision at an international gathering of indigenous peoples at Port Alberni, British Colombia that brought into being the World Council of Indigenous Peoples.[63] There was a lot going on, the Inquiry helped to focus some of it, and as Watkins put it, "The Dene have very properly made use of the Berger Inquiry to further their cause." [64]

And it did: The Berger Inquiry concluded with the recommendation that no pipeline of any kind ever be built through the northern Yukon or the Mackenzie Delta; that a pipeline *could* be built along the Mackenzie further south; but in any case only after a 10-year moratorium to deal with the aboriginal land claims and conservation issues. As it turned out, 25 years would pass before the resurrection of a pipeline project, but this time it was with the significant participation of three of the four Dene nations. As we write, however, no plan has yet been approved by all the parties.

Perhaps of even longer-lasting impact, though, was Berger's endlessly reiterated insistence that Canada had to take its native peoples seriously:

> Euro-Canadian society has refused to take native culture seriously. European institutions, values and use of land were seen as the basis of culture. Native institutions, values and language were rejected, ignored or misunderstood and—given the native people's use of land—the Europeans had no difficulty in supposing that native people possessed no real culture at all. Education was perceived as the most effective instrument of cultural change: so, educational systems were introduced that were intended to provide the native people with a useful and meaningful cultural inheritance, since their own ancestors had left them none.
>
> The culture, values and traditions of the native people amount to a great deal more than crafts and carvings. Their respect for the wisdom of the elders, their concept of family responsibilities, their willingness to share, their special relationship with the land—all of these values persist today, although native people have been under almost unremitting pressure to abandon them.[65]

Equally as important was Berger's articulation of Dene land claims. "Native people desire a settlement of native claims," he wrote, but

> they do not want a settlement—in the tradition of the treaties—that will extinguish their rights to the land. They want a settlement that will entrench their rights to the land and that will lay the foundations of native self-determination under the Constitution of Canada.
>
> The native people of the North now insist that the settlement of native claims must be seen as a fundamental re-ordering of their relationship with the rest of us. Their claims must be seen as the means to establishing a social contract based on a clear understanding that they are distinct peoples in history. They insist upon the right to determine their own future, to ensure their place, but not their assimilation, in Canadian life.[66]

Indigenous Mapping

Clearly this all remains very much more a project for the future than an accomplishment, but the world post *Calder* (1973), post James Bay and Northern Québec Agreement (1975), post *Inuit Land Use and Occupancy Project* (1976), and post Mackenzie Valley Pipeline Inquiry (1977) is very different from the one that preceded it; and in it, the role of indigenous mapping was lost on no one, whether it was the areal model of the Inuit, the harvest-oriented model of the Cree, or the unpublished approach of the Dene. Indeed, all three models have extensive progeny, and copycat mapping projects were immediately initiated among the Inuit, Settlers, and Naskapi-Montagnais of Labrador, the Beaver and Cree along the Peace River in northeastern British Columbia, and the Indians of the Yukon, among others.[67] Without question, the 1976 publication of the *Inuit Land Use and Occupancy Project* was a important landmark; but Brody's publication in 1981 of *Maps and Dreams*—which continues to be in print in a bewildering number of editions—was of *crucial* significance, laying out the methods, as it did, in an evocative and persuasive text that transmitted them throughout the world. Mapping became an indispensible component of treaty negotiations throughout Canada, evolving apace with the legal reforms set in motion by the Supreme Court's *Calder* decision. In particular, maps played a prominent role in the Supreme Court's 1997 decision in the *Delgamuukw* case, which acknowledged oral histories as evidence of aboriginal title.[68] More crucially, perhaps, was the fact that the state—Canada—was increasingly involved in this process, if not through sponsoring the mapping work itself, then certainly through its determination of the standards of accuracy and conventions used. Indigenous peoples may have been making the maps, but ultimately it was the Canadian government that set the limits on what could be mapped and which maps counted when it came to consultation with tribes. This made treaty negotiations controversial, to say nothing of the role that maps played in them.[69] But by the time the 1990s dawned, Canada was far from being the only place where land use and occupancy mapping was occurring. Similar projects were underway in Asia, Africa, and Latin America, as well as in North America.

Crucial to later developments in indigenous mapping—and ultimately to Herlihy's work in the Sierra Juárez—were those that unfolded in Edson's eastern Nicaragua, especially at the hands of Bernard Nietschmann. It is to this history that we turn now.

5

Maps, Guns, and Indigenous Peoples

"More indigenous territory has been claimed by maps than by guns. This assertion has its corollary: more indigenous territory can be reclaimed and defended by maps than by guns."[1] Bernard Nietschmann wrote these lines in 1996, nearly two decades after mapping projects in Canada led to a new round of treaty negotiations with First Nations. He added that "whereas maps like guns must be accurate, they have the additional advantages that they are inexpensive, don't require a permit, can be openly carried and used, internationally neutralize the invader's one-sided legalistic claims, and can be duplicated and transmitted electronically which defies all borders, all pretexts, and all occupations."[2]

Nietschmann clearly wasn't talking about maps that *described* the world. He was talking about maps designed to *change* it, maps that would bring to light the innumerable small wars between states and Fourth World nations festering around the planet. The phrasing came directly from Phoebe Nahanni, George Manuel, and others in Canada, but the militancy was new, and it captured Nietschmann's transformation from a rising star in the field of cultural ecology—and a professor in the Geography Department at the University of California at Berkeley—to an outspoken advocate for indigenous rights.

Nietschmann's transformation had everything to do with his involvement with the Miskito communities of eastern Nicaragua. The relationship began with his dissertation in cultural ecology in the late 1960s, a relationship turned inside out in the early 1980s when armed

Miskito groups were drawn into a decade-long war against the Sandinista government that came to power following the 1979 revolution. Nietschmann's outspoken support for the Miskito earned him a reputation as a "gonzo-geographer" who served as a "scribe to the outside world" for the Miskito during the 1980s.[3] His work also made him a pariah with Sandinistas and their international supporters, who equated Nietschmann's support for the Miskito with an endorsement of the Reagan Adminstration's policy of covertly arming anti-Communist proxy forces in places like Nicaragua, Afghanistan, and Angola.

Self-consciously using maps to transform the world rather than describe it meant hitching mapping to processes of social change, but it also required something more. It required seeing how processes of social transformation create a need for maps in the first place, and tracing out how that in turn shapes approaches to mapping. The map ceases being a tool to be wielded more or less effectively with transcendent outcomes, if it ever was, to provide instead a distinct way of seeing the world and aligning the forces needed to conform reality with maps. It's about maps *and* guns. It's about weaponizing the map.

Modern Times in an Isolated Place

Like Martin Weinstein, Peter Usher, Milton Freeman, and others involved with mapping indigenous lands in Canada, Nietschmann was trained in cultural ecology. The field sought to explain the relationship between people and the environment as a dynamic system of feedbacks and adaptation, merging anthropology with geography. Research in the field focused overwhelmingly on subsistence practices, regarding them as the core of a system "by which man adapts to his environment in order to maintain a viable relationship. In so doing, subsistence strategies may be adjusted and environments modified to assure a group's survival and well-being within tolerable limits of the cultural system and ecosystem."[4] This approach framed the dissertation research Nietschmann launched in 1968, in the Miskito community of Tasbapauni, 175 miles south of the Río Coco and 40 years after Red Mike Edson's patrols in the region.

Like the Dene, Inuit, and Cree territories mapped in Canada, Tasbapauni's apparent isolation belied its historical ties to the rest of the world. Tasbapauni sits on a narrow strip of land separating the flat, brackish water of Pearl Lagoon from the wind-whipped waves of the Caribbean. Less than a thousand feet of land separates the two bodies of water at Tasbapauni, making the community a convenient "haul over" used for portaging boats between the sea and the lagoon. Here residents can access a range of marine, fresh water, and terrestrial environments. They use nets and lines to catch shrimp and fish in the lagoon, and cultivate plots of manioc and plantain along its shore. Coconut palms line the beach, beyond which lie reefs and sea grass beds where Tasbapauni residents search for fish and turtles.

Over the course of 13 months in 1968 and 1969, Nietschmann and his wife Judith, herself an anthropologist, collected information on Tasbapauni residents' subsistence activities. Together they compiled detailed information about where and how Tasbapauni residents grew crops, fished, and hunted. They complemented that work with detailed dietary studies, weighing the amount of food produced and consumed by households. These "yield weights" allowed Nietschmann to construct detailed budgets of the number of calories people consumed and where those calories came from, as Martin Weinstein would do five years later with the James Bay Cree. Along the way, Nietschmann made maps of everything with a fervor that bordered on compulsion. In Nietschmann's hands, the maps did more that flesh out the spatial distributions of the data he was collecting on Miskito subsistence. They became data in and of themselves, revealing the world that the Miskito made through their interactions with the environment.

Maps also provided Nietschmann with a means of scaling up his study beyond the community focus that dominated most cultural ecology work at the time. After earning his PhD in 1970, Nietschmann returned to Nicaragua set on retracing the journey of Ephraim Squier, an American archeologist who had sailed the coast of eastern Nicaragua in the mid-19th century to covertly assess the possibilities for a trans-isthmian railroad.[5] Accompanied by a series of Miskito friends and acquaintances, Nietschmann sailed north from Tasbapauni to Cabo Gracias a Dios, turning up the Río Coco into the heart of the area once patrolled by Edson's Marines.

The trip afforded Nietschmann the chance to corroborate his findings in Tasbapauni and experience, firsthand, the navigational and subsistence practices he had been diligently documenting through interviews and surveys.[6] The experience proved crucial to Nietschmann's sense of the Miskito as a distinct ethnic group whose territorial reach and subsistence practices had been strongly modified by their prolonged "period of intimate market contact with outsiders."[7] During that period the Miskito had neither disappeared nor been assimilated. Instead, their control over their ecosystem and subsistence practices had allowed them to "make extensive cultural adaptations to new economic systems and helped to transform and transfer energy and materials from their ecosystem to overseas systems."[8]

Those relationships to the "overseas" world profoundly shaped Miskito culture and history, most notably during the boom years of the enclave economies from 1920 to 1960. During this period, Miskito people came to rely on intermittent work in the logging camps and banana plantations run by the Standard and United Fruit Companies. They complemented their earnings with the sale of sea turtle meat and shells, logs, and rubber, earning the cash needed to pay for the things that subsistence could not provide: medicine, tools, outboard motors, and clothes. But the enclaves brought more than work and a market for subsistence resources. They brought West Indian migrants and Sandino to the region, and then Red Mike Edson and the U.S. Marines.

By the time Nietschmann arrived in Tasbapauni in 1968, the enclave economies that had once brought a semblance of prosperity were collapsing beneath the weight of the environmental destruction they had wrought. Between 1945 and 1964, U.S. logging companies had felled and exported over 600 million board feet from the enclaves.[9] When laid end to end, this was enough to circle the earth four and half times. Much of it had been extracted from the area patrolled by Edson in the 1920s, between the Coco and Wawa rivers, before being milled and exported from Puerto Cabezas to New Orleans and beyond. The exports built houses in Louisiana, Mississippi, and Alabama, to say nothing of company towns such as Puerto Cabezas. They also decimated the region's forests.

During the 1950s, another New Orleans-based company, NIPCO (Nicaraguan Long Leaf Pine Company), assumed control of Standard Fruit's concession and the mill in Puerto Cabezas. NIPCO expanded its operations into an area north of the Río Coco claimed by both Nicaragua and Honduras, taking advantage of the cartographic indeterminacy of the border.[10] The effort proved short-lived. In 1958, the Honduran Army launched a brief border war with Nicaragua that established the Río Coco as the international boundary, an outcome ratified two years later by a ruling from the International Court of Justice at The Hague. NIPCO shuttered its operations in Nicaragua in 1963, bringing the four-decade-long enclave economy centered around Puerto Cabezas to a grinding halt. One small enclave remained, dedicated to the extraction of turpentine and other resin-based chemicals from the stumps of previously cut pine trees; and Wrigley's Chewing Gum Company continued to buy tree gums and process them for export at their plant in Waspam on the Río Coco. But neither could replace the jobs lost with the enclave economy, creating a pervasive sense of decline and isolation in Miskito communities.

Nietschmann's research in eastern Nicaragua coincided with this decline. It also ran headlong into efforts by the Nicaraguan government to fill the vacuum created by the departure of NIPCO and other U.S. companies. At the top of these was the designation of all of eastern Nicaragua as "state lands" owned by the government due to their lack of occupants or other owners. In the 1960s, the Nicaraguan state was synonymous with a single family: the Somozas. The family had come to power with the departure of the U.S. Marines in 1933 and the assassination of Augusto Sandino a year later by the family patriarch, Anastasio Somoza García. Over the course of their dynasty, the Somozas used the state to enrich their family and friends through, among other things, the sale of natural resources. To consolidate the state's claims over the areas formerly occupied by the enclaves, the Somoza regime contracted with international teams of development experts who mapped the region's natural resources. On these new maps of Zelaya, as the department spanning the former Mosquito Shore was called, the Miskito, if they appeared at all, showed up as little more than isolated dots, villages within the green frontier. The maps reinforced nationalist visions of eastern Nicaragua as a vast empty space barely connected to the rest of the country, an isolation deepened by references to the region as the "Atlantic Coast."[11] There was no room for the Miskito in this vision except to conform to the map . . . whose historical, political, and economic contours had been drawn for them.

The Somoza regime drove the point home by enacting bans on the burning of savannah grasses for pasture and logging without a state permit. These effectively criminalized Miskito livelihoods across the savannah and Río Coco, compounding the sense of regional decline. Anthropologist Mary Helm captured the mood with her description of listening to radio reports about the U.S. war in Vietnam while conducting her research in the Miskito town of Asang on the upper Río Coco: It would take a similar event, her informants argued, something on the magnitude of a Vietnam or Sandino, to bring the Americans back and rescue the Miskito from their isolation and deprivation.[12]

Further south in Tasbapauni, similar dynamics of isolation and decline shaped Nietschmann's research. To accommodate the lack of work in the enclaves, Tasbapauni residents had turned to an older form of cash economy: turtling. Green and hawksbill turtles annually gather in large numbers to graze on the immense sea-grass beds extending off the shore of the Miskito communities into the Caribbean. So large were the migrations that more than one observer, Nietschmann included, likened them to the massive movements of bison on the Great Plains of North America. For as long as anyone could remember, coastal Miskito communities had relied on green turtles for food. Their skill as turtlemen had defined the Miskitos' relationship to the outside world. As early as the mid-1600s Miskito turtlemen were selling a portion of their catch to British pirates and buccaneers who sailed along the coast. It was this trade that had brought Miskito communities into contact with the circuits of goods and people around the Atlantic, and their renown as turtlemen and navigators had earned them berths on the ships of slave traders, buccaneers, and whalers.[13] Throughout, turtle meat had been the single most important source of food in coastal Miskito communities.

It all began to change following World War II when the demand for turtle meat and tortoise shell brought buyers from western Nicaragua, Jamaica, and Grand Cayman to the region. The turtle business, as residents of Miskito communities called it, became all the more important following the end of the enclave economy. Nietschmann's research documented this shift, describing a vicious cycle that led Miskito turtlemen to catch more and more turtles for sale, using ever more lethal methods. The increase was driven by market demands that made turtle meat worth more in the hands of international buyers than on the plates of Miskito families. By Nietschmann's assessment, Miskito turtlemen sold some 10,000 turtles on the market between 1969 and 1972, constituting as much as 90% of their total catch.[14] The subsistence system was on the point of collapse as turtle populations declined and residents of towns such as Tasbapauni relied more and more on cash to buy the "tinned" meats and processed foods once sold exclusively in enclave company stores. The entire subsistence system, built quite literally on the backs of sea turtles, was on the brink of collapse. And "when the turtle collapses," Nietschmann prophesized, "the world ends."[15]

In order to save the Miskito, Nietschmann devoted himself to sea turtle conservation. Over the course of the 1970s, he continued to visit Tasbapauni, regularly collecting data on declining turtle populations that he wrote up in his columns for the U.S. magazine *Natural History*. At the time, the Somoza family owned the largest turtle export business on the coast, Tortugas S.A.; and in 1975 Nietschmann's efforts earned him a personal invitation from Anastasio Somoza Debayle to present his ideas for turtle conservation.[16] Two years later Somoza enacted a total ban on turtling, using conservation to shore up his flagging international reputation as a brutal dictator.

The ban had a devastating effect on the Miskito communities reliant on the turtle trade for buying food. Tasbapauni residents angrily accused Nietschmann personally of "humbugging the turtle business" in spite of their widespread acknowledgment of declining populations. Nietschmann was painfully aware of this predicament. Saving the turtle, Nietschmann wrote, "meant hurting a group of people who had become part of my life." [17]

From Habitat to Territory

Nietschmann's focus on conservation followed the logic of cultural ecology. If the Miskito were to be saved from the coming of modern times, as Nietschmann romantically put it, their environment would have to be protected. Historically excluded from the political and economic life of the nation on the basis of their race, Miskito "rights" to lands and resources "belonging" to the state were not about to be recognized by the Somoza regime. Nietschmann saw resource conservation as a novel way of getting around this rejection, a way that pragmatically negotiated the Miskitos' marginal position in Nicaraguan society. But it hinged on a notion of the Miskito as fully self-sufficient and able to sustain themselves through subsistence; and Miskito responses to the turtle ban argued otherwise, invoking the turtle *trade* as central to their existence. As Nietschmann was actually aware, without turtling, there was little economy. Something had to give, but what?

By the end of the 1960s, in the wake of renewed efforts by states to settle frontiers in the name of national development, anthropologists from around Latin America were beginning to grapple with what they saw as a kind of "internal colonialism." Debates over how to include indigenous people within society had previously hinged on whether land belonged to people who worked it *regardless of race*, or if instead the Indians belonged to the land and were thus the property of its owners. Conservative elites favored the latter feudal approach, which had the unlikely advantage of recognizing more indigenous communities, if only to see them as kept labor. Liberal nationalists, in contrast, maintained that indigenous peoples owned the land that they cultivated, like other citizens. Either way, inclusion in national societies depended on property.

But this emphasis on property rights did not work in the forested frontiers targeted for national development—frontiers perceived by nationalist elites as empty areas inhabited by "wild" Indians living off nature alone. In places such as the Amazon Basin, this meant removing and settling native populations in order to clear the space needed for development. Groups that refused were often killed outright. As late as the 1960s, reports abounded of government representatives hunting "wild" Indians, distributing strychnine-laced sugar, and supporting missionary efforts to "pacify" uncivilized tribes.[18]

In 1971 a group of anthropologists studying indigenous societies in Latin America convened a symposium to denounce state racism towards Indians. Though the group consisted primarily of anthropologists from Latin America, it was organized by an Austrian anthropologist, Georg Grünberg. Grünberg secured funding for the event through his home institution, the Department of Ethnology at the University of Bern. Additional support came from the Switzerland-based World Council of Churches. Spurred by liberation theology, the Council had recently expanded its mission from protecting the rights of Christians living in Communist countries to other forms of discrimination, including racism. Grünberg secured an invitation from the government of Barbados to host the symposium at the University of the West Indies.

The location was crucial to the event's success. With the exception of Grünberg, all of the symposium participants were from Latin American countries with sizeable indigenous populations. Many of these countries were governed by military dictatorships suspicious of anthropologists' political leanings. Barbados, by contrast, had no indigenous population, and therefore no state interest in the "indigenous question."[19]

Over the course of five days in January, attendees hammered out an agenda for "Indian liberation" that they set forth in their Declaration of Barbados. The document denounced Latin American states' ongoing involvement in the "internal colonization" of Indian populations through forced settlement and/or worse. It also called on missionaries and anthropologists to end their paternalist roles as protectors of Indians and throw their weight behind Indians' efforts to represent themselves politically. Toward that end, the Declaration called for full recognition of Indian groups' rights to "territory" and "self-government."[20] The group also released a dossier, in English and Spanish, documenting institutionalized racism against Indians.[21] The English report scarcely received any attention, but the Spanish version was widely denounced, even being burned by the military dictatorship in Uruguay.[22]

In spite of this condemnation, the Declaration became an important resource for Indian political mobilization throughout Latin America. Its denouncement of institutionalized racism led to the reform of Brazil's Ministry of Indian Affairs, and it sharpened critiques of *indigenista* policies in Peru and Mexico that promoted "culturally appropriate" forms of development, forms that allowed limited economic gain while reinforcing racialized hierarchies. The Declaration's emphasis on "territory" and "self-government" further outlined an approach to redefining the debate altogether. Without dismissing the political and economic importance of control over land and resources, *territory* conveyed a sense of broader political rights associated with self-government. At a more pragmatic level it also provided a way to include forest-dwelling groups whose occupancy did not conform to conventional, agrarian notions of property. *Territory* shaped a new understanding of the "habitats" diligently described by cultural ecologists such as Nietschmann as an eminently political space.

This emphasis on territory coincided with growing efforts outside Latin America to establish "indigenous peoples" as subjects of international law.[23] Four years after the Barbados meeting, the World Council of Churches funded a larger event in Port Alberni, British Columbia hosted by the National Indian Brotherhood. As we mentioned in the last chapter, the Brotherhood had been instrumental in calling for recognition of indigenous peoples as *nations*, in keeping with the approach to claiming and mapping land developed by Phoebe Nahanni and the Dene, among others. Groups from around Latin America, including the Miskito, sent delegations to the event. The Brotherhood's founder and president, George Manuel, used the event to introduce his concept of a Fourth World bound by a shared, ongoing experience of colonialism. The concept linked the external colonialism practiced internationally by the Cold War superpowers—the First and Second Worlds—with the internal colonialism practiced domestically by the superpowers and Third World states alike. The meeting culminated with the founding of the World Council of Indigenous Peoples, which subsequently gained status as a nongovernmental organization (NGO) at the United Nations. Two years later, in 1977, the Council organized a second gathering at the United Nations' offices in Geneva to create an international legal framework establishing indigenous peoples' *human* rights to self-determination, self-government, and territory.

The Map, Weaponized

The new emphasis on indigenous rights prompted a critical reassessment of state policies and anthropological studies. Maps like the ones that Nietschmann made were no longer just data. They were *evidence* of use and occupancy that served as the basis for claiming *rights* of ownership and self-determination. The shift was not without precedent. Well before the Inuit Land Use and Occupancy Project, the Indian Claims Commission in the United States had turned to maps made by cultural ecologists such as Julian Steward to financially compensate Indian tribes for lands lost to "gradual encroachment" by settlers.[24] Settling the claims affirmed property titles established by "encroachment" and "adverse possession" in hopes of quieting any future challenges from indigenous peoples. The treaty process in Canada followed a similar pattern, replacing aboriginal titles with property rights and monetary compensation as well as limited recognition of First Nations' political standing.

The situation could not have been more different in Latin America where "state lands" on frontiers were widely viewed as critical to national development. Any recognition of indigenous land rights had to be weighed against the costs to the nation. Nicaragua in 1980 was no exception. Leftist Sandinista guerillas had overthrown the Somoza government a year earlier, bringing the family's four-decade rule to an end. By their own admission, the Sandinista leadership knew very little about the eastern third of the country. In keeping with their nationalist views, they saw it alternately as a vast frontier rich in resources and as a region sacked by a succession of North American interests.

This view of the frontier was challenged by a group of college-educated Miskito youth who had risen up through the ranks of the Miskito political organization, ALPROMISU.[25] Founded with the support of the Moravian Church in 1974, the organization focused almost entirely on economic and political rights for the Miskito as *citizens*, advocating for regionally specific development, education, and health care. The group had also sent delegations to a number of international gatherings of indigenous peoples, including the 1975 meeting in Port Alberni and the 1977 meeting in Geneva. Despite skepticism from Nicaraguan officials, their travel was made possible through Moravian involvement in the World Council of Churches. Through participation in the Geneva meeting and contact with other indigenous political organizations—including a visit by Chief George Manuel from the

National Indian Brotherhood in Canada—the cadre of young, educated Miskito adopted a more radical vision of indigenous rights.[26] In the aftermath of the revolution, the group took advantage of Sandinista suspicions about ALPROMISU's conciliatory stance toward the Somoza regime and dissolved the organization. In its place they founded MISURASATA, whose name was an acronym for the Miskito phrase "Miskito, Sumo, Rama, and Sandinistas Working Together."[27]

In spite of their alliance with MISURASATA, the Sandinistas were still unsure if and how Miskito demands fit within their revolutionary nationalist project. Members of the Sandinista *junta* were clearly divided on the matter, with key leaders continuing to regard the Miskito, along with Afro-Nicaraguan Creoles and other indigenous groups, as a "backward" (*atrasado*) population with questionable loyalties due to their long relationship with Britain and the United States. Others saw the indigenous question as an important part of the revolution, including a number of officials at the Sandinistas' National Agrarian Institute tasked with redistributing property as a key element of the revolution. This was especially true of the person assigned to coordinate policy for the Atlantic Coast, Galio Gurdián. Gurdián had recently dropped out of the graduate program in anthropology at the University of Chicago to return home and support the revolution. One of his first tasks was to coordinate a comprehensive study of land politics on the Atlantic Coast.

Gurdián contracted two fellow anthropologists for the task, Philippe Bourgois, a graduate student in anthropology at Stanford University, and Georg Grünberg, the organizer of the 1971 Barbados symposium. Bourgois was already in Nicaragua, attempting to launch his dissertation research in the Miskito communities of the upper Río Coco. Grünberg arrived in Nicaragua within 72 hours of the Sandinistas' 1979 triumph, seeing it as radical opportunity to advance the cause of Indian liberation. Their report recommended the full decolonization of the Atlantic Coast, recognizing indigenous territorial rights as the basis for a new regional autonomy. Illustrating their recommendation, Bourgois and Grünberg included a 1971 map made by Bernard Nietschmann showing the distribution of Miskito, Sumo, and Rama populations (Figure 5.1). Nietschmann had first published the map in the journal of the Nicaraguan Geographic Institute, later reprinting it in his 1973 book, *Between Land and Water*. There the map had been evidence of the geographic extent and location of subsistence economies; in Bourgois and Grünberg's report, it was repurposed as the basis for indigenous rights to territory. It was too much for the Sandinista

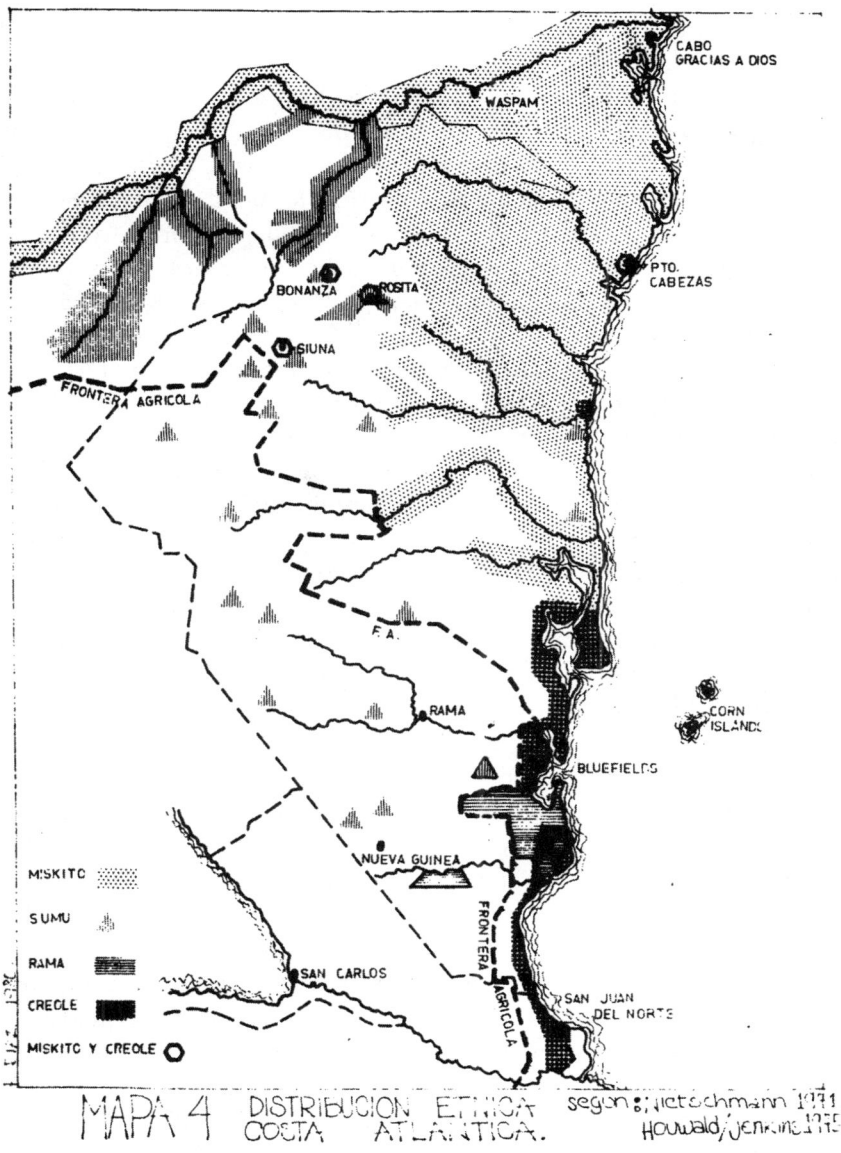

FIGURE 5.1. Map of the Miskito, Sumu (Sumo), and Rama populations from Bourgois and Grünberg's report. Source: Philippe Bourgois and Jorge (Georg) Grünberg, La mosquitia en la revolución: Informe de una investigación rural en la Costa Atlántica Norte, Managua, Nicaragua: Instituto Nicaragüense de Reforma Agraria, Departamento de Planificación, May 1, 1980.

junta, which immediately discredited the report for its "separatism" and expelled Bourgois and Grünberg from the country.

But MISURASATA was not about to let the matter of land die. They organized their own study of indigenous land rights financed by Moravian Church donations and a small grant from Cultural Survival, an indigenous rights advocacy organization founded in the 1960s by Harvard anthropologists.[28] MISURASATA used the money to hire an exiled Salvadoran surveyor, Mauricio Polanco, to make a map showing the areas occupied by indigenous peoples and to compile historical evidence of their ownership.

Polanco's effort coincided with the political mobilization of Miskito communities. Shortly after the inception of the mapping project, MISURASATA demanded that the Sandinistas back a literacy campaign in Miskito, Creole, and Sumu, the prevailing languages in much of eastern Nicaragua. The campaign built on the Sandinistas' tremendous success with a national literacy campaign in Spanish. The earlier campaign had relied on educational materials developed to support the revolution in Cuba. The effort on the coast combined the structure and content of the Spanish campaign with liberation theology-minded documents used by the Moravian Church and Catholic missionaries to empower their congregations, telling the region's history in the local languages.[29] Much of this version focused on the region's status as a quasi-sovereign entity under the rule of a British-recognized Miskito king. MISURASATA's mapping project gave spatial form to this history, defining a territory that sharpened a collective sense of the Miskito as the rightful occupants of the region, owners of its land and resources.

And then, just as the literacy campaign was drawing to a conclusion, Polanco drowned in the Río Coco, leaving his incomplete work in the hands of Steadman Fagoth, the MISURASATA leader accompanying him. Undeterred by his death, MISURASATA made the completion and release of the map a major part of their 1981 agenda.[30] The move heightened Sandinista fears of Miskito separatism, and in February of that same year, Sandinista officials arrested MISURASATA's leadership. All were later released, following protests by Miskito in Puerto Cabezas and along the Río Coco. Some of the newly released leaders rejoined the Sandinistas, but others, including Fagoth, fled the country. Others still, most prominently Brooklyn Rivera, remained in Nicaragua to push MISURASATA's demands for territory. Rivera released yet another land proposal to the Sandinistas in July 1981, framing MISURASATA's demands in terms of international human rights. Once again the Sandinistas rejected the proposal, denouncing it as an imperialist plot by the United States to split the country in two and destroy the revolution.

Unfounded at first, Sandinista fears of Miskito separatism became a reality as the situation in Nicaragua polarized. U.S. President Ronald Reagan made no secret of his opposition to the "Communist threat" posed by the Sandinistas—though the U.S. Congress was less inclined to agree, voicing concerns about a new Vietnam in Central America. In what would soon become a well-established pattern, in November 1981 Ronald Reagan secretly channeled US$19 million to the CIA for the purposes of arming and training anti-Sandinista forces in Honduras and Costa Rica. Most of the money went to the Nicaraguan Democratic Front led by former members of Somoza's National Guard. But they were not the only ones to receive aid. Other counterrevolutionary or "Contra" forces also received U.S. support, including a pair of Miskito groups. One of these was led by Steadman Fagoth. The other was comprised of men from the Río Coco who had broken off from MISURASATA to form a new entity known as Los Astros, named after the seven archangels of the Old Testament. Trained by Argentine and Guatemalan military instructors, Los Astros launched the first attacks on Sandinista garrisons along the Río Coco in December 1981.[31] Fueled by millenarian thinking, they saw themselves as fighting a war for survival that surpassed any broader geopolitical context.[32]

The Sandinistas saw things differently. Convinced that Los Astros and other Miskito groups were U.S. ploys, they unleashed a massive counterinsurgency campaign within days of the Río Coco attacks. Known as *Navidad Roja* ("Red Christmas"), the campaign razed Miskito communities along the length of the Río Coco. Some 10,000 Miskito fled the campaign by crossing the river into Honduras, where many later joined armed groups. The Sandinistas "relocated" as many as 8,500 more Miskito from throughout the northeast to a series of agrarian colonies further south along the sole road connecting Managua with Puerto Cabezas. The Río Coco and the Miskito had once again returned to geopolitical prominence.

Freedom Fighters

From 1977 until 1982, Nietschmann followed events in Nicaragua from afar. The turtle ban had dealt a blow to his relationship with Tasbapauni. Like many others, Nietschmann's initial reaction to the

Sandinista revolution was guardedly optimistic. After Somoza, anything seemed like an improvement. Nietschmann even flew to Managua in 1980 to present his ideas for turtle conservation to the Sandinistas.[33] Conservation was a low priority for the Sandinista leadership then, and with the start of the Contra war, it fell even lower.

At the same time Nietschmann pursued his research interests elsewhere, studying sea turtle and dugong ecology in the Torres Strait Islands between Australia and New Guinea. In the Torres Straits, he encountered a growing land rights movement led by islanders who would later challenge Australia's claim that the islands, like other aboriginal lands, were *terra nullius*—"empty lands"—and therefore property of the state.

Nietschmann continued his transit of the emerging Fourth World with visits to the hill tribes in northern Thailand, the Toradja in Sulawesi, and the Ifugao in the Philippine Cordillera. The latter had been exquisitely mapped by Yale anthropologist Harold Conklin in his *Ethnographic Atlas of Ifugao*. Conklin used his detailed maps of Ifugao terracing systems to argue for a new "ethnoecology" built around ethnographic studies of landscapes, precisely at a time when others in the field were turning to remotely gathered air photos and satellite images to analyze indigenous land use.[34]

But Conklin's *Atlas* was missing something. The maps were still just data—data that failed to capture what was really going on. It was a dilemma Nietschmann faced in his own work. In the preface to his 1979 book, *Caribbean Edge: The Coming of Modern Times to Isolated People and Wildlife*, Nietschmann wrote:

> Academics are accustomed to treat ecological and cultural data in a highly abstract manner several steps removed from the vividness and intimacy experienced during the research. There is nothing wrong with this at all, and the result is usually a more reasoned, dispassionate report. We all do this; it is trained into us and later reinforced. But sometimes much is left out—often the essence and the spirit.[35]

In *Caribbean Edge* that essence and spirit had consisted of efforts by Nietschmann's Miskito informants to convey how they saw the world and themselves in it. By 1983 that essence and spirit were something altogether different. Miskito control over land and resources outflanked conservation concerns. Maps and other research could no longer simply be data or even expressions of a cultural worldview. They had to be weapons useful in fighting for indigenous control over land and resources.

In 1983, Nietschmann returned to Nicaragua through the "back door," traveling with friends of his from Tasbapauni who had joined Brooklyn Rivera's armed MISURASATA group in Costa Rica—friends who began his "new education." "Instead of stars, reefs, winds, and currents, as in the past," Nietschmann wrote, "it is now about informants, security precautions, and who can be trusted."[36] In spite of the difference in approach, land continued to be the fundamental concern. It also signaled MISURASATA's efforts to distance itself politically from the anti-Communist rhetoric associated with the U.S.-backed Contra forces. As Brooklyn Rivera put it, "Right-wing, left-wing, colonialist, capitalist, Marxist, all governments are anti-Indian. They want what is ours: Indian land and resources. They want to make our nation part of their nation."[37]

Nietschmann turned his field research skills to documenting the war between the Sandinista state and the Miskito nation. He mapped Sandinista offensives and Miskito battles, troop movements, and aerial attacks. On film and on paper, he recorded Miskito accounts of Sandinista abuses. Along the way, he dropped his regular column in *Natural History* and started publishing his frontline reports in *Cultural Survival Quarterly,* the *Fourth World Journal,* and *Co-Evolution Quarterly.* Through it all Nietschmann hewed close to the tone set by Rivera, portraying the Miskito as the vanguard of struggles between states and Fourth World nations. It was a withering attack on the racism of the Sandinista leadership with a singular focus on the importance of land.

Like most wars, the Miskito struggle was as much a material reality as it was a rhetorical creation. The Sandinistas had violently displaced thousands of Miskito at the start of the war, and they had subsequently militarized the entire region by placing garrisons in Miskito communities and developing a network of *orejas* or spies. It was classic counterinsurgency that often reprised, with uncanny resemblance, the tactics of Sandino and Edson six decades before. Miskito residents were often caught between the Sandinista Army and Miskito patrols. Abuses by one side, such as torture, the brutal killing of a family member, the razing of a town, or conscription, often led people to join the other. Of course there were many sides to choose from: the Sandinista Army, a litany of Miskito groups, and Contra groups like the U.S.-backed FDN (Fuerza Democrática Nicaragüense [Nicaraguan Democratic Force]), to say nothing of those Miskito seeking ways to advance recognition of their rights within the Sandinista government. It was also not uncommon for fighters to move from group to group for reasons that frustrated every effort to make ideological sense of the war.[38]

But complexity does not win wars so much as prolong them, which is precisely where rhetoric comes into play. Rhetoric allows the separation of forces into sides—the Sandinistas, the Contras, the Miskito—papering over the differences that threaten to split them apart. Rhetoric simplifies the complexity of struggle in a way that enables actors to define goals and objectives, not to mention their ability to reach some form of peace. Both Sandinistas and the Reagan administration had a hard time defining the Contras as a single entity, albeit for different reasons. As one of Oliver North's private contractors put it after visiting Contra forces in Costa Rica that included Rivera's MISURSATA forces, the "whole operation is one big goat fuck."[39]

Miskito forces were hardly immune to divisions either. Fagoth routinely denounced Rivera's negotiations with the Sandinistas as traitorous. At the same time, discontent within Fagoth's own ranks ran high enough to lead to an assassination attempt and a near total loss of confidence in him among his U.S. handlers.[40] Although Rivera's emphasis on land helped him gain moral and political authority among the Miskito, military authority was a different matter. Yet through it all a relatively coherent narrative emerged that the Miskito were fighting the war "to defend our land"—*wan tasbaia dukiara*.[41]

Nietschmann worked tirelessly to prop up that narrative in articles and speaking engagements and on his visits to the various armed groups in Honduras and Costa Rica. In 1987, just as news of the Iran–Contra scandal was breaking in the United States, the Miskito forces succeeded in unifying. Rivera headed the new force, which was called YATAMA, a Miskito acronym translating roughly as "Sons of Mother Earth." The merger redefined the Miskito struggle as the defense of Yapti Tasba, "Mother Earth." Nietschmann stayed on as an advisor to the new organization, helping the faction interested in a political resolution to the war file a proposal for "democratizing" Yapti Tasba with the U.S. government's National Endowment for Democracy.[42]

Nietschmann's willingness to go to the U.S. government was not new. As he put it himself, the Miskito and the United States were "allies who shared only enemies."[43] For the United States, the Miskito were an oppressed minority whose struggle lent badly needed legitimacy to a Contra War run largely by an unholy alliance of ex-members of Somoza's National Guard, drug runners, evangelical Christians, private contractors, and "cowboys" from the Reagan administration.[44] For the armed Miskito groups, the United States was their only hope for maintaining their struggle against the best armed military in Central America *and* resisting the racially charged paternalism of the main Contra forces.

Nietschmann walked a very fine line among the positions. From the very beginning of the war, State Department officials had solicited his participation in "informing" the American public about the plight of the Miskito.[45] At the time, intelligence analysts were widely rumored to be scouring works by Nietschmann and other social scientists for information about a region they barely knew. Nietschmann wrote numerous pieces that portrayed the Miskito as authentic "freedom fighters" in league with other proxy forces battling Communist states with Reagan's help. Nietschmann's most comprehensive and ideologically charged effort to make his case is 1989's *The Unknown War: The Miskito Nation, Nicaragua, and the United States*. It was published by Freedom House, a Washington, DC think tank whose close ties to the U.S. government earned it a reputation during the Reagan years as source of administration propaganda. In the book, Nietschmann rehashed his Fourth World argument that "the Miskito are little understood because almost all descriptions and explanations have had the external goals of making the Miskito fit into European concepts of history, geography, ideology, and loyalty."[46] To get his point across, Nietschmann included a number of maps showing Yapti Tasba, the Miskito motherland, as a nation (Figure 5.2). At the same time, he used his characterization of the Miskito as a nation to directly engage the Reagan Doctrine: "Indigenous nations are a territorial and cultural firebreak to the spread of communism and other totalitarian regimes."[47] Nietschmann took his efforts a step further by appearing on Pat Robertson's evangelical television show, *The 700 Club*, to appeal for support for the Miskito directly from the show's conservative Christian audience, among Reagan's most loyal supporters.[48]

Pulling this off required enormous simplifications: the portrayal of the Miskito as a unified indigenous nation with a defined territory that could stop the spread of Communism. It was a useless representation *in* Yapti Tasba, where nearly every point was contestable; but outside Yapti Tasba it was something else: an image of an indigenous nation fighting for freedom. And nothing *proved* that quite like a map. It's no surprise that among the only maps of Yapti Tasba are those found in the pages of Nietschmann's *The Unknown War*. The maps made it possible for Freedom House's audience to see—with the stunning clarity of clear

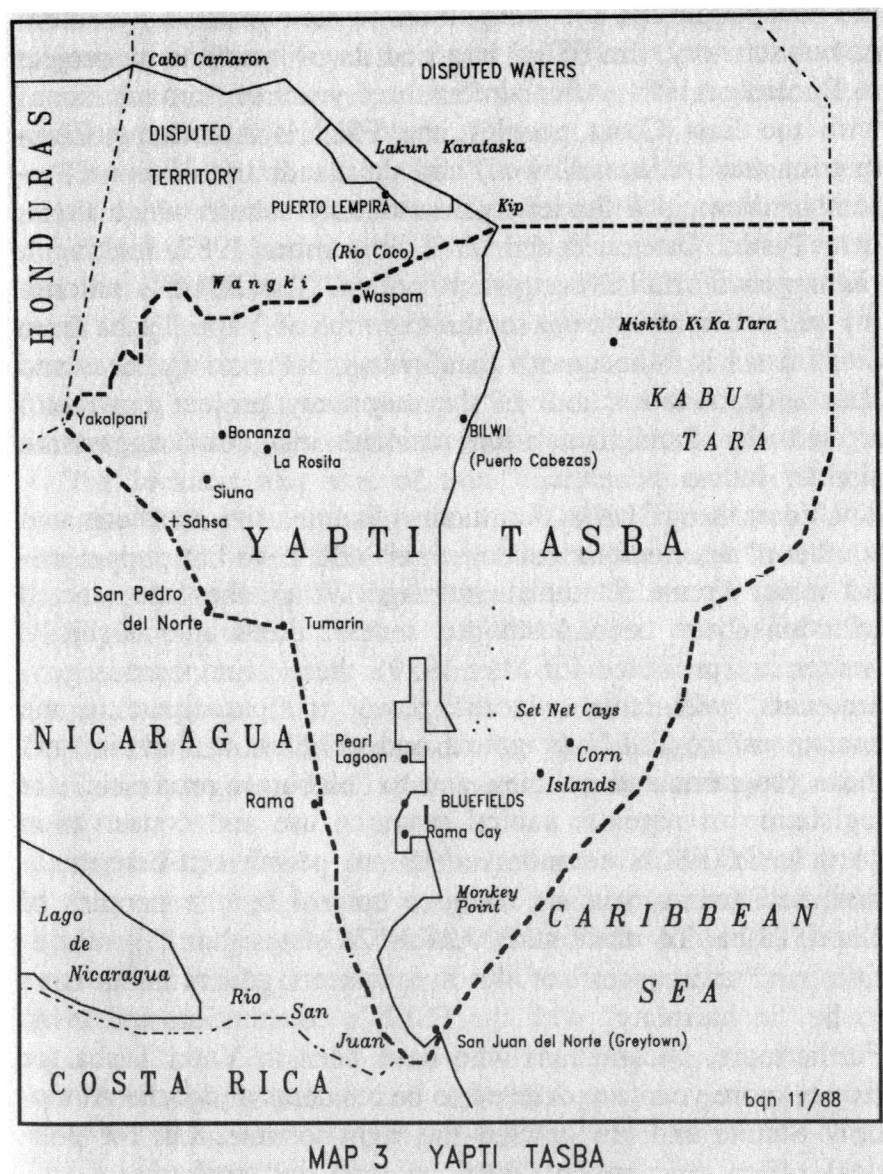

FIGURE 5.2. Yapti Tasba. Source: Bernard Q. Nietschmann, *The Unknown War: The Miskito Nation, Nicaragua and the United States*, Freedom House, New York, 1989, p. 13.

boundaries and unified cultural identity—what was not visible on the ground: an indigenous nation oppressed by a Communist state.

The Fourth World

In 1990 the Sandinistas were voted out of office, losing to a coalition held together by a generous helping of cash from the United States. Their defeat satisfied Contra demands for a "free" Nicaragua. The outcome for the Miskito was less clear, particularly with regard to land. In 1987, in a move that undermined YATAMA's claim to be fighting for land, the Sandinistas had passed a law dividing the Atlantic Coast into two regional autonomies and launching a novel experiment in governance in Latin America. Elections in the autonomous regions were held after the national vote that removed the Sandinistas from office, and they paved the way for YATAMA to transform itself from an armed group to a political party. YATAMA candidates, all prominent veterans of the war, won key positions in the autonomous regional government in the north, fueling the impression that the Miskito had "won the war."

Nietschmann's maps of Yapti Tasba did not clarify the situation so much as they obscured its complexity. By the mid-1990s, eastern Nicaragua, and much of the rest of the country, was mired in overlapping and contentious struggles over land and resources. Once again, Miskito communities found themselves struggling to survive in a context of profound isolation. The Nicaraguan state used the region's autonomous status to justify its abandonment, chronically underfunding the regional governments. Worse still it renewed its claim that all of eastern Nicaragua consisted of national lands. At the same time, scores of former combatants from all sides—Contra, Sandinista, and Miskito—sought to claim land promised them under demobilization agreements. In particular, they singled out "empty" areas of forest where the remaining stands of timber were located. These claims brought the ex-combatants into direct conflict with Miskito and Mayangna communities clamoring for recognition of their indigenous rights. In the shallow, reef-studded waters of the Caribbean, Miskito fishermen faced additional challenges from "pirate" foreign fishing boats illegally searching for lobsters, turtles, and fish. These drove Nietschmann into his postwar indigenous mapping project, whereby he worked with Miskito communities to chart their use of the reefs and secure their ownership of them. It was in the context of *this* project that Nietschmann wrote his lines about maps and guns. Amidst it all, the informal economy in the region was booming, fueled by illegally cut hardwoods and cocaine trafficking. At some point, it all became too much for Nietschmann to bear and he stopped going to Nicaragua. The romantic vision of the Miskito struggle for land that had galvanized resistance during the 1980s was literally torn apart by the political divisions that followed the war. Something new had to take its place.

For his part, Nietschmann went on to expand his Fourth World arguments to a global scale. As he saw it, a "third world war" was already well underway in a series of struggles between states and nations, with the Miskito struggle for Yapti Tasba at the fore.[49] This perspective allowed him to link the Miskito struggle with those of the Kurds and Kawthoolei, the Papuans and Palestinians, the Basques and Baluchis. The bonds between these groups were largely cartographic, each categorized by Nietschmann as a nation with a territory and culture, oppressed by a state. Like so many other visions of social order, Nietschmann presented the Fourth World as a preexisting entity, singled out by states for repression. What the Fourth World lacked in military power, it made up for with its enduring qualities. As Nietschmann wrote, "Fourth World nation combatants don't have to wear uniforms, carry a flag, mint currency, send a team to the Olympics or have nuclear weapons—the mark of the state—to defend their geographies and identities, which have stood the tests of time and bombs."[50] These nations shared one basic need with the states they opposed: They needed maps to bring their vision of a new social order into view. It was a positively Bowman-esque political vision, viewing nations as a kind of organic geopolitical order best positioned to guarantee freedom. And yet, as Nietschmann was aware, it was an order that would never come about without a fight—a fight as timeless as the Fourth World nations themselves.

Nietschmann was already aware of states' efforts to prepare themselves for this new war, redefining the terms of conflict, such as they were, as a justification for a whole new kind of military violence characterized by stunning asymmetry. This new war, Nietschmann promised, would single out Fourth World struggles for an unprecedented level of militarization. "By state definition," he wrote, "state combatants conduct war, counterinsurgency and police actions, and maintain law and order; nation combatants conduct terrorism."[51] To demonstrate his point, Nietschmann pointed to a new unit organized by the U.S. Army, the Small Wars Operations Research Directorate, or SWORD, that aimed to furnish the military with the research and intelligence needed to fight these new wars. "The SWORD unit's vision of future wars is limited by one-sided politics and a blind acceptance of state geographic sovereignty, a sovereignty that the opposing forces do not accept in 72 percent of the world's wars. Most current wars are being fought for territorial goals, not political ones."[52] Maps could make those territorial goals visible, but they could not work by themselves. They had to be backed by guns. The map was now fully weaponized.

6

From Territory to Property

Indigenous Mapping after the Cold War

Nietschmann's crack about maps and guns may have captured the sentiment and ethos of the self-styled "counter-mappers" who emerged in the 1990s, but it didn't begin to get into the *range* of indigenous mapping efforts going on around the world. While the Miskito were mapping coral reefs and coastal lagoons with Nietschmann, forest dwellers in Indonesia were using maps to leverage their role in managing resources. Aboriginal people in Australia, the Ye'kuana in Venezuela, the Emberá in Panama, and the Guaraní in Bolivia were all using maps to formulate land claims filed with state officials. Black communities in Colombia were doing the same. In Canada, the controversial reopening of treaty negotiations with First Nations kicked off a profusion of mapping projects in British Columbia. There were mapping projects in Papua New Guinea, Kenya, Namibia, Nepal, and the Philippines, not to mention Bangladesh, Brazil, Thailand, and the United States. Having previously thought of themselves as Sirionó and San, as *indios*, or even as no more than residents of a particular region or community, *through mapping* many of these groups were becoming indigenous in a novel political sense. Mapping helped foreground common struggles over land and resources and cast these in terms of "territory" and "autonomy" recognized as rights through newly developing legal standards being drafted at the United Nations and the International Labour Organization (ILO).

The emphasis on legal rights to territory and autonomy echoed calls for a more sweeping decolonization. In the run-up to the 1992 quincentenary of Columbus's arrival in the Americas, indigenous organizations across Latin America made plans for a countercelebration commemorating their 500 years of resistance. One of the biggest planning meetings produced the 1990 Declaration of Quito, which called for the decolonization of the Americas:

> The existing nation states of the Americas, their constitutions and fundamental laws are judicial/political expressions that negate our socio-economic, cultural and political rights.
>
> From this point in our general strategy of struggle, we consider it to be a priority that we demand complete structural change; change which recognizes the inherent right to self-determination through Indians' own governments and through the control of our territories.
>
> Our problems will not be resolved through the self-serving politics of governmental entities which seek integration and ethno-development. It is necessary to have an integral transformation at the level of the state and national society; that is to say, the creation of a new nation.[1]

The Fourth World envisioned by Nietschmann and George Manuel seemed close at hand, its contours focused in a profusion of maps depicting indigenous rights to territory and autonomy.

But the wave of indigenous mapping that began in the 1990s did not usher in an unprecedented era of decolonization. Instead it heralded the mainstreaming of indigenous mapping as, above all, a means of governing what might otherwise have been an unruly bunch of claims and counterclaims for territory and autonomy. Instead of being driven by the aims and goals of Fourth World nations, these projects were, by and large, funded by NGOs. They were joined by an even more powerful set of players: the Ford Foundation, the U.S. Agency for International Development (USAID), and the World Bank.[2] In every case, mapping provided a means of nominally recognizing indigenous peoples' rights, while at the same time assimilating them into a territorial order whose lines were codified by the law.[3]

To understand how this happened, it's necessary to see how new mapping technologies—many of them initially developed for military applications—facilitated a proliferation of mapping projects. At the same time it's important to see how indigenous mapping fit into the rather Bowman-esque challenge of fashioning a new global political order following the demise of the Soviet Union and the end of the Cold War. But it's equally important to see how lines drawn by Cold War proxy battles in places such as Nicaragua gave shape to the order that was still to come. This was particularly true for places where counterinsurgency tactics had broadened beyond armed conflict to engage peoples' normative desires for legal protection of their rights, for development, and peace. All of those desires laid the groundwork for an order that not only required states to be implemented; they laid the very basis for the state's claims to legitimacy as the guarantor of stable property rights, of economic growth, and of security. Counterinsurgency engages these desires, as demonstrated throughout this book, as a means of imposing a social order that facilitates distinguishing civilians from combatants and friends from enemies . . . in the name of more efficiently fighting wars where the perimeters of the battlefield are socially rather than territorially defined.

As Isaiah Bowman had made clear in his work with The Inquiry during World War I, organizing space politically according to ethnic identity, common history, and language provided a means for transforming the boundaries of war into a blueprint for governing. In much the same way, the World Bank, USAID, and the Ford Foundation, along with countless other organizations, picked up the techniques of indigenous mapping by way of making space *governable*.[4] Along the way they transformed the *object* of mapping from "territory" to "property," replacing the emphasis on autonomy with the language of markets and the rule of law.[5] In short, indigenous mapping was swept into the dominant refrain of capitalism's triumph at the end of the Cold War.

Counterrevolution by Other Means

Central America proved to be pivotal for the mainstreaming of indigenous mapping. In all accounts, the wars of the 1980s had left the region in shambles. Their conclusion in the early 1990s did not bring about peace, but did mark a shift from the violence of military conflict to that of political and economic restructuring. This change itself was a legacy of counterinsurgency in the region. Rebranded by the U.S. military as "low-intensity warfare," counterinsurgency lowered the emphasis on decisive military victory. Instead, it aimed for "a level of internal security that permits economic, political, and social growth through balanced development programs."[6]

The doctrine proved effective at fighting conflicts in which the battlefield included all of society, weaponizing everything from social divisions to the procurement of basic needs for food, shelter, and water. Of course, counterinsurgency did fail to achieve decisive military victories. Indeed, many of the wars in Central America ended as much because of cuts in military aid from the United States and the Soviet Union as from military stalemate and the exhaustion induced by attrition and terror. The end of armed conflict no more solved the problems than it resulted in peace. Instead, it opened the door for ways of fighting war by other means, by planning for peace through ongoing efforts to forge an enduring social order. From a U.S. policy perspective, this way of waging war by peaceful means was already at hand in the form of development. Long used to buffer security during the Cold War, development's goal of economic growth provided a guide for remaking Central America in line with U.S. economic and security interests.

Nicaragua was the most egregious example of this trend. As a 1991 USAID report summarizing policy objectives for Nicaragua put it, "the strategy presented in this document is an extremely ambitious one. It is difficult to overemphasize the degree of change in the Nicaraguan economy and Nicaraguan society which it envisions."[7] With former Contra leaders and their allies in control, this transformation promised a "slow-motion counterrevolution" executed through technocratic reforms to property rights and the privatization of state assets.[8] In Guatemala, Honduras, and El Salvador, USAID officials faced a different task: one of rehabilitating the legitimacy of states through development. But in spite of the difference, the goals were similar to those in Nicaragua: making use of market-oriented reforms to contain and neutralize political opposition while pursuing regional unification and integration with U.S. markets. International financial institutions such as the World Bank and the Inter-American Development Bank reinforced the approach, urging states to privatize nationally owned assets, cut social spending, and stabilize national economies. Unsurprisingly, none of those efforts guaranteed any semblance of peace. Instead, they merely forced the region to remake itself in accordance with U.S. security and economic interests.

This market-oriented vision for rebuilding Central America compounded the threats of dispossession and displacement faced by indigenous peoples throughout the region. One of the mandates of structural adjustment programs directed states to auction off state-owned assets, including access to natural resources found on so-called *national lands*. Historically regarded as *national frontiers*, states claimed ownership to these regions due to their lack of identifiable property rights, often with blatant disregard for indigenous peoples living there. Violence in these same regions during the 1980s had compounded the displacement of indigenous peoples, as insurgent groups and proxy forces sought cover in the dense forests that dominated frontier areas; and the conversion of these regions to battlefields accelerated the displacement of indigenous peoples as refugees, emptying them of their occupants and strengthening perceptions of these areas as "empty" national lands.

In addition to the pressure from extractive industries and the strain of rebuilding lives displaced by war, indigenous peoples throughout the region faced a third threat from efforts to settle frontier areas through converting them from forests to farmland. Marching at the head of this wave were newly demobilized soldiers from insurgent groups and national armies whose efforts to claim frontier lands brought them into direct conflict with the indigenous communities living there. As such, so-called national lands quickly became the focal point for struggles between competing attempts to bring a sociospatial order to post-war Central America, pitting newly militant demands for indigenous rights against state-led development plans. In short, it marked a return to the standing war that had dominated Central American societies since colonial times. If there was any hope for a peaceful way out of the predicament, it was going to require an alternative way of valuing resources and conceiving of social order. And that is precisely where Nietschmann's call for indigenous maps came in.

Conservation through Cultural Survival

In 1992 the National Geographic Society's *Research and Exploration* journal published a special, bilingual map supplement that proposed just such an alternative plan for ordering Central America. *The Coexistence of Indigenous Peoples and Natural Ecosystems in Central America* demonstrated the coterminous relationship between forests and the areas in which indigenous peoples lived, the boundaries of indigenous homelands coinciding with those of the region's remaining forests. In keeping with National Geographic's house style, the map was a two-sided affair. One side (see Figure 6.1) presented the cartographic "facts" of the argument, laying natural ecosystems in a range of greens and blues over a beige-colored "base" map with the boundaries of the seven Central American states: Guatemala, Belize, El Salvador, Honduras, Nicaragua, Costa Rica, and Panama. On top of both were the locations, in shaded black boundaries, of what the map called "indigenous territories." A trio of inset maps floating over the blue of the Caribbean

FIGURE 6.1. Detail from *The Coexistence of Indigenous Peoples and Natural Resources*, showing Miskito Territory and the forests of eastern Honduras and northeastern Nicaragua. Reprinted with permission from The National Geographic Society.

chronicled the deforestation of the isthmus, showing forest coverage for the region in 8000 B.C.E., the 1940s, and from 1987 to 1992. The point of the map, as its title made clear, was the relationship between indigenous territories and natural ecosystems, especially forests. The map's flipside reinforced the argument with National Geographic-quality photographs and brief overviews of the indigenous peoples and environment of each state.

The map was produced by Mac Chapin, at the time the Central America Program Director for Cultural Survival. During the 1980s, Cultural Survival had emerged as the leading outlet for advocacy-oriented work by academics involved with indigenous rights in the Americas, including Nietschmann. Through the pages of its journal, *Cultural Survival Quarterly*, scholarly contributors wrote about the various struggles over territory and autonomy that increasingly defined "indigenous peoples" as a *political*, as opposed to an anthropological, category. Maps figured in that conversation, with Cultural Survival directly contributing to Miskito land tenure efforts in the early 1980s as participants in the Polanco map effort we described in the last chapter.

Chapin knew this history personally, having visited communities throughout Central America in his official capacity at Cultural Survival. Two decades earlier, he'd also completed a PhD under cultural anthropologist Keith Basso, whose interest in place had led him to experiment with mapping methods.[9] But the idea of making a map himself didn't strike Chapin until 1989 while reviewing a submitted manuscript on "Mapping the Distribution of Indians in Central America."[10] As Chapin tells it, the article's charting of indigenous populations made an intriguing match with a National Geographic map that hung on his office wall showing deforestation in Central America. Chapin admits he knew next to nothing about how to transform his intuition into a map, so he turned to Nietschmann for help. Nietschmann quickly latched on to the idea. In his 1989 book on the Miskito struggle, he'd already produced a map showing the "states and nations of Central America" (Figure 6.2).[11] Nietschmann gave Chapin a crash course in cartography and helped to secure funding for the project from the National Geographic Society's Committee on Research and Exploration.

With the National Geographic money in hand, Nietschmann helped Chapin organize the field teams tasked with gathering data for the map in each of the seven Central American states. In-country experts headed the field teams in Guatemala, Belize, and El Salvador, and Nietschmann personally directed the work in Nicaragua. In Panama and Honduras,

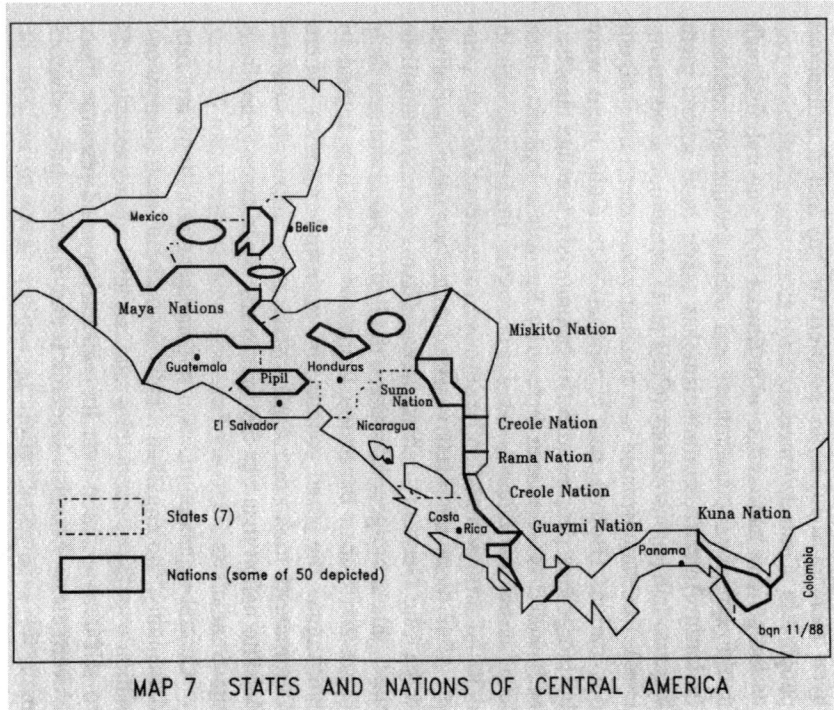

FIGURE 6.2. States and Nations of Central America. Source: Bernard Q. Nietschmann, *The Unknown War: The Miskito Nation, Nicaragua and the United States*, Freedom House, New York, 1989, p. 51.

Chapin turned the task of leading the field teams over to a PhD student trained by the lead author of the very article on the distribution of indigenous populations that had sparked his interest. The student was Peter Herlihy, the very Herlihy we met mapping the Sierra Juárez in Chapter 1.[12]

In spite of Chapin's role in producing the map, Nietschmann's intellectual influence on the message was unmistakable. In 1987 Nietschmann published an incendiary article in *Cultural Survival Quarterly* that kicked off a two-issue series on "Militarization and Indigenous Peoples." Nietschmann's article drew together the worldwide struggles of indigenous peoples into what he called "The Third World War."[13] Fought continuously since the end of World War II on "every continent except Antarctica," these "full-sized little wars" pitted *state* armies against the *nation* peoples who comprised the Fourth World. These wars were not to be confused with conflicts between states and insurgents like the Contras, battling over the control of states. Insofar as the Contras, like the Sandinistas before them, were bent on state control, from the perspective of the Fourth World their efforts amounted to "colonialism by non-Europeans."[14] Fourth World struggles involved, quite literally, *remaking the map of the world* and challenged the dominant geopolitical order imposed by states. Since these wars were not fought between state forces, they were waged "without rules," outside the limits of the Geneva Conventions regulating state conflict.

Nietschmann's arguments captured the imaginations of fellow Fourth World advocates. It also caught the attention of former Cold War warriors scrambling to rethink geopolitical security following the demise of the Soviet Union. Their interest led Nietschmann to reprise and amend his arguments. In a return to his background on cultural ecology, Nietschmann added an environmental element to his argument. With his "Rule of Indigenous Environments" Nietschmann asserted that "where there are nation peoples with an intact, self-governed homeland, there are still biologically rich environments." In contrast, "state environments" were "almost always areas of destructive deforestation, desertification, massive freshwater depletion and pollution, and large-scale reduction of genetic-biological diversity."[15] The 1992 *Research and Exploration* map helped solidify Nietschmann's argument by highlighting the coterminous relationship between indigenous peoples' territories—nations—and natural ecosystems. Indeed, Nietschmann repeatedly referred to the map as demonstrating his rule and holding it out as a blueprint for decolonization.

But no matter how much the map seemed to demonstrate Nietschmann's rule, its text and images equally lent themselves to a far less radical reading. Over and against historical readings of indigenous peoples' struggles to secure a place for themselves in the world, the emphasis on their relationship to the environment presented the territories as essential, ecological truths. As much as both Nietschmann and Chapin recognized this tension, the text that accompanied the map emphasized the relationship between indigenous peoples and the environment as an essential one. Instead of pointing to their rights, Chapin's text portrayed indigenous peoples as victims of colonial aggression. "Until recently," he wrote, indigenous peoples in Central America "had lived in isolation, in regions of refuge, largely protected from the incursion of the outside world by the thick stands of forest, the heavy rainfall, and generally inhospitable environment" that dominates the region's Caribbean lowlands. "Now all this has changed," he continued:

> Since the 1940s, exponential population growth throughout Central America, capitalist schemes utilizing new technologies, and advances in public health have combined to open up these areas to exploitation by loggers, cattle ranchers, and swarms of landless peasants. Multinational companies are extending their tentacles into the last remaining hinterlands in search of cheap timber, oil, and precious minerals. As all of these forces advance, the forests are cut down and burned off at a steadily accelerating pace, and the native inhabitants are deprived of their resources, displaced, and driven into cultural extinction. [16]

Chapin's text transformed the map from a blueprint for decolonization to a proposal for better governing Central America through conservation and development. It put indigenous peoples at the forefront of this struggle, but not as nations, as Nietschmann advocated. Instead it emphasized their role as conservationists. Chapin underscored this point, calling for international conservationists "to form alliances and to work with local people as a means of solving common problems and reaching common objectives" [17]

The map's intervention was both carefully scripted and well timed. In 1992 the polemics about international conservation were near the crest of their mania for "saving the rainforest," the Amazon Basin in particular. Chapin's inset maps chronicling deforestation, together with his text, portrayed the forests of Central America as vulnerable to the threats faced by rainforests everywhere. The idea of the rainforest was a powerful icon for mobilizing international concern about the environment, but it was an idea that concealed as much as it revealed.[18] Conservationists' representations of rainforests as isolated regions untouched by modern social and economic forces drew directly on state classifications of these areas as frontiers, but substituted an emphasis on conservation against the historical importance of settlement. But these areas were no more untouched by modern forces than they were seas of green devoid of humans. Instead they were thoroughly inhabited by indigenous peoples—among others—whose long history of struggle with states, resource extractors, and settlers shaped their understanding of the forests as *indigenous territories*.

The concept of territory knit together particular histories of struggle with newly emerging international human rights standards. As we said in the last chapter, during the 1970s indigenous movements and their allies used the concept of *territory* to critique the historical emphasis on *land* as a means of production characteristic of agrarian reforms. This was particularly true for groups living in lowland forest areas and made extensive use of forested areas beyond agriculture. At Barbados in 1971 and again at Port Alberni in 1975, the concept of territory was further expanded as fundamental to any meaningful exercise of rights to self-determination. That formulation was reinforced by efforts to develop international human rights standards for indigenous peoples in the 1980s at the United Nations and the ILO. In 1989 the latter adopted the first legally binding statement on indigenous rights, Convention 169 on the Rights of Tribal and Indigenous Peoples, defining territory as "the total environment of the areas which the peoples concerned occupy or otherwise use."[19]

The *Research and Exploration* map subtly acknowledged those efforts with its cartographic rendering of "indigenous territories," yet the map was conspicuously silent on the struggles behind the term. The legal reference was no doubt deliberate, offering a technical definition that was more palatable than alternative terms such as "homeland" or "nation." Tellingly, the map did not include the boundaries of the autonomous regions in eastern Nicaragua, nor did it make any reference to the recently concluded armed struggle there or anywhere else in the region. Instead, it opted for a distinctly 19th-century view, labeling the Nicaraguan littoral the "Costa de Miskitos." The map's portrayal of natural ecosystems took an equally technical approach, relying on land cover data gathered by satellites rather than engaging the long history of environmental destruction in eastern Central America. Indeed, the forests so thoroughly sacked by U.S. logging companies in northeastern Nicaragua appeared as a mosaic of natural ecosystems.

And yet this vision was not patently false. As on all maps, the vision was a proposition. It remained to be seen how the newly elected U.S.-backed government in Nicaragua would accommodate Miskito demands, and whether the idea would spread elsewhere. The emphasis on conservation helped defuse that politically charged debate, recasting indigenous peoples and forests as objects in need of saving by a decidedly international force. In Nicaragua, conservationists were already hard at work lobbying to upgrade the status of two large areas of forest located respectively on the borders with Costa Rica and Honduras, converting former battlegrounds into "peace parks." In the Miskito communities immediately south of the Río Coco, Nietschmann was lobbying for another reserve, the Miskito Coast Protected Area, comprised of coastal lagoons, mangrove forests, and coral reefs in northeastern Nicaragua. As Nietschmann saw it, such efforts engaged conservation as a critical first step toward creating the spaces in which indigenous peoples could exercise their rights to self-determination.[20] Indeed, it was in the course of this work on the Miskito Coast Protected Area that he penned his (in)famous line on maps and guns.

Further afield, indigenous peoples were forcefully making the case for recognition of their territorial rights as essential to effective conservation, showing up in force at global events such as the United Nations' 1992 Earth Summit in Rio de Janeiro. Two of the key documents drafted at the Rio Earth Summit, Agenda 21 and the Convention on Biological Diversity, specifically acknowledged the importance of including indigenous peoples in sustainable development and conservation. Instead of being false or misleading, the *Research and Exploration* map made a proposition, namely that the best route to peace in Central America lay in integrating conservation with the protection of indigenous territories to guide a form of sustainable development.[21]

The proposition of "conservation through cultural survival," to use another phrase that Nietschmann helped popularize, elicited . . . considerable skepticism.[22] Conservationists—Chapin's target audience—were at the head of the list. The map started a conversation that quickly shifted in emphasis from indigenous peoples to conservation and development. In place of recognizing the vast differences in indigenous communities, conservationists read the map in terms of its viability as part of a global solution for environmental problems.[23] For them, indigenous control of land and resources scarcely guaranteed conservation. If anything, it diminished the degree to which conservationists working through governmental and nongovernmental channels could dictate the terms of resource use and management. From their perspective, indigenous territorial claims introduced a degree of unwelcome uncertainty, politicizing questions of resource control that conservationists thought should be dealt with scientifically, if not through direct state intervention.[24]

One of the more immediate uses of the *Research and Exploration* map was its inclusion in plans for the USAID-funded Paseo Pantera (Panther's Path) project. Developed with principles of conservation biology, it proposed the creation of a biological "corridor" that linked southern Mexico with northwestern Colombia through a chain of parks and protected areas. Biosphere reserves in Honduras and Nicaragua formed the heart of the project, encompassing a complex of forest and marine ecosystems vital to its success as a conservation project. The two U.S.-based NGOs directing the project, the Wildlife Conservation Society and the Caribbean Conservation Corporation, had no trouble using the *Research and Exploration* map as a source of data on land cover. They were far more skeptical of indigenous peoples. Although they acknowledged that the "rules of conservation" had no "regard for the ways of life of indigenous peoples," the project organizers saw them as potential threat.[25] "Yet stresses on these peoples, modernization of their lives, and growing population pressure have meant they [indigenous peoples] no longer live sustainably within their regions, creating another source of conflict." Not surprisingly, the project dropped all reference to indigenous rights, in spite of recommendations made to the contrary by one of its hired experts: Peter Herlihy.[26]

The view not only demonstrated conservationists' skepticism of indigenous rights. It also established their support for USAID's broader effort to achieve a degree of regional integration that would, in time, pave the way for Central America's inclusion in the then-proposed North American Free Trade Agreement (NAFTA). Conservation and regional integration thus became the cartographic bricks for building "peace" in the region. Stumping for the Paseo Pantera project, former Panamanian President Jorge Illueca succinctly captured the spatial convergence:

> Border development and integration is a priority of the Central American peace plan. In general, frontier areas are politically and economically marginalized and poor in infrastructure. They usually have the highest incidence of poverty. At the same time, because of lower population densities, they are often areas dominated by forests and other natural ecosystems. In the past, the combination of marginalization, poverty, and dense vegetation cover have made some of these areas fertile grounds for armed insurgency.[27]

Conservation and development were fast becoming security strategies, broadening the "slow-motion counterrevolution" U.S. policymakers were trying to establish in Central America.

Property and Peace?

Illueca's reference to security pointed to another problem faced by indigenous peoples and conservationists. In the erstwhile frontiers of Central America, the wars of the 1980s had scarcely ended. In fact they continued in these areas, refracted through innumerable, contentious struggles over land and resources aided and abetted by the withdrawal of state armies. The infrastructure of roads, refugee camps, and training bases that had been built to support the proxy wars in Central America had created unprecedented access to frontier areas previously inaccessible. USAID recognized the problem this infrastructure posed, and following the war launched reforestation projects in places such as the Honduran

Mosquitia.[28] The effort proved woefully inadequate at stemming the settlement of these areas by landless and land-poor peasants, many of whom were newly demobilized soldiers looking for a place in civilian society. In eastern Nicaragua and the Honduran Mosquitia, settlement put the newcomers in direct conflict with the indigenous communities living there and was accompanied by an immediate risk of deforestation.

These problems were particularly acute in Nicaragua's Bosawas Protected Area on the upper reaches of the Río Coco. The area fell at the outermost margins of the zone patrolled by Red Mike Edson and the Marines in the 1920s, though it was traversed repeatedly by Sandino's forces, which often came into conflict with the indigenous Mayangna communities there. Miskito and Contra forces brought war back to the region in the 1980s, relying on its thick forest canopy to provide cover for patrols crossing the Río Coco from Honduras. Following the war, former combatants familiar with the area returned to stake claims to land promised them by the terms of their demobilization, claims fanned by the ongoing perception of the region as "national lands." Former high-ranking Contras and supporters of the Somoza regime returning from exile in Miami applied the same rationale and pressured the Nicaraguan government to open the area to commercial logging and mining.

Both the former combatants and resource speculators ran into direct conflict with the Mayangna and Miskito communities living in the core of the protected area, many themselves newly returned from refugee camps in Honduras. By 1993 the conflict had become increasingly volatile, with groups of Contra and Sandinista veterans taking up arms—sometimes together—and demanding state recognition of their rights to the land. At the same time these veterans were being drawn into the burgeoning illegal timber trade, trafficking in tropical hardwoods such as mahogany cut without state permits. Similar dynamics were unfolding across the Río Coco in the Honduran Mosquitia, and in the Petén in northern Guatemala.

Off the coast of northeastern Nicaragua, Nietschmann's efforts to help create a Miskito Coast Protected Area were running into similar trouble.[29] Boats from Honduras were illegally entering Nicaraguan waters, taking advantage of the rapid demilitarization of eastern Honduras and Nicaragua to buy lobster from Miskito divers. Drug traffickers from Colombia were also taking a keen interest in the Miskito Cays at the heart of the proposed protected area to refuel the high-powered "fast boats" that were all but invisible to radar. Soon Miskito divers were involved in trafficking, moving small amounts of cocaine inland to transshipment points and even consuming crack. Ex-combatants from the Miskito communities where Nietschmann was working initially tried to confront the traffickers head on, dusting off AK-47s kept carefully hidden after the demobilization to detain boats, but the efforts proved no match for the traffickers, who used cash and low-grade crack cocaine to buy support in the communities. By 1995 traffickers had all but taken control of Sandy Bay, the largest community at the center of the proposed protected area.[30]

Prized by conservationists, the forests and reefs of Central America were fast becoming hotspots of postwar violence. USAID intervened in the forests in the Petén, along the Río Coco, and in the Miskito Coast Protected Area, using the Paseo Pantera project to shore up conservation efforts in all three areas. In the latter, Miskito involvement in drug trafficking provided all the excuse USAID needed to give international conservation organizations full control over the project. The move perpetuated the very kind of top-down, colonial conservation that Nietschmann so adamantly opposed. Further north in the Petén of Guatemala, conservationists solidified this approach by appropriating counterinsurgency tactics as a blueprint for policing resource use.[31]

In the Bosawas Reserve, USAID took a markedly different approach. There USAID joined forces with the Nicaraguan Ministry of the Environment to fund a land tenure study in the core of the protected area. The contract for the project was awarded to The Nature Conservancy, which in turn hired anthropologist Anthony Stocks, a professor at the University of Idaho and a Cultural Survival affiliate, to conduct the study. Stocks transformed the project into a comprehensive mapping of indigenous land use and occupancy, drawing direct inspiration from the Inuit Land Use and Occupancy Project as well as the work of Nietschmann and Chapin. As in the Inuit study, Stocks's goal was not to promote indigenous "nationalism" nor to curry favor with conservationists, but rather to determine the basis for organizing ownership of land and resources in the area. The convergence of conservation and security in Bosawas proved pivotal to the mainstreaming of indigenous mapping.

Stocks's approach distanced itself from both the documentation of indigenous knowledge and from Nietschmann's Fourth World ideas, instead emphasizing the delivery of concrete legal results.[32] He was not alone in doing this. Across the Río Coco Peter Herlihy was taking a similar approach to mapping Tawahka and Miskito lands in Honduras, the latter in conjunction with Mac Chapin's newly formed Center for Native Lands. Stocks's work went further, though, by using maps to delineate boundaries for dividing the reserve area into six territories "based on common history, ethnicity, geography, and land use."[33] It was Bowman all over again, with the twist of subdividing states into mutually exclusive ethnic territories. But as with the Bowman Inquiry, Stocks's maps did not depict a preexisting reality. Instead they projected an *ideal* configuration of people and resources modeled on state approaches to governing, and so they drew battle lines for what was going to remain a very active struggle.

Like Chapin's *Research and Exploration* map, the rhetorical efficiency of Stocks's maps leant them to alternative readings. Chief among these was the spin given the maps by officials at the World Bank, whose officials were struggling for approaches to development that could incorporate the critiques leveled against market-oriented policies by both indigenous peoples and environmentalists.[34] As we know, these critiques weren't new. In the 1980s critics had taken the bank to task for the destructive effects of its support for road and dam construction in environmentally sensitive areas such as the Amazon Basin. Concerns were compounded by the bank's turn toward neoliberal policies favoring the growth of markets over infrastructure. Among other things, these policies called for opening up state lands to private investment through resource concessions and the roll-out of new private property rights. These policies led the World Bank to support state efforts to auction off the logging and mineral rights on so-called "national lands." The idea was that this would stimulate efficient economic use of natural resources and revenue growth that would, in turn, help alleviate the Latin American debt crisis.[35] It was no surprise that indigenous peoples vehemently opposed these policies as violations of their human right to territory. This opposition resonated with program staff at the World Bank.

Although a minority voice within the World Bank's vast bureaucracy, key staff in the institution's Latin America program had been arguing for several years that recognition of indigenous rights to land and resources was entirely compatible with the bank's market-oriented approach. They grounded their analysis in a scientific appreciation of the efficacy of indigenous peoples' environmental knowledge as a basis for successful resource management. The problem was that existing forms of recognizing indigenous land rights through protected areas, reserves, and agrarian communities all continued to impose state categories orthogonal to indigenous practices of use and occupancy. Recognizing indigenous territories, in contrast, offered a means of successfully engaging indigenous peoples' environmental knowledge to create more inclusive and efficient forms of development.[36]

The most vocal supporter of this approach was Shelton "Sandy" Davis, an anthropologist in the World Bank's Latin America program. Davis argued that the concept of "indigenous territories" provided "a new model of land *tenure* and resource management."[37] In the lines on the maps delimiting indigenous territories, he saw the potential for a new approach to governing conflicted areas, accommodating indigenous demands while advancing World Bank interests in expanding the markets through property rights.

By way of supporting his point, Davis offered two examples: the Kuna in Panama and the *resguardo* system in Colombia. In Panama, the Kunas' de facto control over land and resources had been a boon to conservation efforts and resulted in one of the more successful examples of indigenous-led conservation.[38] In Colombia, World Bank officials had begun to support state efforts to designate extensive *resguardos*, or reserves, in the Amazonian lowlands. There recognition of indigenous land rights proved to be an effective, if temporary, means of advancing conservation and defusing armed conflicts sparked by the land settlement programs funded by the World Bank in the 1970s. As far as Davis was concerned, both cases demonstrated the fact that recognition of indigenous tenure rights was essential to development strategies that could merge conservation with resolution of ethnic conflict. Tapping into a long, if contested, history of finding culturally appropriate forms of development for indigenous peoples, Davis termed this approach "ethnodevelopment."

Given the emphasis of this approach on land tenure, it's no surprise that mapping figured prominently in the plans for its implementation. In 1989 Davis contracted Canadian geographer Peter Poole to produce a report exploring the virtues of indigenous mapping for resolving land tenure disputes. Though focused on Latin America, Poole's report drew from his dissertation research on Inuit hunting in the Canadian North and the Inuit Land Use and Occupancy Project. In his introduction to the report, Davis made the importance of that experience clear:

> Originally, we had hoped that the study would result in the production of a manual; a sort of "how to" book which would draw upon the experiences of native peoples in the Canadian North (where the chief consultant had prior experience), Latin America, and other parts of the world and provide policy

makers and project planners with a guide on how to design areas which combined traditional land-use practices and conservation objectives.[39]

The report fell short of Davis's expectations of a "how-to" manual. Whereas mapping efforts in Canada were rapidly attaining a level of scientific rigor, Poole found that much of the work in Latin America was "still at an experimental stage and not ready to be generalized as universal models."[40] Still, the widespread use of mapping demonstrated its potential. As Poole concluded: "For indigenous societies, security of tenure or settlement of land claims is a crucial precondition for the pursuit of self-determined social and economic development, which may include a conservation-type project."[41] And yet, although indigenous participation was crucial to making these projects work, Poole was clear about the overall objective: making conservation work, economically and socially. "Conservation projects must make local sense in social and economic terms," he wrote, advancing the World Bank's concern with "global ecosystem conservation."

By the early 1990s, projects led by the likes of Chapin, Stocks, and Herlihy were gaining attention as models for mapping indigenous land tenure. All they were missing was the legal piece that would create a basis for recognizing the rights illustrated on the maps. Chapin had perfected a means of addressing this problem by ensuring that state mapping agencies participated in the production of the maps to stave off challenges to the technical accuracy of the final product.[42] Stocks and Herlihy took a different tack, emphasizing the importance of tenure and preparing the way for legal recognition of indigenous rights to property. World Bank officials such as Sandy Davis sympathized, emphasizing the importance of "regularizing" indigenous lands through formal demarcation and registration with state agencies.[43] The World Bank had already begun to integrate regularization with projects in the Amazon Basin, using the recognition of indigenous land rights to clear the way for development projects and mitigate displacement of indigenous peoples in projects there and in Peru.

But it was the World Bank's involvement with efforts in Colombia that proved transformative. As Davis was well aware, regularizing indigenous land rights through the *resguardo* system had effectively merged conservation with conflict resolution in the lowland Amazonian region of Colombia. The 1991 Colombian Constitution expanded this approach, formalizing the *resguardo* system as a form of communal land tenure for indigenous peoples. It also adopted parallel language for black, or Afro-Colombian, communities. The reforms were not solely the product of state officials' efforts. They came about as part of efforts to address demands from indigenous and Afro-Colombian (black) groups for recognition of their rights to territory and autonomy. For Afro-Colombians, these demands were driven by the fact that they made up almost half of all those displaced by decades of armed conflict.[44] Indigenous organizations had, in contrast, used demands for territory to carve out a space of relative autonomy for themselves in war zones around the country. By the 1980s, their efforts had become increasingly militant, with groups such as the Quintín Lame Armed Movement defending indigenous territories against incursions by Colombian military and right-wing paramilitaries on one side, and left-wing guerrilla armies on the other.[45] The militancy of the Quintín Lame as well as that of other indigenous organizations across Colombia made them indispensible parties to any meaningful peace negotiations.[46] For these indigenous organizations, an emphasis on territory and autonomy provided a means of forging a coalition among organizations, whose demands varied considerably due to their specific history and location, to say nothing of their interpretations of what it meant to be "indigenous."[47] The emphasis on territory further helped the indigenous organizations to find common cause with Afro-Colombian groups. Territory drew both groups together, allowing them to use the constitutional process as a means of reimagining the Colombian state as a pluralist society. As utopian as that sounds, their efforts came to fruition. The 1991 Constitution included language formally recognizing communal rights to land, setting in motion a flurry of legal reforms and mapping efforts. Davis and his allies at the World Bank hailed the reforms as a model for land regularization, combining recognition of ethnic demands with conflict resolution.[48] Other World Bank officials followed suit, supporting initial efforts to title and demarcate indigenous *resguardos* and black communities.[49]

In 1993, the World Bank expanded its approach in Bolivia by providing support for a comprehensive overhaul of agrarian reform laws that included recognition of indigenous community titles.[50] As in Colombia, the approach promised a technical solution to land tenure conflicts, focusing on the "regularization" of property rights through formal demarcation and registration in a national cadaster. To World Bank officials, both were essential elements for building markets in land and resolving tenure disputes that discouraged foreign investment. As a "key asset for the rural and urban poor," land as property provided a way to include the poor within markets and help build civil society.[51]

Under the guise of land regularization, the World Bank was now increasingly committed to the limited recognition—and even creation—of indigenous territories.

The World Bank was hardly alone. By the mid-1990s, community mapping was happening everywhere. Though the projects routinely described themselves as led by communities, they were widely supported by the Ford Foundation and a growing network of NGO intermediaries. In 1995, Poole reprised his effort for the World Bank, this time working under contract with the World Wildlife Fund's Biodiversity Support Program, funded by USAID, to survey 63 mapping projects worldwide.[52] Nearly all of the projects surveyed had been launched since 1990 and were making use of GPS and other "geomatic" technologies to map community lands in Brazil, the Dominican Republic, Namibia, Indonesia, the Philippines, Thailand, Indonesia, Papua New Guinea, and the United States. As far as Poole was concerned, the geographic reach of the projects, combined with their use of new mapping technologies, demonstrated the widespread applicability of community mapping for conservation and development planning.

World Bank staff such as Sandy Davis wholeheartedly agreed, championing mapping as a pivotal means for regularizing land rights and establishing a "new relationship among indigenous peoples, scientists, national governments and international organizations for the conservation and sustainable use of the world's tropical forests."[53] More importantly, from the perspective of World Bank officials and other market-oriented policymakers, this new relationship was "a contractual one, whereby indigenous peoples are provided with juridical recognition and control over large areas of forest *in exchange for a commitment to conserve the ecosystem and protect biodiversity*" (emphasis added). Clearly, indigenous mapping was headed for the mainstream.

Although this mainstreaming engaged counterinsurgency's concerns with stability, social order, and economic growth, its focus was markedly different. Instead of fighting a war where the battlefield included the entire population of a region, the World Bank, Ford Foundation, and conservation groups' interests lay in making society governable.[54] Economic concerns dominated understandings of how to govern efficiently, but market-based reforms alone could not achieve this. Instead, they sparked indigenous movements that threatened to bring about the great wave of decolonization that Nietschmann and others had claimed was coming. Mapping provided a means of engaging indigenous demands for land and resources head-on, but in order to neutralize the powerful challenges against state claims to territorial sovereignty, the mapping objectives had to change. An emphasis on *tenure* went a long way toward effecting this switch. Instead of mapping territories, with their connotations of self-determination and self-government, tenure shifted the focus to *ownership* of land and resources as rights registered and guaranteed by states. In a word, it focused on *property*. The application was arguably new to Central and South America, but the approach was not. Recall that in the Canadian North, land use and occupancy mapping was premised on the state's interest to extract resource wealth from deposits of minerals, oil, and gas found there. At roughly the same time in Alaska, the United States had been prompted to settle native land claims as part of opening up new oil fields and securing a right-of-way for the Trans-Alaska Pipeline.[55] Indeed, the very first tribal council in the United States was created not for the purposes of advancing tribal self-determination but rather to facilitate a deal gaining access to Navajo land for Standard Oil.[56] What's more, each of these developments was bound up with the militarization of their respective regions, whether through direct military presence or mobilization of resources for military need. In each case, property rights facilitated incorporation of land and resources into markets, advancing state presence in those same regions as the ultimate guarantor of those rights.

Eager to advance this approach, World Bank officials actively sought to expand on its experience in Colombia and Bolivia, integrating mapping with property regularization projects in new places. But to do so the bank could not just impose its agenda; instead it needed to find willing collaborators among state officials and indigenous groups, a convergence that they found in—of all places—Nicaragua! Nicaragua sat at the intersection of multiple agendas within the World Bank. By the mid-1990s, bank officials were warming to the idea of "conservation corridors," such as the Paseo Pantera, which combined protected areas with buffer zones where limited, sustainable land use was allowed. Indigenous peoples fit into the World Bank's idea of the latter, and the corridor idea looked like a panacea for resolving the conflicts between people and parks. In 1998 the World Bank channeled upward of $90 million to the failing Paseo Pantera project, rechristening it as the Meso-American Biological Corridor.[57] The forests flanking the Río Coco constituted the heart of this project, magnifying their value to conservationists. Bank officials were equally aware that the region was densely inhabited by indigenous Miskito, Mayangna, and Tawahka peoples as well, thanks in part to their mapped claims to the regions' forests. Lastly, Nicaragua's system of property rights, thrown into disarray by Sandinista-era reforms in the eyes of many at the World Bank, lent themselves to efforts to regularize property rights. The convergence made Nicaragua the perfect place to expand the model for combining community property rights, reforms, and conservation already tested in Colombia and Bolivia.

Many in Nicaragua shared World Bank officials' interests, at least superficially. Though former Contras and their allies dominated the Nicaraguan government in the early 1990s, there were still pockets of Sandinista influence within key ministries bent on saving what was left of their revolution. There was also a new class of Miskito politicians, many of whom had led armed groups in the previous decade, actively looking for ways to continue their struggle for land and resources by other means. Both groups ran into opposition from landless populations, particularly those comprised of demobilized soldiers or "excombatants." Their competing political projects converged on the question of land—who owned it and controlled its resources—inciting contentious struggles across rural Nicaragua. In places such as Bosawas, those conflicts threatened to become small wars. Elsewhere in western Nicaragua, state efforts to return lands expropriated by the Sandinistas to prerevolution owners had escalated into violence. To remedy this development, the World Bank leant financial support to a series of programs aimed at building the national cadaster and overhauling existing property regimes. In eastern Nicaragua, this effort meant first figuring out who owned what land. It was a problem uniquely suited for mapping. In 1996 the Nicaraguan Ministry of Agriculture used World Bank funds to contract a comprehensive study of land tenure in indigenous and black communities.

Miskito political leaders were aware of the stakes. Played one way, the study could affirm the findings of the mythical Polanco map that documented indigenous control over the entire Caribbean Coast. But the study also posed the risk of identifying unoccupied lands owned by the state and adding to the wave of dispossession already fueled by nonindigenous settlers and excombatants. The study was controversially awarded to a team led by researchers from Nicaragua and the University of Texas at Austin, several of whom had been prominent supporters of the Sandinistas in the 1980s.[58] By the team's own admission, "to be offered, much less accept, World Bank funding for our research during the 1980s would have been inconceivable." And yet:

Contradictions notwithstanding, the opportunity to carry out this study opened an important space, which we could not pass up. It offered financial support for what we viewed as a crucial step in the ongoing struggles of coast peoples for legally validated rights to their lands. More generally, it was the main type of space within which progressive political initiatives in Nicaragua currently occur; that is, opportunities to pursue incremental, negotiated change from within, pushing at the limits of what is conceded from above, while avoiding all-out mobilizations for utopian transformations.[59]

To fit indigenous and black tenure rights within the state system, the team carried out mapping exercises in 128 black and indigenous communities. The maps were backed by ethnographic reports documenting the "collective memory of struggle" that shaped communities' understandings of the importance of land rights.

The study came back with a number of surprises. First, most of the communities surveyed elected to document their tenure with other communities, consolidating their tenure areas into 17 multicommunity blocs. Only 12 of the 128 communities opted to document use and occupancy individually. Then, when mapped, the blocs and individual communities covered the entirety of the two autonomous regions created by the Sandinistas, affirming communities' contention that there were no "national lands" in the region (see Figure 6.3). This second finding led Nicaraguan officials to suppress the distribution of the report, but it was too late for that. The team that prepared the report had left copies of the maps with each participating community. The World Bank was also eager to see the report distributed, as its findings lent empirical substance to their parallel efforts to draft legislation recognizing indigenous and black community property rights, using the Colombian laws as a guide. In fact the bank had already funded an exchange of experts from both countries.[60]

The Rule of Law

The World Bank's efforts to support the new law received an unexpected boost from a 2001 ruling by the Inter-American Court of Human Rights in a legal complaint brought against Nicaragua by the Mayangna community of Awas Tingni. At the height of the postwar resource rush, Nicaraguan officials had granted exclusive rights to log on lands claimed by Awas Tingni to a Korean-financed company. Aided by lawyers brought in by the World Wildlife Fund as part of the organization's involvement

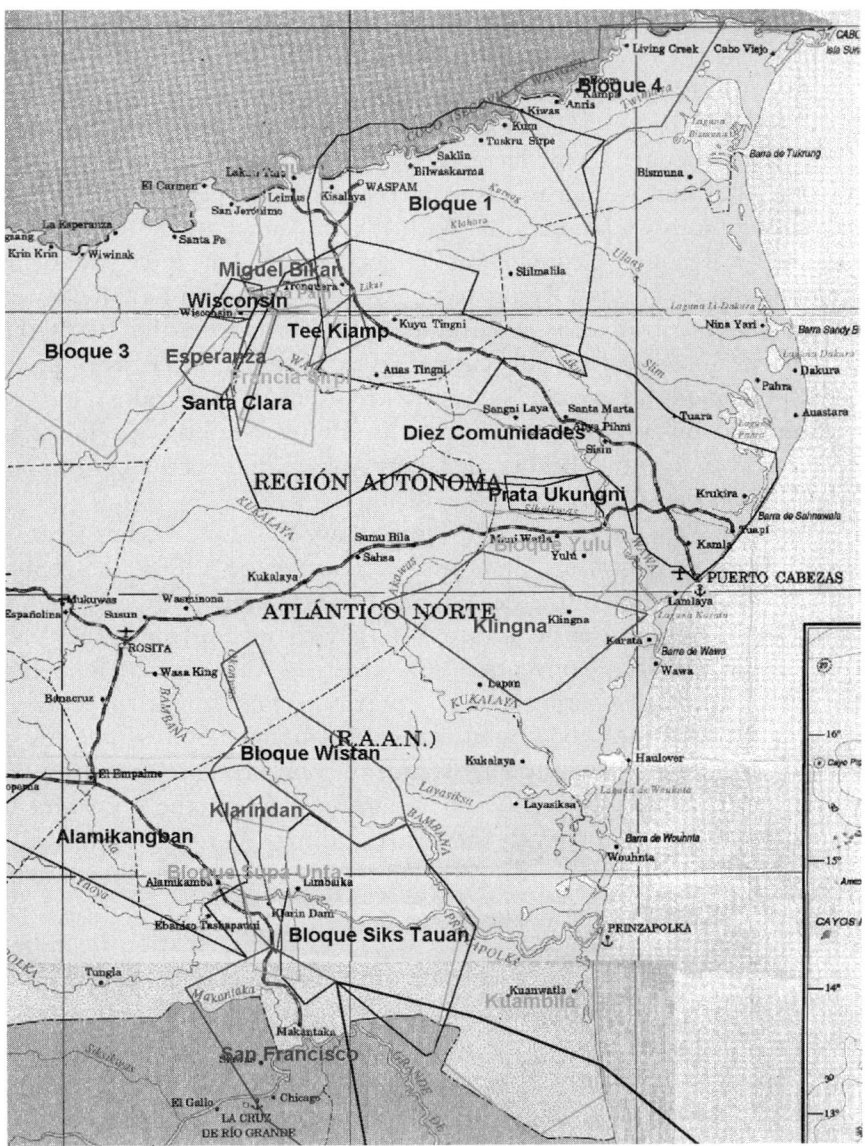

FIGURE 6.3. Multicommunal tenure blocs, North Atlantic Autonomous Region (RAAN), Nicaragua (Caribbean Central American Research Council, 1998). Source: Edmund Gordon, Galio Gurdián, and Charles Hale, "Rights, Resources, and the Social Memory of Struggle: Reflections on a Study of Indigenous and Black Community Land Rights on Nicaragua's Atlantic Coast," *Human Organization* 62(4), 2003, pp. 369-381; Charles R. Hale, "Resistencia Para Que?: Territory, Autonomy and Neoliberal Entanglements in the 'Empty Spaces' of Central America," *Economy and Society* 40, 2011, pp. 184–210.

in the Paseo Pantera project, Awas Tingni filed a lawsuit challenging the concession in 1995.[61] The community won a favorable ruling from a Nicaraguan court a year later, though state officials refused to comply with the court's order to suspend the concession pending resolution of the community's land claim. With the assistance of the U.S.-based Indian Law Resource Center, Awas Tingni petitioned the Inter-American Commission on Human Rights to intervene. The Indian Law Resource Center was no stranger to Nicaraguan politics. During the 1980s the center was an outspoken advocate of Miskito rights, often working in collaboration with Bernard Nietschmann. The Harvard-trained lawyer handling the Awas Tingni case and future United Nations' Special Rapporteur on Indigenous Rights, S. James Anaya, had been a confidante of Miskito leader Brooklyn Rivera during the war.

Neither were Anaya and the Indian Law Resource Center strangers to the Inter-American Commission. During the 1980s, the center had played a key role in the development of international human rights standards for indigenous peoples, assisting with the drafting of the U.N. Declaration and ILO Convention 169, and they were actively lobbying the Inter-American Commission to contribute to a similar process in the Americas through the Organization of American States. The Awas Tingni case underscored the importance of committing to that process. In 1997 the commission issued a report instructing the government of Nicaragua to comply with the court's ruling. Again, state officials refused and the commission forwarded the case to the Inter-American Court of Human Rights.[62]

The case proved to be another occasion for innovation in indigenous mapping. Shortly after learning of the logging concession, Anaya and the Indian Law Resource Center contracted with Cultural Survival's Ted Macdonald to map the community's use and occupancy of the area. Applying the methods developed by Chapin and Nietschmann, Macdonald produced a map later submitted to the commission to illustrate Awas Tingni's claim.[63] The map itself was fairly basic. Using GIS, it superimposed ethnographic information from the community, georeferenced by field teams that had documented the areas used for hunting, fishing, and agriculture, along with old village sites and place names. As occurred in Canada, maps helped the judges *see* a technical resolution to Awas Tingni's claim in terms of titling and demarcation.[64] In its ruling, the court accordingly directed the state to "adopt in its domestic law . . . the legislative, administrative, and other measures necessary to create an effective mechanism for delimitation, demarcation, and titling of the property of indigenous communities, in accordance with their customary law, values, customs and mores."[65] The court mandated the state to create not only new legal mechanisms for applying Sandinista-era laws recognizing community property rights, but it also directed the state to create another map that would make those reforms tangible (Figure 6.4).

The court's ruling spoke directly to the World Bank's concerns with regularizing indigenous land rights in terms of property. Above and beyond indigenous rights, the bank's concern with property regularization continued to be a function of its interest in regional economic integration. To that end in 2001 the bank had collaborated with the Inter-American Development Bank to launch the Plan Puebla–Panamá to integrate energy production, transportation, and telecommunications from southern Mexico across the seven Central American states to Colombia. Like the Mackenzie River Project in British Colombia and the James Bay Project in Northern Québec, much of the proposed infrastructure either passed through indigenous territories or relied on resources found there. Recognition of their property rights would mitigate any potential conflict, recognizing indigenous ownership of lands and resources but requiring their cooperation in securing the access to the land and resources. The Awas Tingni case made property into a human right, recognizing indigenous claims while drawing them into the institutional matrix of property reform and regional integration. Indeed it was the World Bank that pushed for implementation of the ruling, over and against the wishes of state officials, by making compliance a precondition for loans Nicaragua badly needed. Within a year of the ruling, the Nicaraguan National Assembly took the first step toward compliance, approving the community property rights legislation drafted with the bank's help. But the new law no more guaranteed protection of community property rights than the maps themselves had.

In addition to Nicaraguan officials continued reluctance to title lands belonging to indigenous and black peoples, the sheer number of maps produced as part of the World Bank's efforts posed its own problem. In keeping with the bank's emphasis on tenure, the team contracted to complete the study had drawn provisional boundaries around the area claimed by each bloc, using dashed lines superimposed on base maps compiled from government sources. When the blocs were put together, they didn't fit like jigsaw pieces. Instead the blocs overlapped, often extensively, creating an unruly tangle of lines. State officials insisted that prior to titling, the communities would have to resolve the tangle

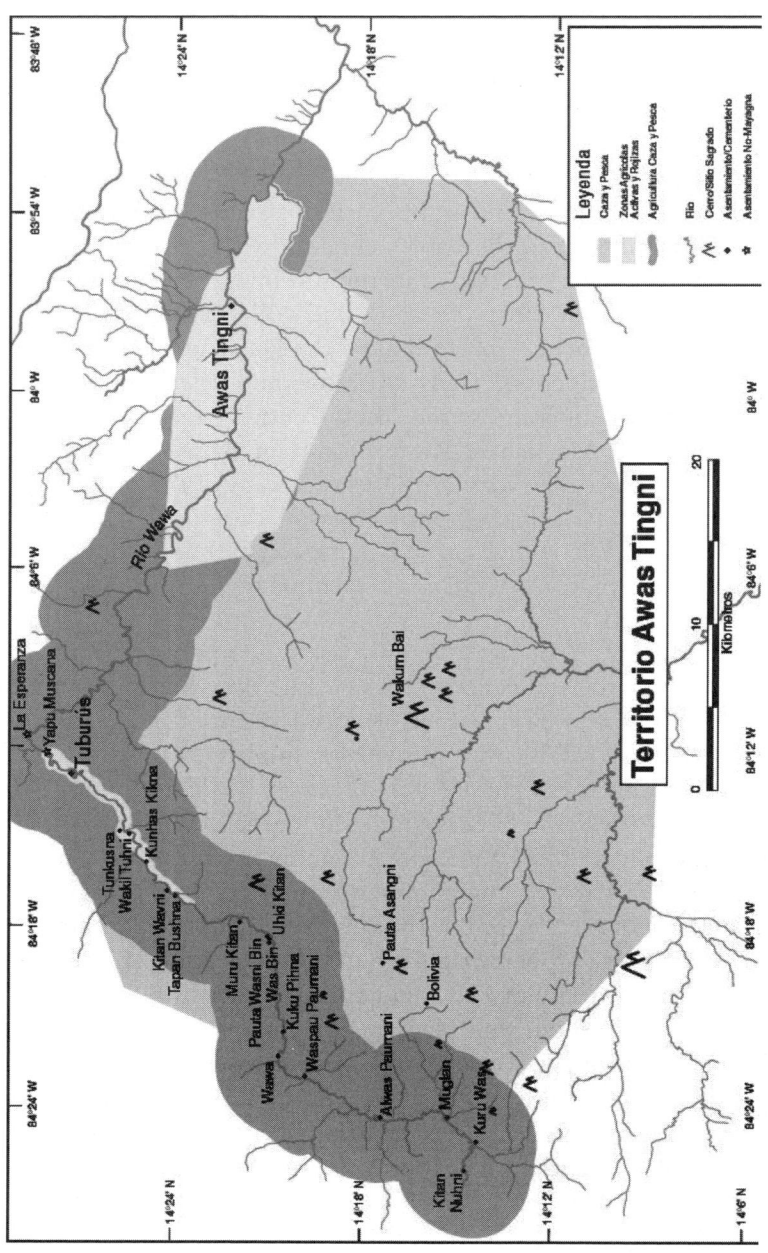

FIGURE 6.4. Map showing Awas Tingni's use and occupancy of land claimed by the community, 1995 (Theodore Macdonald). Source: S. James Anaya and Claudio Grossman, "The Case of *Awas Tingni v. Nicaragua*: A New Step in the International Law of Indigenous Peoples," *Arizona Journal of International and Comparative Law* 19(1), 2002, pp. 1–15.

of lines into boundaries delimiting mutually exclusive spheres of ownership—ownership commensurate with prevailing notions of property. This requirement was reproduced in the legislation passed with World Bank support, effectively converting a striking feature of on-the-ground tenure arrangements into a cartographic problem.

This situation set the stage for a series of bitter struggles over land and resources that has yet to be completely resolved. In the case of Awas Tingni, the overlaps contributed to a seven-year delay in the state's issuance of a title, during which the people of Awas Tingni lost over half their claim to neighboring Miskito communities, exacerbating tensions between groups who otherwise appeared equally "indigenous." This problem of overlaps was reproduced when the team of anthropologists that did the Nicaragua study completed a Ford Foundation-funded study in Honduras. There too Honduran officials used the overlaps as a pretext for delaying titling while complying with bank-supported efforts to draft community property rights legislation. Some of the Honduran communities had a measure of success in resolving problems amicably, though in other areas the overlaps exacerbated racial tensions among black, indigenous, and mestizo groups.[66]

The problem of overlaps did not, however, diminish the importance of mapping. Indeed, as a uniquely cartographic problem, it all but demanded a cartographic solution. Whether by design or not, the effect was to reinforce the importance of maps in producing a governable space crafted from the underlying rationales that structure colonialism: culture and economy. Far from ushering in a wave of decolonization, maps tightened its grip. In their lines and polygons, judges at the Inter-American Court and officials at the World Bank could see the possibilities for a geographical order that obscured the lives of people living there. This order required boundaries, clean lines that would denote mutually exclusive areas of ownership, whose security could be guaranteed by the state. Instead of protecting indigenous territories, boundaries made dispossession easier by making the state the ultimate arbiter of who owned what land.

Curiously one of the more concerted efforts to challenge this approach was also one of the more widely celebrated—and critiqued—Central American mapping projects of the 1990s. This project too took shape around a petition filed with the Inter-American Commission on Human Rights. The case involved a legal challenge filed by Mopan and Q'eqchi' Maya communities against Belize, contesting state plans to log and build roads that were intended to integrate the southern half of the country within the national economy. The Maya communities were legally represented by the Indian Law Resource Center, the same organization that represented Awas Tingni. Led once again by S. James Anaya, the lawyers pursued a strategy of producing maps to demonstrate the communities' use and occupancy of the lands in question.[67] After first approaching Mac Chapin to produce the maps, the lawyers turned to a familiar friend and ally: Bernard Nietschmann.

Between 1996 and 1998, Nietschmann's team of geography students from the University of California, Berkeley worked closely with Maya communities to produce a series of 23 maps later published as *The Maya Atlas*. One example can be seen in Figure 6.5. Instead of relying on GIS to reinforce their legitimacy, Nietschmann used hand-drawn maps made from aerial imagery and field surveys. The techniques marked a forceful return to Nietschmann's cultural ecology training, producing ethnographic and demographic evidence to bolster the Mayas' claim. Though the *Atlas* was submitted to the commission as evidence, it had a much larger effect circulating among U.S. and European environmentalists. Its international success had the effect of rehabilitating Nietschmann's reputation among those academic colleagues dismayed by his "political" work of the 1980s. Bernard Nietschmann, famed cultural ecologist, was back, this time in the guise of the "pragmatic post-colonialist."[68]

But *The Maya Atlas* only reinforced a broader reading of indigenous mapping as fundamental to the production of governable space. With their territories mapped as property, their rights guaranteed by an apparatus of domestic and international law, indigenous peoples found themselves bound tightly within spaces thoroughly defined, if not actually dominated, by states. The much hoped for wave of decolonization that maps were supposed to bring about had stalled out on the most basic question about what an anticolonial geography would look like.[69] That question fueled academic debates as much as it led indigenous peoples' organizations to reappraise mapping as a tactic. Whatever else could be said about this predicament, it made it clear that anticolonial struggles would have to fashion their own geographies or replicate colonialism's spatial grammar. The revolution, in other words, would not be mapped. Indeed, mapping seemed little more than a slow-motion counterrevolution. Over and over, mapping ran into the same problem: There was no preexisting matrix of territories, state, indigenous, or otherwise available for the maps to document. Instead, mapping soldiered on as the initial gloss on a new social order.

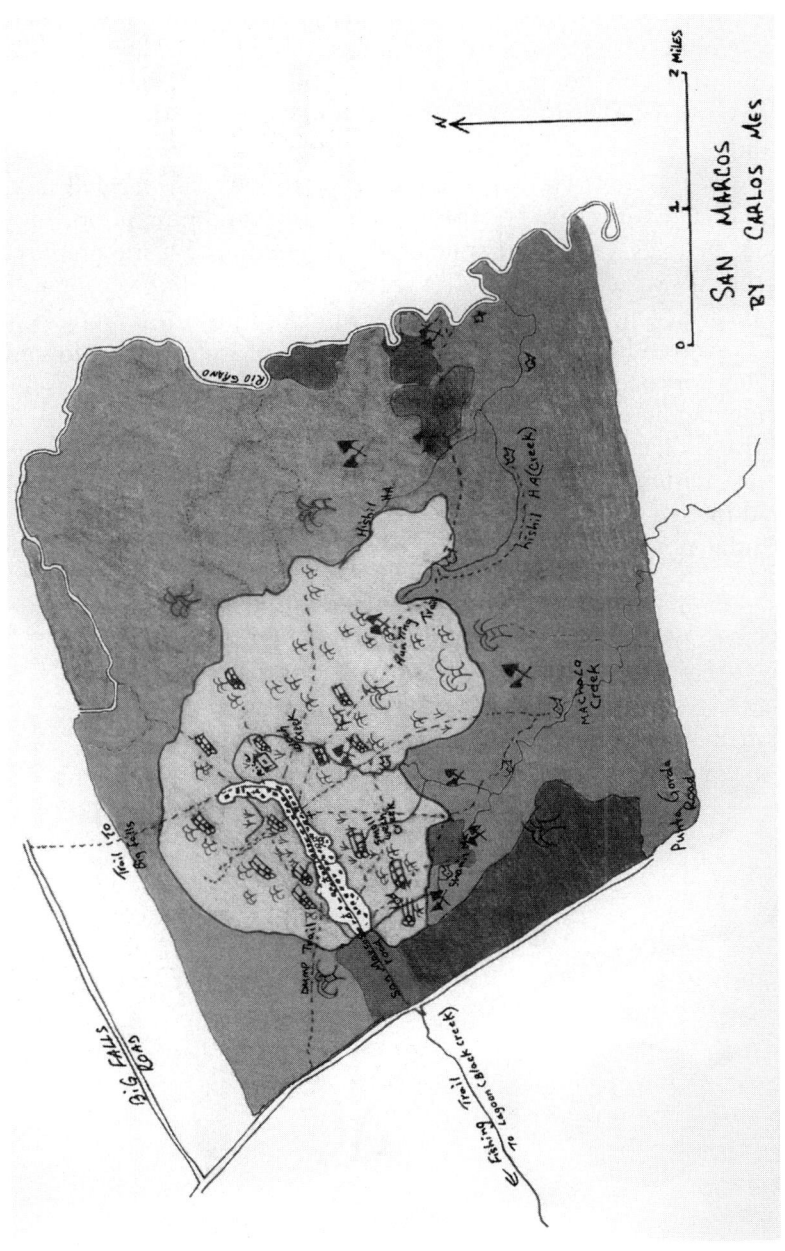

FIGURE 6.5. Village of San Marcos, *The Maya Atlas*. Used with permission from the Toledo Alcaldes Association (authors).

In this regard, Nietschmann and the legion of self-described "counter-mappers" he inspired were correct to claim that anticolonial struggles would have to take on the power of maps wielded by states and, before them, by colonial authorities. But in their commitment to maps, they missed the importance of actually creating the space that would undo colonialism's twin logics of culture and economy. Chapin's *Research and Exploration* map and those that followed clearly fell short of that challenge for a variety of reasons, political and pragmatic.[70] For indigenous peoples, taking the state on directly required being prepared to have the full weight of the state's security apparatus fall on them. It also meant forging alliances with environmental groups (among others) more invested in the cartographic grammar of states, one allowing "workable" approaches to conservation. Nietschmann's *Maya Atlas* has come under withering critique in this regard, "transferring the duty to catalog the indigenous culture from the anthropologist to the native informant" without interrogating the logic that structured the task.[71] Left unaddressed, that logic could easily turn mapping into a weapon *against* indigenous peoples.

7

Counterinsurgency and the Rise of the "Warrior Scholars"

Property, property, property: It was becoming a hot topic in the military during these years too. Especially at Fort Leavenworth's Foreign Military Studies Office. Especially for a retired Army officer working there as a senior researcher. Geoff or Geoffrey Demarest—he publishes under both names—spent 23 years in the Army in multiple positions in Latin America: Guatemala in the late 1980s and Colombia in the 1990s as commander, staff officer, attaché, intelligence analyst, instructor, and ... frequent student. In addition to a JD from the University of Denver law school in 1980, and a PhD from the University of Denver's Graduate School of International Studies in 1989, Demarest is a graduate of the U.S. Army War College, the U.S. Army School of the Americas, the Defense Attaché Course, the Foreign Area Officer's Course, the Defense Strategy Course, the Defense Language Institute, the Spanish Language Institute, and the John F. Kennedy Special Warfare Center. Others too, probably. As we type this chapter, he's finishing a PhD, in geography, at the University of Kansas.

He's a well-educated man, is Geoff Demarest. He's a self-professed libertarian too, and a *prolific* writer on a range of topics, including, preeminently, the importance of thinking through property—that is, thinking through the lens of property about ... just about anything.

As he wrote a few years ago: "The success of a society depends on construction of formal, liberal property regimes."[1]

Where he came up with this notion is hard to say, but an essential component of his style of "libertarian pragmatism"[2] could easily have come from Robert Ardrey's *The Territorial Imperative*. Ardrey's second book—the first had been the notorious *African Genesis*—came out in 1966 with the subtitle *A Personal Inquiry into the Animal Origins of Property and Nations*, and it prepared the ground for Demarest's thinking about the significance of property. During the later 1960s *The Territorial Imperative* was a Book-of-the-Month Club selection, editorial fodder, and a water-cooler conversation topic. Written by a playwright—*Star-Spangled*, 1936; *Thunder Rock*, 1939; *Sing Me No Lullaby*, 1954, among many others—and screenwriter (many movies, from 1946 to 1966), Ardrey's book had flair. And Ardrey's undergraduate education *had* included work in anthropology (a University of Chicago BA in 1930), so he was not unfit for the writing either. What Demarest got from Ardrey was the root idea that "a territorial species of animals, therefore, is one in which all males, and sometimes females too, bear an inherent drive to gain and defend an exclusive property."[3] For Demarest, territory constituted a "conflict threshold": "Almost every territorial trespass is subject to spatial mapping, and those territorial violations that imply war are often the most ardently mapped."[4]

Property. As Demarest would write in the preface to his 1998 *Geoproperty: Foreign Affairs, National Security, and Property Rights*, where he recommended that students of conflict might want to think about rereading Ardrey's *Territorial Imperative*:

> It is curious that the notion of property has been so lightly regarded and so under-used outside the dense confines of Marxist theory. Curiouser still considering the surprise, indignation, and sanctimony that often accompanies the uncovering of some secret interest in land. While investigating processes and systems, levels of conflict, game theory and motivators of violence, we presumed the object of human struggle into invisibility. An intellectual fig leaf was dangled over a powerful word. Now it is time to peek back under there; and that is the mission of this book. With it the reader is invited to consider a property-based appraisal of human struggle.[5]

Geoproperty is Demarest's best-known book. Standing on the ground Ardrey had cleared for him, Demarest found orientation in C. Reinhold Noyes's *The Institution of Property: A Study of the Development, Substance and Arrangement of the System of Property in Modern Anglo-American Law*, and in John Powelson's *The Story of Land:*

A World History of Land Tenure and Agrarian Reform.[6] From Noyes, Demarest took the idea of property as a collection of rights and obligations, what he refers to as "a contract with the State for the recognition and support of specific preferential rights."[7] What he got from Powelson was the idea that "customary land tenure, nonliterate society, trend migrations, slavery, and continuous warfare (conflicts not expected to be resolved) all go together; they contrast with fixed tenure (land registration), literacy (written contracts), settled existence, free wage labor, periodic peace, and the expectation that contracts will end."[8] Power, here, "becomes the ability to gain or protect property—and strategy the rational use of available resources to gain or protect property."[9] From this perspective, rights are clearest where land ownership is transparently registered and privately held. Therefore, register *private* property (which is a bundle of preferential rights and obligations). In a separate work ominously titled *The Peasant Betrayed,* Powelson described how recognition of private property rights was the only way to meaningfully include the poor in the economy and, by extension, give them political power.[10] All other approaches risked simply expanding the power of the state as the ultimate aribiter of land, creating antidemocratic societies such as those found in Mexico, Bolivia, Peru, and Nicaragua, not to mention Tanzania, Iran, and Egypt, among others.

Demarest published *Geoproperty* while the United States and Colombia were negotiating what would become Plan Colombia, a U.S.-backed counterinsurgency plan for fighting drug cartels and left-wing guerilla armies. Demarest turned to the application of his arguments to the Colombian situation, the very case that had influenced his thinking about private property and its role in counterinsurgency. Demarest had been in Colombia in a variety of capacities many, many times. Among other things he had led the effort to create a proposal for the "Feasibility of Creating a Comprehensive Real Property Database for Colombia," which the Foreign Military Studies Office issued in 2002. This opens with a statement of Demarest's operating assumptions, largely derived from Powelson:

1. There is a positive correlation between formal land ownership and material progress;
2. There exists a complementary relationship between formal land ownership and social peace;
3. Informally owned and unregulated land ownership favors illicit land use and violence;

4. Property manipulation is an overlooked dimension of outlaw political and military strategies and behaviors;

5. Collection of property information on a strategic scale is made feasible by new technologies; and

6. Property information can be used in support of law enforcement, military, developmental, economic, and diplomatic decisions and programs.[11]

Therefore, Demarest concludes, some effort should be mounted for obtaining and organizing property data to support U.S. decision making. In support of his position, he attached as Exhibit A his paper, subsequently published as a booklet, *Mapping Colombia: The Correlation between Land Data and Strategy*.[12]

By this time Hernando de Soto had published *The Other Path: The Invisible Revolution in the Third World* and his even more popular *The Mystery of Capital: Why Capitalism Triumphs in the West and Fails Everywhere Else*.[13] De Soto added another string to Demarest's libertarian lute. *The Other Path* is a detailed analysis of the success of Peru's "extralegals"—who inhabited Lima's massive slums—in acquiring land, running businesses, and providing transportation, along with a plea for their official legitimation. Near the end of his book, de Soto concludes that

> what Peruvians want, first and foremost, is firm property rights, reliable transactions, and secure activities. They want facilitative legal instruments, which they do not now have. Second, they want to avoid obstructive legal norms as far as possible. Third, they want to replace the state with informal and private organizations in many areas.[14]

Through formal recognition of the extralegals' de facto property rights, de Soto insisted, the Peruvian state could defeat the Maoist Shining Path, drying up the guerrillas' sea by offering stability to its base.

But the lessons were much broader than Peru. The "transition from dispersed, informal arrangements to an integrated legal system" was foundational to the success of capitalism. In a chapter from *The Mystery of Capital*, de Soto argues that this transition was crucial to the transformation of the United States from "a Third World country" into a modern state in the mid-19th century. That's when U.S. politicians recognized that they could not impose, through legislation, an order on the frontier that spread west from the Mississippi; they could only legislate *after* settlement occurred. For de Soto, the 1862 Homestead Act was a stroke of political genius, transforming de facto property rights into the basis for a legal system. "Historically," writes de Soto, "the Homestead Act does have great symbolic value, symbolizing the end of a long, exhausting, and bitter struggle between elitist law and a new order brought about by massive migration and the needs of an open and sustainable society."[15] Altogether missing from de Soto's account was any mention of the Indian wars of the 19th century, wars that created the space in which homesteading could even occur, much less make sense. Nor does he discuss the equally disastrous effect that the 1887 Dawes Act played in offering up "Indian land for sale" by allotting ownership of tribal lands to private individuals. The wars that created that space, as we know, became the paradigm for the strategies and techniques that would, in the 20th century, turn into counterinsurgency.

De Soto's highly selective reading of U.S. history mirrors his attempt to make sense of urban Peru in the 1980s where, he blandly states, "in the past forty years, indigenous migration has quintupled the urban population and forced cities to reorganize."[16] Like Enlightenment philosophers of industrializing England, de Soto wasn't concerned with the violent upheavals that had produced this migration. He was content to take its end result—*pueblos jovenes*, or slums, teeming with "informals" used to living outside the law—as a simple problem of population growth. He'd reduce a complex history to a question of security.[17] His approach allowed him to treat informality as a technical problem, one resolvable through measurement by using the law to structure practices of surveying and titling the "dead assets" created by informals into "living capital."[18] De Soto's ideas drew praise from Richard Nixon to Margaret Thatcher, George H. W. Bush, and Bill Clinton, reinforcing neoliberal claims that the free market was the best institution for creating freedom, and that private property rights were essential to its functioning. World Bank officials took note. Buoyed by lucrative World Bank consultancies, de Soto became "the global guru of neo-liberal populism," hiring out his services as a consultant to countries around the globe.[19]

Free-market acolytes at the World Bank were not the only ones to take notice of de Soto's arguments. Military personnel such as Demarest were paying attention too, drawn to de Soto's interweaving of economic growth, security, and individual freedom that fell in line with their own ideas about the world. Demarest's 2007 *Property and Peace*, which opens with a quote from de Soto ("How could something so important have slipped our minds?"[20]), rethinks security along these lines. "First,"

Demarest insists in his opening pages, "there must be formal property, *and then* there can be security and peace."[21] Demarest's assessment was informed by his identification of a series of security challenges that defied modern military organization. These "postmodern" threats proliferated in open, unregulated spaces characterized by a lack of private property rights. Indigenous territories constituted one such space occupied by everyone from the Zapatistas in southern Mexico to coca growers in Colombia, Peru, and Bolivia. The Internet furnished another example, a space where "indigenous rebels, feminists, troublemakers, and idle wireheads could participate in a cause with a limitlessly undefinable and complex population of sympathies and sympathizers."[22]

Demarest wasn't entirely alone in his assessment of security risks. A 1999 bibliography published by the Foreign Military Studies Office identified "insurgencies, terrorist groups, and indigenous peoples" as the leading threats to security in Latin America.[23] What Demarest added to the debate was a singular emphasis on private property as the means of engaging these threats. His take-home message from the first chapter is "Property is an elemental approach to conflict resolution"; for the second chapter, "Power, property, and wealth can be the same." The third chapter applies these ideas to a case study of Colombia, and the fourth turns them into a critique of "obsolete, big-space geopolitics." "What is to be done?", the concluding fifth chapter asks. Demarest's answer? "Study social contracts. Improve property."[24]

His more recent *Winning Insurgent War: Back to Basics*, while busily praising classical ideas of strategy and attacking French-influenced thinking in the social sciences, has property concerns at its core: "Property as the heart of a peaceful social contract; Property analysis as a way to understand power."[25] Expanding a little on the first, Demarest says, "I would promote the general construction of a sustainable property ownership system, which we can also call the social contract."[26]

In the "Restatement" concluding *Winning Insurgent War,* Demarest writes: "Create property. Property is the social contract, the contract of contracts that serves to resolve conflicts before they become violent. Property is where economics, law and geography meet. Know who owns what land."[27] The necessity for mapping ownership in areas without property records that's implied by this doctrine, especially as a counterinsurgency technique, has driven Demarest since the mid-1990s. As former U.S. Ambassador to Panama, Ambler H. Moss, Jr., puts it in his preface to Demarest's 2003 *Mapping Colombia*, "to succeed in both counternarcotics as well as the suppression of lawlessness, an indispensable starting point is the knowledge of ownership and the value of land."²⁸ There is nothing general, nothing theoretical about the proposition here: At stake is the U.S. war on drugs *and* peace in Colombia.

From the get-go, Demarest has had the implementation of these ideas in view. In his 2002 proposal, "Feasibility of Creating a Comprehensive Real Property Database for Colombia," Demarest was already writing:

> Modest funding of participation in the forum by the Colombian Geographical Society (CGS) is highly recommended. It is further recommended that said participation be done in partnership with a suitable United States counterpart, possibly the American Geographical Society (AGS). To these parings, IGAC/USGS and CGS/AGS, could later be added a counterpart pair of independent think-tanks, a pair of universities and, perhaps, a pair of news organizations. ²⁹

But why the AGS rather than the AAG, the far larger, far more prestigious Association of American Geographers? Perhaps the diminished AGS, with its small number of members, minimized the possibility for debate while conferring a veneer of academic legitimacy. Or was it because of the AGS's history of working for the military? The AGS was also under new leadership at the time. In 2001, geographer Jerome Dobson left his 26-year career at the Oak Ridge National Laboratories in Tennessee to take a position at the University of Kansas, 30 minutes from Fort Leavenworth. One year later, in 2002, he became president of the AGS. Why Dobson retired to the University of Kansas is anybody's guess, but by 2005 Dobson had secured Foreign Military Studies Office funding for the AGS's first Bowman Expedition, which, through Kansas geographer Peter Herlihy's México Indígena project, was designed to take the first step toward Demarest's dream of that universal, transparent cadaster capable of telling him "who owns what land." But we'll be detailing this history in the next chapter . . .

Meanwhile . . .

Meanwhile the United States's 21st-century wars in Iraq and Afghanistan were not going well, not with the massive deployment of troops and arms that had been the way the United States had thought about fighting wars since . . . World War II. The invasion of Iraq may have been shocking, but it did not awe Iraqis into submission, and when their insurgency began in the summer of 2003—barely after the invasion—the United States was wholly unprepared to deal with it. As the 1st Marine Division prepared to return to Iraq in the spring of 2004, its commander, Major General James Mattis, suggested that his officers reread Red Mike Edson's *Small Wars Manual*.[30] About the consequences of this, Nicholas Schlosser wrote:

> One can see the manual's impact on how the Marines approached operations in Iraq. Upon returning to Iraq in 2004, Marines focused on engaging the population, stressed the need to respect the local culture, and undertook concerted efforts to build local police forces similar to the constabularies described in the *Small Wars Manual*. Most counterinsurgency operations in Iraq have focused on company- and platoon-sized patrols designed to both destroy the insurgency and protect Iraq's civilian population. The axiom that political and cultural understanding are as important to achieving victory against insurgencies as defeating the enemy on the battlefield has also been a hallmark of Marine Corps operations in Iraq, especially during the Second Battle of Fallujah. As Marines prepared for battle in November 2004, they used information operations to encourage civilians to vacate the city before the impending battle and conducted the operation in cooperation with the Iraqi National Army and interim government. This approach can be seen throughout the *Small Wars Manual*, drafted almost 70 years before Operation Phantom Fury.[31]

The following year, Mattis joined then Lieutenant General David Petraeus at Fort Leavenworth's Combined Arms Center to oversee the drafting of a new (2006) U.S. Army/Marine Corps *Counterinsurgency Field Manual, FM 3-24/MCWP 3-33.5*.

Like Mattis, Petraeus too had a reputation as a reader, even as an intellectual—his PhD in international relations is from Princeton—and he had used *Small Wars Manual*-style counterinsurgency measures to build a sense of security and stability in the area of Iraq under his command. He used force judiciously, worked to rebuild the economy, immediately staged local elections, and in general took a *Small Wars Manual* approach to the occupation. That is, Petraeus and Mattis were on the same page of the same book when it came to the production of *FM 3-24*.

Neither Edson nor the *Small Wars Manual* are directly referred to in this new text, though the *Small War Manual* is included in the bibliography under "The Classics"; and there's a boxed treatment of the Vietnam-era Combined Action Program, which was a direct and successful outgrowth of Edson's Nicaraguan experiences.[32] Beyond these there seems to be a distinct effort to avoid referring to the older manual in any way, though Edson's predecessor Callwell's *Small Wars* is referenced, together with the works of T. E. Lawrence and Napoleon. Beyond these, references reach back not much further than Vietnam, and there's a sense that the thinking advocated by the new manual is supposed to be, well, new.

The manual was greeted with great acclaim on publication and has been downloaded literally millions of times, but there were also immediate reservations, and not just from antiwar elements. One of Demarest's students, for example, Major Jon-Paul N. Maddaloni, authored a monograph in 2009 that concluded: "The argument has been made that *FM 3-24* was largely written for the nuances of Iraq and is narrowly focused. The frame of *FM 3-24* fails to understand the fundamental nature of insurgencies and comprehensively examine counter-strategies."[33] As if acknowledging the validity of these sorts of criticisms, the services in question released a new edition in 2014.[34]

> Robert Jones, a retired Army Special Forces colonel and an adviser to the U.S. Special Operations Command who attended a conference of experts involved in the latest field manual rewrite, said the version produced under Petraeus provided an overarching view of counterinsurgency at a time when such a perspective was needed.
>
> However, the document doesn't fill the traditional mission of a field manual, which should be to focus on tactical operations at brigade level and below, he said.
>
> "It did not . . . work well to guide the day-to-day tactical actions of the Army brigades and Marine regiments sent into the fights in Iraq and Afghanistan," he said.
>
> The new field manual will describe efforts that were tried out in Iraq and Afghanistan without having a specific approach for future conflicts.[35]

Whether this new version of the manual will satisfy the critics from *outside* the services, however, is a very different matter.

Most serious among these critics are the anthropologists who have risen almost en masse to complain, object, revile, and make fun of the manual, a product they too read as more directed toward a U.S.-armchair audience than battlefield commanders, especially since its 2007 republication by the University of Chicago Press. But as Marshall Sahlins has written about the *Manual*, "the truly good news is that the military's appropriation of anthropological theory is incoherent, simplistic, and outmoded—not to mention tedious—even as its ethnographic protocols for learning the local cultures amount to an unworkable fantasy."[36] And that's a mild assessment. Nor is it far from the military's more recent appropriation of "human geography." Catherine Besterman, Andrew

Bickford, Greg Feldman, Roberto González, Hugh Gusterson, Kanhong Lin, Catherine Lutz, David Price, David Vine, and many, many others have raised dissenting voices, some raucous. But, because the *Manual* isn't all they're objecting to, perhaps we should introduce Petraeus's other Leavenworth project.

For *FM 3-24* wasn't the only thing Petraeus was overseeing during his tenure at Fort Leavenworth. There was also the Human Terrain System (HTS). Like the new field manual, the HTS was another response to the situation in Iraq that had begun deteriorating in 2003. In the summer of 2004 Major General Robert Scales testified before the House Armed Services Committee that

> more than a year after the Iraq war began soldiers are rotating home with a sense of unmet expectations. Consensus seems to be building among them that this conflict was fought brilliantly at the technological level but inadequately at the human level. The human element seems to underlie virtually all of the functional shortcomings chronicled in official reports and media stories: information operations, civil affairs, cultural awareness, soldier conduct . . . and most glaringly, intelligence, from national to tactical.[37]

The lessons were clear, Scales insisted in his testimony, that "computers and aerial drones are no substitute for human eyes and brains," and this led him to propose emulating the late-19th-century British practice of immersing bright officers in the cultures of, for example, China (Charles George "Chinese" Gordon) or Arabia (T. E. Lawrence). "At the heart of a cultural-centric approach to future war," Scales concluded, "would be a cadre of global scouts, well educated, with a penchant for languages and a comfort with strange and distant places."

Iraq, for instance, or Afghanistan.

A year later, in 2005, Montgomery McFate and Andrea Jackson published "An Organizational Solution for DOD's Cultural Knowledge Needs." McFate was an anthropologist, at the time with the Office of Naval Research, and Jackson was the Director of Research and Training for the Lincoln Center, a military contractor. In their paper they identified the importance of having social scientists join soldiers in the field, and they proposed a $6.5 million pilot program.[38] Within the year the Foreign Military Studies Office had sketched out the HTS as a collection of specially trained "human terrain" teams—including social scientists—embedded in the field, connected to "reachback" research cells, subject-matter "expert networks," a "tool kit," assorted other techniques, and "human terrain information."³⁹ The proposal fit squarely with one of Petraeus's own lasting conclusions from his experience in Iraq, namely that *"cultural awareness is a force multiplier."*⁴⁰ Continuing on, Petraeus writes:

> Knowledge of the cultural "terrain" can be as important as, and sometimes even more important than, knowledge of the geographic terrain. This observation acknowledges that the people are, in many respects, the decisive terrain, and that we must study that terrain in the same way that we have always studied the geographic terrain.

In 2006, BAE Systems, the global defense and security contractor hired to run the HTS, began posting advertisements for positions. In 2008 Michael Bhatia, Nicole Suveges, and Paula Loyd became the first social scientists killed in action.⁴¹ It was also the year Don Ayala, a former U.S. Army Ranger then working for BAE Systems, was charged with murder in the death of Loyd's assailant.

By 2010 the program's budget was $150 million.⁴²

Given that anthropologists with the program could earn up to $300,000 a year while posted abroad, the paltry number who've signed up is startling: Fewer than eight of the HTS's well over 400 employees have advanced degrees in the field.⁴³

This "no show" could be the result of the rapid opposition to the HTS that began to coalesce. In 2007 the American Anthropological Association issued a "Statement on the Human Terrain System Project" that concluded with, "The Executive Board views the HTS project as an unacceptable application of anthropological expertise," having argued that, "in the context of a war that is widely recognized as a denial of human rights and based on faulty intelligence and undemocratic principles, the Executive Board sees the HTS project as a problematic application of anthropological expertise, most specifically on ethical grounds. We have grave concerns about the involvement of anthropological knowledge and skill in the HTS project." The board followed this up with a series of statements, sessions, letters, reports, and other forms of condemnation, all of which garnered extensive media coverage. Since anthropology was what the HTS was mostly supposed to be, this was all pretty damning.

Then in 2009 Roberto González published *American Counterinsurgency: Human Science and the Human Terrain*, the same year that the Network of Concerned Anthropologists came out with *The Counter-Counterinsurgency Manual: Or, Notes on Demilitarizing American*

Society.⁴⁴ In 2010 González's substantially heftier *Militarizing Culture: Essays on the Warfare State* appeared, followed in 2011 by David Price's *Weaponizing Anthropology: Social Science in the Service of the Militarized State*.⁴⁵ The books are serious critiques of both the HTS and FM 3-24 from a variety of professional perspectives, and they seriously raised the stakes of the game here at home. So did Marine Corps Major Ben Constable's piece in the *Military Review*, "All Our Eggs in a Broken Basket: How the Human Terrain System is Undermining Sustainable Military Cultural Competence." Constable questions the very need for the HTS—at the expense of the existing indigenous foreign area and civil affairs officer programs—while simultaneously addressing its many problems, including the way "the practice of deploying academics to a combat zone may undermine the very relationships the military is trying to build, or more accurately rebuild, with a social science community that has generally been suspicious of the U.S. military since the Viet Nam era."⁴⁶ Yeah, how did McFate overlook that problem?

HTS Meets the Bowman Expeditions

Another 2009 publication, independent journalist John Stanton's *General David Petraeus' Favorite Mushroom: Inside the U.S. Army's Human Terrain System*, pulled together the two themes of this chapter: Demarest's property mapping and Petraeus's Iraqi and Afghan initiatives.⁴⁷ Most of Stanton's book is concerned with the HTS, but there's a chapter called "HTS Meets the Bowman Expeditions" that starts off, as we do in this book, in the Sierra Juárez of Oaxaca: "On January 14, 2009 the Union of Organizations of the Sierra Juárez of Oaxaca (UNOSJO) issued a press release accusing the principal researchers/managers of México Indígena—a program in the larger Bowman Expeditions—of unethical conduct for not fully disclosing that the U.S. Army is a sponsor of the Bowman Expeditions. They also accuse the principals of geopiracy. According to a member of the anthropology community, 'This is a nasty little story.' " ⁴⁸

"What," Stanton asks, "is U.S. Army TRADOC thinking?"

TRADOC, the Army's Training and Doctrine Command, is located all over the United States, but its G-2 (intelligence) unit is at Fort Leavenworth. Within TRADOC's G-2 Enterprise Intelligence Support Activity section, you'll find the Foreign Military Studies Office (where, of course, Geoff Demarest has his lair); and in TRADOC's G-2 Enterprise Operational section you can find the HTS. The two exist in parallel positions in TRADOC's G-2 intelligence function. Field manual writing is also, as we've said, centered at Fort Leavenworth, in the Combined Arms Doctrine Directorate of the Combined Arms Center. But Leavenworth is a big place, with some 15,000 residents and employees, and relationships are complicated; though with the Foreign Military Studies Office launching the HTS, it's unlikely that relations with Petraeus were anything but warm. It's even possible that Petraeus knew about Demarest's interest in property registration, but certainly they met, at least once, when the board of the AGS, shepherded by Demarest, spent a couple of hours with Petraeus on a 2006 field trip to Leavenworth. We love the way Dobson introduces this visit in his report printed in *Ubique*, the AGS newsletter, "Each year the AGS Council visits an institution where geography is practiced as an integral part of one or more major missions directly affecting science and society. . . . This year we chose Fort Leavenworth, Kansas, home of the Command and General Staff College, Foreign Military Studies Office, and other important institutions."[49]

Dobson's report does acknowledge the Foreign Military Studies Office's support of México Indígena—and Herlihy was part of the AGS tour group—as well as a then forthcoming expedition to the Antilles, but it's *lavish* in its attention to Petraeus, who, Dobson assures us, "At the conclusion of our visit, he said he had a new appreciation for geography as a source of such understanding" (of cultural landscapes). Heady stuff! Not to mention having the prime mover of the new *Counterinsurgency Field Manual* and the HTS in friendly conversation with the prime movers of the Bowman Expeditions and the México Indígena project.

No matter the degree of cordiality, however—and the accompanying photo of Petraeus leaning across a table to shake the hand of AGS Executive Director Mary Lynne Bird (Dobson grinning in the background) is, oh, *so* cordial—none of this should be allowed to gainsay Stanton's contention that

> while UNOSJO's claim of direct linkages between the Bowman Expeditions/ México Indígena (BEMI) and the HTS remains unsubstantiated at this time and may, indeed, be incorrect (a call to Lieutenant George Mace, PAO, of the HTS program could not be returned in time for release), the fact is that U.S. Army TRADOC owns the troubled HTS and a $500,000 chunk of the controversial BEMI—and the data that goes with both. As reported in prior pieces on the HTS, sources state that data from Human Terrain Mapping (HTM) for HTS does not remain compartmentalized but is shared with other U.S. Army intelligence related databases. There is no reason to expect

the BEMI data has been treated any differently. Furthermore, the BEMI appears to have accomplished what the HTS program promised but could not produce: a useful deliverable in the form of a user-friendly geographic information system (GIS).

This, of course, is Dobson's and Herlihy's claim: The AGS *can* deliver the goods! To underscore their pitch, the Executive Summary PowerPoint presentation on the México Indígena project opened with a slide featuring a picture of Petraeus, accompanied by his statement on the value of understanding the cultural terrain.[50] The quote as much as the AGS's promised deliverable (a "multiscale database") supports the claims of UNOSJO and other detractors of the Bowman Expeditions that although the work may be open-source, it's nonetheless military intelligence.

David Price writes about the importance of this sort of open-source nonclassified intelligence in *Weaponizing Anthropology*. Having rehearsed Robin Winks's account of how, in 1951, a small group of Yale historians, using no more than declassified materials in Yale's library, came up with 90% of the material in a competing report by CIA analysts with their access to classified data, Price points to the way that American scholars recently outperformed the CIA's best estimates for dangers in postinvasion Iraq, again using only publically available sources. "I would suspect," Price continues, "that a repeat of the Yale Report experiment focusing on the Middle East would find another 10% intelligence gap, but with the academy now winning due to the deleterious effects of generations of CIA intellectual inbreeding."[51]

The point is that intelligence doesn't have to be secret to be intelligence. This is what Demarest's demand for transparent property records ultimately comes down to. In his 2011 "Urban Land Use by Illegal Armed Groups in Medellin," Demarest makes clear how he imagines these records being used. He concludes by saying, "Looking at IAG [illegal armed groups] presence through the lens of land use can contribute positively to the construction of an urban counter-crime, counterinsurgency, and sustainable development method (probably in Geographic Information System format)."[52] Detailing the effective strategies exploited by the Medellín government, he writes: "Perhaps most intriguing, the government assured that streets were clearly named, addresses specific, residents registered by location, and ownership rights formalized." Successes of the Colombian security forces in Medellín, Demarest stresses, are due to their adoption of city planning language and methods. The assertion built on the outcome of a 2002 counterinsurgency campaign, known as Operation Orion, launched by the Colombian military against urban guerrilla groups in Medellín's Comuna 13. Often regarded as a decisive success, Operation Orion was allegedly carried out with CIA involvement as part of Plan Colombia.[53] In Demarest's assessment, the key to the operation's success was the decision by "military and civilian authorities to maintain a permanent, physical presence [in] all parts" of Comuna 13, the sector of the city where fighting was concentrated.[54] Prior to the operation, guerillas had relied extensively on the control of commercial property to maintain their strategic control over the sector, hollowing out state presence much as the Shining Path had done in the slums of Lima, Peru in the 1980s. As further proof of that success, by 2008 Demarest already had at his disposal the findings of another Bowman Expedition, this one led by a geography professor from Western Kentucky University, David Keeling, to . . . *Medellín's Comuna 13*. Under Keeling's direction, that Bowman Expedition used GIS to record property rights in Comuna 13 and document incidences of violence. The Bowman Expedition maps, along with Operation Orion, pulled together the elements of success that Demarest first identified in 2003's *Mapping Colombia:*

> Winning an internal war requires that the whole gamut of government enterprises and every aspect of a nation's power be used to a common end. In the context of an internal war, when it comes to the cartographic function of the state, there can be no logistical distinction between military and nonmilitary effort. Cartography supports the whole range of state interests, from the most immediate combat needs to long-term pacification and development, but it must in any case be comprehensive and current.[55]

Nothing more, he might add, was applied in Herlihy's mapping of land in Tiltepec and Yagila than such simple planning methods, and it's to that project that we now turn.

8

The AGS, the Bowman Expeditions, and the México Indígena Project

In retrospect, it's easy—perhaps too easy—to see the México Indígena project as an inevitable weaponization of indigenous mapping. The pieces were all there, from Herlihy's mapping experience to Demarest's interest in counterinsurgency and property, along with the AGS's ambitions to "bring geography back" to relevance among U.S. policymakers through . . . *exploration*. The two areas visited by the project, the Huasteca Potosina in San Luis Potosí and the Rincón de Ixtlán in Oaxaca furnished the rest, offering up "indigenous communities" whose collective rights to property, recognized in the aftermath of the 1910 Mexican Revolution, were actively slated for conversion to individual titles under the auspices of sweeping land reforms introduced in 1993. Rationalized in terms of neoliberal economic policies favoring private property and free markets, the reforms played a direct role in mobilizing indigenous communities in defense of their lands, most notably in the 1994 Zapatista uprising. Add to that, U.S. security analysts' general interest in Mexico as the site of a running drug war and a potential failed state and the story quickly takes shape.

Or at least the conspiratorial version of events does. We say *conspiratorial* deliberately, not to cast doubt on the plausibility of that account but rather to underscore the difficulty of ever really knowing what went on. The only people able to do that are Demarest, Herlihy, Dobson—the México Indígena team—and the communities of the Rincón themselves. At different moments each has tried to distill the truth of the project from its own perspective. Their accounts take twists and turns of logic that suggest there was no master plot to target the Sierra. Instead, they describe a convergence of related but disparate strands of indigenous mapping, counterinsurgency, and geographical research. Indeed if there's any conspiracy, it's the one that *denies this convergence of interests* by pretending the México Indígena project was nothing more than routine research, geography as usual. Any military interest in the project, as Herlihy and Dobson repeatedly claim, was purely circumstantial. Still, that's no excuse for ignoring the political and intellectual implications of that convergence—implications that, when seen from the Sierra Juárez, were all too clear.[1]

As a reminder, here's what Tiltepec, one of the communities mapped by Herlihy's team, had to say in March 2009:

> They [the México Indígena team] never told us that the data they collected in our community would be turned over to the Foreign Military Studies Office (FMSO) of the United States Army, nor did they inform us that this institution was one of the sources of funding for the project. For this reason, we believe that the researchers deceived our General Assembly in order to take information from us that served their own purposes.[2]

Two years later, in 2011, municipal authorities from five other Rincón communities affirmed Tiltepec's declaration, saying:

> We do not agree with the way in which the México Indígena team conducted its research in the communities of San Juan Yagila and San Miguel Tiltepec between 2006 and 2008. The team did not inform the communities of the source of the funds used to support the project, expressly hiding the United States Army's role in the project, in violation of our right to free, prior, and informed consent recognized by the United Nations' Declaration on the Rights of Indigenous Peoples. We also support both communities in their efforts to confront any problems they may encounter as a result of the completed research.[3]

Herlihy and his team have yet to reply to the particulars raised by Tiltepec, much less the other Rincón communities.

Instead, Herlihy has steadfastly reiterated his belief that "the AGS Bowman Expeditions program, and the México Indígena project in particular, again show how the military and other government funding can support valuable geographic scholarship."[4] But affirming the concerns raised by the Rincón communities, Melquiades Cruz, a resident of Yagavila, has written:

> Aside from the rumors of inappropriate conduct or the lack of informed consent from the communities, this situation worries us because we have always seen a strong link between geography and the interests of the military industrial complex, especially in the recent attempts to create worldwide property databases. The *México Indígena* project subscribes to a political-military strategy. We cannot forget that this mapping occurs in the midst of the debate over a package of military financing from the United States known as the Mérida Initiative. The control and displacement of indigenous communities is intended to prevent potential conflicts in "hot spots," contribute to the military control of the region, and finally free up natural resources for the benefit of the government and its transnational allies.[5]

In short, if there's any dirt to be dug up about the México Indígena project, it's not in the Sierra Juárez (or the Huasteca Potosina). Instead it's in the constellation of ideas and forces brought to Mexico by the Bowman Expedition.

After all, the maps that Herlihy's team produced weren't really for the communities. They *knew* where they were. The maps were for Demarest, the Foreign Military Studies Office, and the AGS and the professional reputations of its latest corps of explorers. The circulation of the maps within these realms brings into view a larger complex of institutions and individuals—an *apparatus*, if you will—organized in the name of defending a social order, codified in maps and the law, and backed by force against any and all threats to its security. It is to this military–industrial–academic complex that we now turn our attention.

The Mapmakers

Let's start with Herlihy and the AGS. As detailed in Chapter 6, Herlihy was a product of the boom in indigenous mapping that occurred in the 1990s. Following the completion of his dissertation on the Emberá and Wounaan in Panama in 1986, Herlihy built his career as an authority on what he calls "participatory research mapping" through work with other mappers in Honduras and Panama. In spite of these collaborations, Herlihy claims—as on his faculty website—to have "developed the first participatory research mapping (PRM) methodology in Latin America in 1992, since pioneering its research and applied use in geography, other disciplines and development work, particularly conservation work in Central America."⁶ No Fourth World-ist, Herlihy consistently emphasized the importance of mapping for tenure rights, through consultancies with groups such as The Nature Conservancy and the German NGO, Gesellschaft für Agrarprojekte.

In 2003 Herlihy's research trajectory took a different direction when he received a U.S. Department of State Exchange Grant to teach geography in Peru, followed a year later by a University of Kansas faculty grant to study "Communal Land Titling, Natural Resource Conservation and Social Conflict in the Neotropics"; but the grant that really changed the course of his research came in 2005, when Herlihy was awarded a Fulbright to spend a year at the Autonomous University of San Luis Potosí in Mexico. In addition to teaching courses on mapping, Herlihy proposed to study "Communal Land Titling Systems in México" in conjunction with a Mexican colleague, Miguel Aguilar Robledo. With Robledo, Herlihy secured a second grant to conduct a participatory mapping project in the Huasteca Potosina focused on natural resources and conservation. In that project, Herlihy and Aguilar began mapping Teenak and Nahua communal lands. Again, research as usual for Herlihy, except this time it drew the attention of the U.S. Army's Foreign Military Studies Office.

AGS President Jerome Dobson helped guide the Foreign Military Studies Office to Herlihy's work. Beginning in 2000, Dobson had guided the AGS's efforts to "reestablish exploration as a key mission of the Society."⁷ This return to the AGS's 19th-century roots came with a 21st-century twist: "Through technology," Dobson wrote in the AGS newsletter, "exploration can now be extended to places where humans cannot go, to phenomena that cannot be observed directly by human senses, and to macroscopic processes so large they can be observed only though remote sensing and geographic information science (GIS)."

But technology alone could not float such ambitious plans. It would also take money. To that end the Society sought "to promote exploration by government, business, and academia." The AGS took its first step toward that goal in December 2000, when it revived its old tradition of having explorers sign its globe. This time around, the Society used the ceremony as a way of soliciting donations from private companies such as Esri, the privately held GIS software giant. But Dobson wanted more and he set out to find funds for actual exploration, eventually leading him to the Foreign Military Studies Office.

When and where Demarest met Dobson is not (yet) known. Over the course of his 26-year career at the government-run Oak Ridge National Laboratory, Dobson did extensive research for the Defense Department and other government agencies.[8] His last major endeavor at Oak Ridge was a $2 million project to develop the LandScan Database for the Department of Defense, a project that mapped the global distribution of population at one-kilometer resolution. Dobson continued to work on the database after leaving Oak Ridge, developing applications for it listed on his curriculum vitae as "Casualty Estimation" and support for NATO's "Operation Able Ally" during 2002 and 2003.[9] One of the database's chief functions is to identify *where* people live, and that was of interest to Demarest.

At the Foreign Military Studies Office Demarest had devoted himself to researching and developing counterinsurgency strategies based on his experience in Guatemala and Colombia. Both wars were disastrous for indigenous peoples, culminating in the genocide of Maya people in Guatemala and the displacement of millions of indigenous and black communities in Colombia. But Demarest's concern lay less with indigenous peoples than with their land. Hence his tight focus on property. Property information allowed states—and therefore anyone with access to the registry—the ability to know not just where people lived but also *who* they were. Through registration of property rights, Demarest envisioned a means of controlling populations whose lives remained otherwise unintelligible to him, using their desire for rights to try and win their allegiance. This motivation fueled Demarest's 2002 proposal to create a comprehensive property database for Colombia as part of revamped counterinsurgency efforts against leftist guerrilla forces, the proposal in which he identified the AGS as a "suitable counterpart."[10] Though as far as we know the project did not proceed, Demarest must have known that Dobson, by then the AGS president, sat in an office less than an hour's drive away.

Though the AGS hadn't worked with the Army for some time, the University of Kansas most certainly had. In 2001 Anthropology Professor Felix Moos made a bid for turning the university into a training ground for intelligence professionals, helping to set up, with a former CIA director, the Pat Roberts Intelligence Scholars Program at Kansas.[11] That relationship received an institutional boost in 2004 when Kansas and Fort Leavenworth signed a memorandum of understanding aimed at strengthening "both institutions' abilities to train future military and civilian leaders, particularly in international relations and national security."[12] A year later Dobson landed his first Foreign Military Studies Office contract (for $185,949) to create an "open-source" database of property rights in Mexico, building on both his LandScan experience and the Geography Department's strengths. Herlihy was the co-principal investigator. The same year, 2005, Dobson and Herlihy signed a second contract with the Foreign Military Studies Office for a "Mexican Property Development Survey" worth $96,264. Herlihy put that money toward his work in the Huasteca Potosina, comparing community tenure arrangements gathered through participatory mapping work with government property rights data. By the end of 2005, Herlihy, Dobson, and Robledo combined the mapping project funded by the Mexican government with the two Foreign Military Studies Office contracts to form the México Indígena project.

The new combined project cemented Dobson's efforts at the AGS to bring exploration back to the Society (and geography, more generally). As with Demarest, Dobson's primary interest lay less with property, per se, than with security. In the February 2006 issue of *Ubique*, Dobson hailed Herlihy's work in Mexico as a prototype for future Society-backed expeditions. Lamenting the lack of basic geographical knowledge among U.S. policymakers and building on Isaiah Bowman's work with the Inquiry during World War I, Dobson summarized a proposal he'd written "suggesting that the AGS send a geography professor and two or three graduate students to every country in the world for a full semester each year, with teams rotating on a five-year cycle so that each country is understood by five separate teams."[13] Dobson estimated the program would cost $125 million per year to implement, "a pittance compared to what the intelligence community typically pays for far less effective information." Dobson continued, "I circulated the proposal and found allies at Ft. Leavenworth, Kansas. They marketed the idea and funded a prototype for the larger concept that, ideally, would reach every country in the world." "Allies" undersells what in Dobson's hands quickly became a very lucrative relationship between the Foreign Military Studies Office and the AGS. After securing $281,213 from the Foreign Military Studies Office for the México Indígena project in 2005, the following year Dobson landed another contract for an expedition to the Antilles worth $520,598. In 2007, the Foreign Military Studies Office contracted with the AGS for two more expeditions, to Colombia ($176,761) and Jordan ($544,116). Demarest had a hand in helping secure Foreign Military Studies Office funds, with Dobson's prototype expedition, México Indígena, paving the way.

A "Silent Revolution"

Still, why start with Mexico and not Colombia, as in Demarest's earlier work? Part of the answer had to do with Herlihy's own research interests in indigenous mapping and, to a lesser extent, Dobson's GIS background. Mexico provided the context for merging these concerns, with Demarest's focus on property in a setting that was, arguably, more accessible than Colombia and yet still relevant to security experts. Mexico was also in the process of turning against the trend toward communal titling found elsewhere in Central and South America with an aggressive program of property reform.

Known as PROCEDE, an acronym for the Program for Certification of Ejidal Rights and Titling of Urban Lots (*solares*), this program aimed to privatize collectively owned lands in the name of bolstering the security of property rights. The program further set its mark on undoing the "social sector" of collective property rights established following the 1910 Mexican Revolution. Most of these lands were formally recognized as *ejidos*, created through the restitution of lands to agricultural workers upon the dissolution of large estates or *haciendas*. Another category of communal property, the *comunidad agraria* or *comunidad indígena*, recognized existing forms of communal land tenure and often closely followed colonial-era arrangements. Both categories worked well enough on paper, accommodating revolutionary demands that rights to land belonged to those who worked it, and establishing a regime of collective property rights that functioned alongside private property rights.[14] In practice they proved problematic for state officials who, among other challenges, were notoriously vexed in their efforts to map these lands and properly register their ownership. Their indeterminacies and ambiguities created what historian Raymond Craib has termed a "fugitive landscape" that at once escaped official control and abounded with opportunities for malfeasance and illegality.[15]

In the wave of the market-oriented reforms of the 1990s, Mexican officials targeted this this fugitive landscape for "regularization" that would straighten out questions of ownership and improve the efficiency of land markets. Echoing Hernando de Soto's claims, Mexican officials determined that the pivotal first step in this regularization process hinged on dividing communal lands into individually owned pieces of private property. Once titled, individual owners could use their property as collateral to access credit needed to shift production away from local consumption and toward national and international markets. The reforms cleared the way for further integrating agricultural production in the farthest flung corners of rural Mexico, as envisioned by the 1994 North American Free Trade Agreement (Figure 8.1).[16]

To achieve these goals—of dissolving communal lands and registering private property—Mexican officials created PROCEDE. As with land regularization programs elsewhere, PROCEDE required the production of new surveys and maps as a crucial first step toward regularization. These surveys and maps did not necessarily reflect on-the-ground tenure arrangements so much as create, in the abstract, an imagined order on which regularization could be modeled. The program had immediate and far-reaching implications for rural Mexico that went well beyond the simple titling of land. As the México Indígena team put it, "For thousands of indigenous communities in Mexico, the PROCEDE program represents a silent revolution, undoing social property and changing communal ownership patterns that in some cases date back to pre-Colombian times."[17] Much as allotment policies were used to dissolve tribally held lands in the United States in the late 1800s, PROCEDE actively sought to incorporate indigenous lands in Mexico into the market through registration of individual rights and subsequent privatization. The effort amounted to assimilation by economic means, relying on the market to overcome centuries of political and economic marginalization frequently justified in terms of race.

The México Indígena project claimed to help make this "silent revolution" audible by mapping the impacts of PROCEDE on indigenous communities in two respects. Participatory mapping projects such as the one underway in the Huasteca Potosina could be used to gather data on community tenure—data that could be compared with PROCEDE titles and further analyzed for errors and discrepancies. Herlihy's approach to mapping would, in turn, allow the database to reach into those pockets where property rights were most obscure, if not altogether nonexistent. As far as Demarest was concerned, it was a convenient opportunity to test out his property-based approach to counterinsurgency. In 2006 and 2007 the Foreign Military Studies Office provided an additional $435,150 for the project on top of the $281,213 committed in 2005.[18]

Into the Sierra

With over $700,000 in military contracts in hand, the México Indígena project clearly had to expand into other parts of the country. But

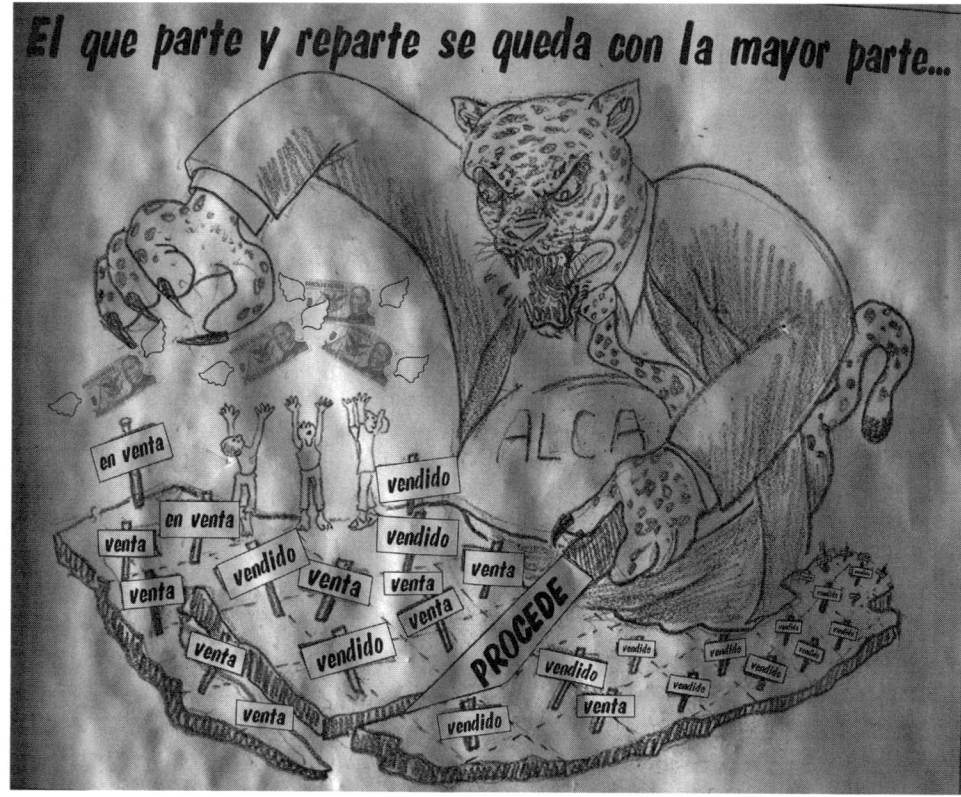

FIGURE 8.1. Editorial cartoon from the January 2006 edition of the newsletter *Lekil K'optik* and displayed on the walls of UNOSJO's offices in Guelatao. "ALCA" is the Spanish acronym for the 2003 "Free Trade Area of the Americas" proposed by U.S. President George W. Bush. The Spanish proverb at the top (*El que parte y reparte se queda con la mayor parte* . . .) translates roughly as "He who cuts and shares the cake keeps the best piece for himself." An image of the same cartoon appears on the México Indígena website hosted by the University of Kansas.

where? PROCEDE concluded in December, 2006, having mapped all but the most conflicted agrarian areas in rural Mexico as well as those that had expressly rejected PROCEDE.[19] These appeared in Herlihy's and Dobson's property rights database as unsurveyed pockets. It was these property-less areas that the México Indígena project targeted for expansion. The position was entirely in keeping with Demarest's assessment that property-less places provided conditions conducive to illegal activities, posing security threats that could be neutralized by mapping property regimes.

Still, there were practical problems to consider. Some areas, such as the Huichol communities of the western Sierra Madre, the mountainous forests on the boundary between Chiapas and Oaxaca, and the Lacandon lowlands, were simply too conflicted to allow for research. Other areas, such as the Mixteca in Oaxaca and Guerrero, had long histories of armed resistance to state incursions. Ultimately, Herlihy settled on two areas: the Raramuri communities of the Sierra Tarahumara in Chihuahua and the Zapotec communities in the Sierra Juárez of Oaxaca.

To get access to either locale, Herlihy needed an invitation from communities or organizations in the region. The invitation was more than a formality. One of the more strident aspects of the rights-oriented approach to mapping that Herlihy helped develop in the 1990s was its emphasis on only making maps at the request of an indigenous organization or group of communities. The need was born out of indigenous mapping's self-styled anticolonial approach to making maps "of, by, and for the peoples of Latin America," as the title of one of Herlihy's articles has it.[20]

While attending the Conference of Latin Americanist Geographers in Morelia, in 2005, Herlihy serendipitously received what amounted to an invitation to visit the Sierra Juárez from a biologist named Gustavo Ramírez. At the conference, Herlihy had reported on his mapping work with Teenek and Nahua communities in the Huasteca Potosina. There his mapping work had helped communities identify errors in the completed PROCEDE surveys. The maps produced by the México Indígena project empowered the communities not only to point out these problems to government officials, but to propose solutions to them.[21]

Herlihy's presentation spoke to a problem that Ramírez knew well. Since graduating from college with a degree in biology in 1997, Ramírez had returned to his hometown of Ixtlán de Juárez where he'd found work with UNOSJO. Ramírez fit in well with the leadership of UNOSJO, many of whom had combined their upbringing in the Zapotec communities of the Sierra with university training to organize against destruction of the region's resources and way of life. Most recently, UNOSJO had led efforts to refuse PROCEDE's entry into the Sierra. By all accounts, the Sierra constitutes one of the few areas in Mexico where state-sanctioned property does not exist (Figure 8.2).[22] As we

FIGURE 8.2. Sign greeting visitors to Ixtlán de Juárez declaring that "private property does not exist in this community" and prohibiting the sale of lands. A similar photo appears in one of the articles published by the México Indígena team.

know, this status is not an accident, nor a product of the region's isolation from the rest of Mexico. As is true for many indigenous territories, it is a legacy of struggle.

How far one wants to trace that struggle back is a matter of choice, but one important stop is certainly the agrarian reforms introduced by the Mexican Revolution, the very ones targeted for dismantling by PROCEDE. The *serrano* communities had never bought the logic of agrarian reform used to identify *ejidos*. In contrast to the agrarian mantra that "the land belongs to those who work it," the *serrano* communities had maintained their own, distinct form of communal tenure. Indeed, although the state nominally recognized their status as *comunidades indígenas*, the configurations of communal lands were not mapped by state officials until the mid-1990s. By then PROCEDE was well under way, and the effort to map the community boundaries offered something of a compromise. On the one hand, if PROCEDE was going to enter the region, state officials needed to know what the communal system was in order to privatize it. On the other, the agrarian titles extended formal legal recognition to the communities who had fought for decades, quite literally, to stave off extractive industries. The state's efforts to subdivide communal lands, coupled with the communities' own history of struggle, had led UNOSJO to develop an interest in having its own, internal capacity to produce maps.

UNOSJO's interest in mapping intertwined the history of the Sierra with that of the organization.[23] Beginning in 1955, the Mexican company Fábrica de Papeles de Tuxtepec logged extensively in the Sierra with the blessing of a presidential decree, converting the Sierra's forest into paper products. In the absence of community titles, state officials justified the concession in part due to the region's status as state lands. That era came to halt in 1986, when communities successfully blocked renewal of the concession. Mexican officials responded by proposing that the remaining unlogged areas of the Sierra be reserved for biodiversity conservation, once again threatening the Sierra communities with displacement and dispossession. UNOSJO itself was founded in 1992 in the midst of these struggles, and immediately took a lead role in opposing efforts by "biopirates" to take genetic samples, for agribusiness interests, of the hundreds of varieties of corn cultivated by Sierra residents (Figure 8.3).[24] Since then a spike in global gold prices led Mexican officials to renew their efforts to auction off state-held subsurface rights to transnational mining companies.

These struggles reinforced the importance of the communal assemblies as the basis for residents' collective way of life, a concept that *serrano* scholar and UNOSJO co-founder Jaime Martínez Luna terms "*comunalidad*," a term he defines as "an historical experience and a vibrant, present day set of behaviors, which is constantly renovated in the face of the social and economic contradictions generated by capitalist individualism."[25] *Comunalidad* wasn't something that could easily by mapped, since territory was only one element of its collective life,

FIGURE 8.3. Street art protesting biopiracy in Oaxaca.

indivisible from governance (the community assemblies), labor (sustained by communal tenure and work), and the annual fiestas that bring together residents from around the region. And yet, as UNOSJO saw it, maps could be used to defend elements of *comunalidad* from outside threats. Herlihy's presentation in Morelia offered a compelling opportunity to explore that potential, and so Ramírez invited Herlihy to make his case at UNOSJO's offices in Guelatao. With new money heading his way from the Foreign Military Studies Office, and pressure to expand the México Indígena project into new regions, Herlihy's invitation from Ramírez to the Sierra Juárez could not have come at a more fortuitous moment.

Here is where things begin to get murky. What we do know is that Herlihy went to Oaxaca in August 2006 to pitch the México Indígena project to UNOSJO. We also know from Herlihy's project status report to Radiance Technologies—the contractor hired to manage the AGS contract with the Foreign Military Studies Office—that he was initially impressed with UNOSJO. Herlihy described the organization as an

> authority on such social issues in Mexico, one sought out by other indigenous leadership from other indigenous regions, including the Zapatista movement (Comandante [sic] Marcos himself came to consult with indigenous leadership in Ixtlán de Juárez and Guelatao during 2006), with the Zapatistas and UNOSJO having a long standing relationship.[26]

But did he tell UNOSJO that he was passing all this information along to the project's funder, the U.S. Army? Not surprisingly, Herlihy says he did and UNOSJO representatives claim he did not. Both claims are, nonetheless, made in retrospect. At the time, UNOSJO deemed Herlihy's project sufficiently legitimate to recommend that he pitch the project to the community assemblies in the Rincón, where he hoped to map.

But what was Herlihy's team going to map? "Everything" is what one man trained by Herlihy in Yagavila as a community researcher, Gregorio Urbano, says.[27] But even "everything" was ambiguous. Would it be the boundaries of the *comundidad agraria*? Place names? Herlihy gave a preliminary answer while training Urbano to use a GPS: Herlihy was planning to map parcels worked by individual families, the very thing that PROCEDE had wanted to map. Well aware of PROCEDE's efforts, Urbano reported Herlihy's interests back to the Yagavila authorities, who promptly withdrew from the project. Or, as Herlihy put in his report to Radiance Technologies, "some individuals in Yagavila voiced concerns about the research activities, so we temporarily suspended research there."[28] Herlihy's team had better luck in Zoogochi, where they trained a community field assistant and mapped "four parcels" after meeting with the assembly. In Yagila, they also met with the assembly and mapped one parcel. Outside the Rincón, Herlihy mapped areas where families had built homes on communal lands, or *solares*—another PROCEDE target—in the urban parts of Guelatao and Ixtlán. For a project that claimed, in part, to be studying the effects of PROCEDE, the México Indígena team seemed to be bent on carrying out PROCEDE's unfinished business.

Meanwhile, things were heating up in Oaxaca. In May and June of 2006 a teachers' strike escalated into an occupation of the television and radio stations in the capital city of Oaxaca. By August the movement had convened the Peoples' Popular Assembly of Oaxaca (APPO) as an alternative governing body to a state government they deemed corrupt. Two months later, in October, APPO's supporters and allied groups managed to all but shut down the city of Oaxaca with their occupation. On October 29, Mexican President Vicente Fox sent 7,500 federal and military police to put down what was increasingly hailed as a popular uprising. The police counteroccupation culminated in a two-day crackdown on protestors, bringing their occupation of the city to an end. The crackdown did not end at the city limits, however, even though most of the Rincón communities stayed out of the fray, listening to events unfurl on the radio. But UNOSJO saw the uprising as an important political opening and joined APPO. Their involvement, which sent at least one of the organization's leaders into hiding for more than a year, diverted the organization's attention from Herlihy's mapping project.

Herlihy's attention, however, did *not* waver. After the uprising ended, he returned to the Sierra in January 2007, and again in June, July, and August, to complete mapping work in the Rincón communities of Yagila, Zoogochi, and perhaps Teotlaxco.[29] He also sought to expand México Indígena's reach, sending a team to pitch the project in Lachixila, a Zapotec-speaking town on the far northern edge of the Sierra Juaréz. By extending the effort to include Lachixila, the México Indígena project could map one of the more biologically diverse ecosystems in the country and make good on Herlihy's continued interest in conservation. Lachixila refused entry for the project, however, forcing Herlihy's team back into the Rincón.

Things did not go much better there. In July, Herlihy reported to Radiance that Zoogochi had dropped out of the project due to "the help of local APPO sympathizers."[30] Yagavila continued to refuse the project, leaving only Yagila. Complicating matters was the apparent withdrawal of Guelatao and Ixtlán from the project. Worse still for Herlihy, his team appears to have been shut out of the other main area where they proposed to work: the Sierra Tarahumara in the northwestern state of Chihuahua. Visiting the region in 2007, Herlihy's team found "a different indigenous problematic" involving extensive militarization and "trafficking in many forms."[31] Against that backdrop, no Raramuri community or organization appears to have invited Herlihy's mapping project in. After their success in the Huasteca Potosina, the México Indígena project looked to be falling far short of its ambitious goals.

Whether because of the APPO uprising or their growing suspicions about the project, UNOSJO was of little help, so Herlihy turned again to his initial contact, Gustavo Ramírez. Ramírez had been working as a consulting biologist with one of the more remote communities in the Rincón, San Miguel Tiltepec. When Ramírez first started working with Tiltepec in the late 1990s, the community was on the verge of selling rights to log on community lands in exchange for building a road to the community. Nearby Santiago Teotlaxco had already done the same, only to find its forest all but destroyed by logging. Ramírez urged the community to forgo logging and hold out for ecotourism instead. The community assembly agreed not to log, and in 2003 the Mexican government finally built a road to the community. But Tiltepec also faced

The AGS, the Bowman Expeditions, and the México Indígena Project 157

another problem. In the 1960s, Mexican officials had helped a group of land-poor farmers from the lowlands resettle at the edge of Tiltepec's lands. Lacking a formal map of their community, Tiltepec's claims had been considered but largely ignored by state officials. In the 1990s, this agrarian settlement, La Luz, began to seek regularization of their community boundaries under PROCEDE. Tiltepec could see the value in having its own map with which to counter La Luz's claims, as well as in staking a claim to forested areas that were gaining value as ecotourism destinations and in providing "ecosystem services." With Ramírez's help, Tiltepec agreed to join the México Indígena project in July 2007.[32]

Still two communities, Yagila and Tiltepec, did not make for a successful project, given the amount of money and resources Herlihy had at his disposal. Herlihy must have known this, as by 2008 he was, once again, making the rounds of the Rincón communities using the "success" of the Yagila and Tiltepec projects to fend off concerns. As part of that effort, Herlihy coordinated "The First Bowman Conference" in Oaxaca in August 2008, including a day-long visit to Yagila.[33] Two staff from the Foreign Military Studies Office made the trip, as did the U.S. State Department Geographer and a representative from the National Geospatial-Intelligence Agency team supporting the Army's newly formed African Command, AFRICOM.[34] Dobson crowed about the success of the meeting in the pages of the AGS newsletter, *Ubique*, stating that

> what the villagers appreciate most is Peter's gift of "participatory research mapping-GIS." . . . They learned to map their indigenous community's boundaries, farm plots, and resources and now use the resulting data to protect what they own and develop their land in responsible ways. This Bowman Expedition has lived up to its billing and exceeded expectations many times over.[35]

But outside the pages of *Ubique*, a different story was unfolding.

Sometime in late 2008, Herlihy made a visit to Yagavila, the first community to reject the México Indígena project. By chance, Herlihy ran into UNOSJO's Aldo González. According to González, Herlihy was continuing to use UNOSJO's initial support for the project in his attempt to convince Yagavila to join the project. González was furious! And understandably so: Following the APPO crackdown, communities throughout Oaxaca had been caught up in a wave of suspicion on the assumption that the police intervention had been aided by infiltrators; and González was also facing a political challenge from Ramírez, who was, at this point, firmly aligned with Herlihy. Once back in Guelatao, González started searching the Internet for more information on Herlihy. What he found in his research only made matters worse.

To borrow from Demarest's own assessment of the Internet, the search made "finding causes and identifying enemies easy for the online individual."[36] To support the México Indígena claim to be doing "open-source" research, Herlihy's team had put together a website chronicling its activities. The site included maps from the Huasteca, photographs of the communities, and copies of the various reports and presentations made by the team. Among the latter were the project reports to Radiance Technologies and a series of PowerPoint presentations pitching the military relevance of the Bowman Expeditions, complete with Petraeus's quote about the value of the "cultural terrain." It became to clear to González that the México Indígena project was more than a mapping project of dubious qualifications. It was a military project, bought and paid for by the U.S. Army's Foreign Military Studies Office. Other staff at UNOSJO and friends of the organization helped complete the picture, turning up Internet links to Demarest's voluminous writings on counterinsurgency.

The first thing UNOSJO did was to question Ramírez, Herlihy's primary ally in the Sierra. Ramírez apparently knew very little about the project's funding, though he was promptly expelled from UNOSJO for his role in helping it along (though by Ramírez's account, he was expelled as part of González's campaign against him). González's next move was to take the revelations directly to the communities. Yagila's Assembly was convinced, and agreed to stop working with Herlihy's project in spite of the fact that the maps were nearly complete. Tiltepec was less convinced; they wanted the map for their own purposes and regarded Herlihy and González with equal suspicion. The division worked in Herlihy's favor. He desperately needed to finish the project in at least *one* community.

González, however, returned to Guelatao to write the first official statement denouncing the México Indígena project, the one released in January 2009. Issued by UNOSJO, the press release quickly circulated on the Internet. In it UNOSJO denounced Herlihy and Ramírez and voiced its concerns about the Foreign Military Studies Office's support of the project as a Bowman Expedition. In UNOSJO's estimation, the Foreign Military Studies Office was responsible for creating a "global database instrumental to the functioning of the Human Terrain System," the Army's better-known, and far more controversial, counterinsurgency initiative. UNOSJO backed up its concern by citing the México Indígena project's reports to Radiance Technologies that discussed "newly forming team[s] working in the Antilles, Colombia, Jordan, and Iraq."[37] UNOSJO declared the entire project an "act of geopiracy" that paralleled previous attempts at *bio*piracy. Once again, outside interests were coming to the communities of the Rincón, using the communities' own desires for food, shelter, land, and now maps to gain access to resources that could be stolen and used to advance outside interests.

UNOSJO's press release sparked a minor controversy in the United States. Dobson and Herlihy were bombarded with e-mails asking for clarification, most of which they dismissed as the work of "cyber bullies."[38] Instead, Dobson implored: "Let the indigenous people of Oaxaca speak for themselves," going on to accuse González of trying to destroy the AGS's "noble effort" to combat the "scourge of geographic ignorance."[39] Dobson also took aim at a "bravely anonymous campus revolutionary [who] slipped broadsides, denouncing me and my colleagues and everything we stand for, under doors all over campus except my own." Dobson's reply was further emboldened by the Army's support for the Bowman Expeditions, claiming that the military had given the program $2.5 million. Dobson qualified the sum as "a good 'down payment' but far less than what's needed to make a sizeable dent in the American scourge of geographic ignorance."[40] The AGS followed up Dobson's missive with a press release reiterating the goals of the Bowman Expeditions. The press release insisted that Tiltepec had endorsed its efforts over and against the critique of an unnamed "activist from outside the municipality"—presumably González—"and implored us to continue helping the community with future projects. We seek no higher endorsement of our work or the AGS Bowman program."[41]

The endorsement would not last. In March 2009, Tiltepec issued its own sharply worded statement denouncing the México Indígena project. By then Tiltepec already had the three copies of the map produced by the project and a GPS unit that Herlihy had given them. What they didn't have, however, was the computer with GIS software and the training Herlihy allegedly promised them. Nor did the map he gave them show everything that community residents had helped Herlihy's team document. Instead, the map Herlihy gave them only had the boundaries long since surveyed by Mexican officials as part of the community's agrarian title, and basic topographic information from the national mapping agency. The community's sole contribution to the map consisted of Zapotec place names and the route of a pre-Hispanic "road" connecting the region with the Caribbean lowlands in Veracruz. Conspicuously absent from the maps' credits was any mention of Demarest and the Foreign Military Studies Office. Tiltepec's authorities felt they'd been had. Herlihy had failed to tell them about the military's role in the project and had failed to deliver the promised training and equipment. Instead he'd given them three copies of a map and a GPS. Tiltepec's statement picked up on UNOSJO's accusation of geopiracy, demanding, among other things, the "return of all information obtained from our community."[42]

Herlihy's team partially complied, removing Tiltepec's map from the project website shortly thereafter. Herlihy replaced the community's data with a statement of his own, accusing González of being an outside agitator and denying UNOSJO's accusation of geopiracy as false.[43] Herlihy also claimed to have informed González (at least) of the Foreign Military Studies Office's role in the project, though he devoted far more space to distinguishing the Bowman Expeditions from HTS than explaining that role. Instead, he claimed that the Foreign Military Studies Office's role had been a "hands off" one "that allowed our scholars the freedom to do their own research projects." That may well be an accurate characterization of Herlihy's relationship to the Foreign Military Studies Office. But it doesn't come close to accounting for what he handed over to the military via Radiance Technologies. Nor does it address who might have that data now, and what it might be used for in the future.

Given all this, it's not hard to fathom why figuring out the "truth" of what happened is beyond anyone's immediate grasp. But there are a few points worth noting. None of the maps Herlihy gave to Yagila and Tiltepec bear the Foreign Military Studies Office logo or any indication of the military's role in the project. Nor does it appear that the Huasteca communities were informed either, though, with Dobson, Demarest had visited the region in July 2005. Herlihy may well have told people at UNOSJO and even in the communities that the Foreign Military Studies Office was funding the project. What he likely did *not* do was explain to them what the Foreign Military Studies Office was (a U.S. Army research and analysis center) nor discuss its mission (to provide military and security experts with information about the "Operational Environment"). He was correct in stating that the Bowman Expeditions were not part of the HTS, though the HTS was also created by analysts at the Foreign Military Studies Office, and one of its backers, Karl Prinslow, was familiar with Herlihy's work.

The Bowman Expeditions were significantly different from the HTS.[44] As we pointed out in the previous chapter, the HTS's explicit mission was to embed anthropologists and other social scientists with combat units in well-defined war zones such as Iraq and Afghanistan. In keeping with the University of Kansas's memorandum of understanding with Fort Leavenworth, by 2007 anthropology professors were training HTS personnel under contract with the Foreign Military Studies Office.[45] In contrast, the Bowman Expeditions take a more "open" approach by drawing on research done by civilians and compiling their findings into a database accessible to military personnel. Instead of focusing on war zones, this effectively turns all of society—all of the planet—into a potential battle zone. As befit the new thinking about global security during the "War on Terror," the Bowman Expeditions helped elevate counterinsurgency from a tactic to a strategy.

But the Bowman Expeditions' militarization of academic research was hardly unique in this regard. Instead it fit within a growing military interest in the social sciences, and human geography in particular. If culture was indeed a force multiplier, what was to stop geography, anthropology, or any other social science from being fully weaponized? Especially when there was ample money to be made from military contracts?[46] Not critiques from fellow geographers. Not calls for new ethical standards. And certainly not the Association of American Geographers, the primary professional society for the discipline.[47] Much like maps made by indigenous peoples to challenge the state, the efforts could only call attention to the problem but not solve it. The problem was simply too big to map.

CODA

Kill the Insurgent and Save the Man

Indigenous Peoples and Human Terrain

The United States Indian Industrial Training School opened in Lawrence, Kansas in 1884. Fated to grow into Haskell Indian Nations University, the school was founded on the principle of saving Indians from the death and isolation created by the Indian wars of the 19th century. Through education the school promised a less violent way of assimilating Indian people into American society, certainly one more humane than the wars that preceded it, many of which had been launched from nearby Fort Leavenworth. It was an awkward impulse born out of the best of (white) intentions, whose complexity is concealed by Pratt's infamous call to "kill the Indian and save the man."

Pratt was himself a veteran of the Indian wars, serving as a second lieutenant in the 10th Negro Calvary, better known as the Buffalo Soldiers. Pratt firmly believed that Indians, like freed slaves, were not essentially different from whites.[1] What distinguished them was not biology but culture. Pratt's advocacy for boarding schools rested on this belief. An opponent of segregation, Pratt believed that the future well-being of Native Americans lay in relocating them from reservations to boarding schools like Haskell. Through learning English, education in basic subjects, and training in trades such as home economics, agriculture, and industrial arts, Native Americans could "quit being tribesmen and take upon themselves the obligations and advantages of American citizenship."[2] It was a task for educators and the Army alike, the latter sworn to uphold the U.S. Constitution that extended the rights of citizenship to all persons born within the territorial United States. Pratt used the Fourteenth Amendment to argue that, as with freed slaves, the future for Native Americans lay with their civilization. If there could be no justice for the oppression inflicted by conquest, leading Native Americans into civilization and later citizenship was the second best. Through education and training, Native Americans could be settled into the sociospatial order of property rights and labor, race and nation that defined the emerging United States. It was a debate as old as the conquest itself.

The boarding school system did not end the Indian wars but created a new battlefield for fighting them.[3] Some children thrived in the new schools. Many more simply survived, going on later to forge their own lives. Others died in the schools from illness, poor medical care, and worse. The cemetery at the edge of the Haskell campus perpetuates the names and tribal affiliations of 103 children who died at the school.[4]

We've already told you this, but it bears repeating. It bears *spelling out*: Half of these kids died within five years of the school's founding, thanks to the lack of heating, sanitation, and medical care. The youngest person buried in the cemetery is a six-month-old baby boy named Henry White Wolf, born to Cheyenne parents temporarily housed at the school. Most of the children died in their teens, with a handful passing into their 20s. They died of malaria, tuberculosis, and pneumonia, from poor diet, exposure, and neglect.

The names of their tribal affiliations are engraved in the Army-issue white marble headstones. Digger, Caddo, Delaware, Wyandotte. Potawatomi, Ponca, Ukie, Modoc. Assiniboine, Osage, Northern Cheyenne. Sioux, Cheyenne, Comanche. Eastern Cherokee, Chippewa, Quapaw. Pawnee, Arapahoe, Kaw, Kickapoo. Navajo, Seminole, Hopi, Mission. Paiute, Miami, Omaha, Papago. Winnebago, Wichita, Ute, Oneida. Mojave. Shawnee. Peoria. The names of the tribes map the Indian wars through which the territorial United States came into being.

Those are just the marked graves. Wooden crosses in the cemetery commemorate at least nine unmarked graves. Countless other bodies were buried in unmarked graves elsewhere on the campus. Still others lie in the wetlands extending south of the main campus, where the sedges, red cedars, and pawpaw trees offered escape from the gaze of school officials.[5]

Coda

In 2011, 100 years after a "Mission" Indian from California, Antonio Prieto, age 16, was buried in the cemetery, the AGS convened the World Conference on Human Geography.[6] In a nod to the festering controversy over the AGS's México Indígena project, the conference was supposed to focus on "community research and ethics." The setting at Haskell underscored the AGS's need to demonstrate its commitment to indigenous peoples—a commitment questioned by an ever-growing list of geographers and indigenous peoples. Haskell's president, Dan Wildcat, hosted the event, and students from the school moderated the panels. And then there was that eagle staff, brought in with a U.S. flag by a Color Guard of Native American veterans. Wrapped in otter skin and tipped with eagle tail feathers, the staff was there, as we've said, to remind all in attendance to speak truthfully and freely.

But it only took a glance at the conference agenda to understand the event's real purpose. Hosted by Haskell in conjunction with the University of Kansas Department of Geography and the AGS, as we've said the event was wholly funded by the U.S. Army Research Office ("developing and exploiting innovative advances to insure the Nation's technological superiority").[7] Army and State Department personnel topped the list of presenters that included the geographer of the State Department and the dean of the Army's Command and General Staff College. There were representatives of the HTS and of the newly formed Cultural Knowledge Consortium that aimed to compile social science expertise to support U.S. government and military programs. These were joined by the leaders of the Bowman Expeditions to Mexico, the Antilles, Colombia, and Jordan, along with a smattering of Native American scholars, many affiliated with Haskell, reporting on mapping projects of their own. Last, but hardly least, were the handful of geographers such as ourselves attending at the invitation of the AGS. Discussion of concerns raised by the México Indígena project were few and far between, overwhelmed by the AGS's main objective: the trumpeting of the relevance of applied geographical research for security and military purposes. As Lee Schwartz, the geographer at the U.S. Department of State, put it, "classic human geography" was essential to guiding a new "3D" approach to security that integrates "development, diplomacy, and defense."[8] The visceral lesson of the Haskell cemetery—and the legacy of the Indian wars—could not have been more distant.

This was not the first time that Dobson's and Herlihy's efforts to pitch the Bowman Expeditions to military officials had traversed the legacy of the U.S. Indian wars. During their 2006 visit to Fort

Leavenworth to meet with General Petraeus, Herlihy, Dobson, and the AGS Executive Committee had visited the quarters of Lieutenant Colonel George Armstrong Custer before posing for a photo in front of the Buffalo Soldier monument. The stops were not extraordinary per se. They are regular features on tours of Fort Leavenworth that highlight its historical role in the Indian wars, nor did Herlihy or Dobson see any connection. Defending the Buffalo Soldier photo and the reputation of the AGS Bowman Expeditions, Herlihy later wrote that their actions revealed nothing "whatsoever about the attitudes and values of those in the photo."[9] True enough. Neither Herlihy nor Dobson has condoned the Indian wars. And yet their actions reveal a lot about their attitude toward the history the Bowman Expeditions selectively engage. In his presentation at Haskell, Herlihy argued that what set him apart from those who fought the Indian wars was his desire to do right by indigenous peoples by giving them maps. Herlihy refused to see any parallel between his work and previous forms of colonization. Like Pratt, Herlihy claimed to be helping indigenous peoples by securing a place for them in a world drawn by conquest, refusing to acknowledge any role in reproducing that geography. In this respect, the Bowman Expeditions were hardly an aberration. Instead they offered a variation on a long running theme of advancing imperial interests through the twinned militarization of geography and saving the Indian.[10]

* * *

Maps no more simply document an existing reality than expeditions simply explore. Both are driven by a desire to discover, *to create* a world shot through with political ambition. This is not to say that the trees and mountains and houses visible on satellite images with their locations plotted by latitude and longitude don't exist. They do. It *is* to say that the order they're *supposed* to reveal—a forest, a mountain range, a community—is very much a creation of the maps. Similarly there are no indigenous territories out there just waiting to be documented, no more than at the end of World War I there were territorially defined nations submerged in the continental empires awaiting Bowman's articulation. Instead there is only a world to create.[11] Mapping is an irrevocably political task.

One thing the mapping projects we've described have brought into being is "indigenous territory." The kernel of this idea lies within indigenous peoples' own deep knowledge of and experience in particular places. The importance of territory comes from their lives. But if their knowledge, traditions, and *lives* are place-specific, their precise location and extent are dynamic and changing, immersed in political and economic change. More fixed definitions of territory conceal that dynamism. Their approach can scarcely be regarded as exclusively indigenous. Instead, they bear the traces of many hands, especially considering the many ways in which the idea has been mapped and defined by everyone from military personnel to the anthropologists who penned the 1971 Barbados Declaration and led the Inuit Land Use and Occupancy Project, from the lawyers and judges tasked with adjudicating claims in Canada and at the Inter-American Court of Human Rights to the land tenure experts at the World Bank. Not to mention geographers such as Nietschmann and Herlihy.

Their collective efforts capture the multiple and often opposing forces that have produced areas variously known as "tribal areas," "Indian Country," "indigenous territories." Haskell's position charts one aspect of these efforts. The HTS's focus on tribal areas tracks another. The Denes' and the Miskitos' efforts to portray themselves as nations points to a third. Chapin's connection between natural ecosystems and indigenous homelands frames yet another. Each of these—and all the others—have endeavored to translate the complexities and nuances of peoples' everyday lives into the grids of latitude and longitude, of property and nation, of ecology and economy ... *for people otherwise unfamiliar with the terrain*. For outsiders, if you will, for strangers. En route they transformed areas formerly regarded as "frontiers" into spaces governed by laws and recognizable principles of government.

Yet as it stands, indigenous territories remain inordinately vulnerable to the violence visited upon them by settlers, extractive industries, state officials, drug traffickers, and, yes, armies that continue to see these areas as empty, as lacking a state-enforced order. Building on Hernando de Soto's claims about the lack of property rights, Demarest has come to explain this violence in terms of missing state institutions. Herlihy and Dobson make a related yet different argument. Here, for example, is Herlihy explaining to a reporter the Bowman Expeditions' interest in indigenous peoples: "We just don't understand these areas that well, and those areas are where you have a lot of violence."[12] But violence no more corresponds to the lack of secure property rights than it does the presence of "bad guys." Violence has more to do with the role that frontier areas—Indian Country—have played in the economic, political, and military formation of states. By claiming that he is simply mapping territories that are already there, Herlihy gains a powerful alibi for obscuring the active role that he plays in transforming them into . . . *intelligence for the U.S. Army*. It also lets Herlihy preserve his role as the "good guy," the benevolent mapper, whose work is driven by a scholarly interest in documenting the world as it is rather than by the strategic interests of his Army benefactors.

The U.S. military's interest in indigenous peoples—if you can fairly call it "interest"—has never reached to understanding the nuances of their day-to-day lives.

Historically "Indian Country" has served as a foil against which the United States defined itself politically and economically. In the immediate aftermath of the American Revolution the term referred loosely to areas claimed by the United States but not controlled by it. The ensuing wars—a great many of them, again, launched from Fort Leavenworth—transformed Indian Country into a de facto free-fire zone of open warfare. The outcome was the displacement of indigenous peoples to reservations and boarding schools like Haskell . . . or their outright extinction. It also redefined Indian Country into what it is today: places governed by federal Indian law. This law was to preserve these places for tribes and to preserve the tribes themselves from the violence visited upon them by the very entity tasked with enforcing the laws: the U.S. government.[13]

This emphasis on the law—now widely replicated at a global scale through the recognition of territory as a right of indigenous peoples—cannot obscure the violent struggle through which these spaces came into being. From the Indian wars through the Philippine–American War of 1899–1904 through the current wars in Iraq and Afghanistan, and drone attacks in Yemen and Somalia, the identification of "Indian Country," of "tribal areas," has been used to demarcate zones where a particular kind of war can be fought.[14] As Nietschmann aptly put it, since wars fought in these areas are not fought against states, they can be waged without rules.[15] This allows the state armies fighting in them to claim the establishment of order and stability as a military objective; and to do so effectively requires understanding not only *where* people live but also *how* they live, the better to enlist their help in fighting the war. This was the idea "Red Mike" Edson so perceptively grasped in the 1920s in his battles with Sandino's forces in Nicaragua; and his idea continues to shape counterinsurgency's strategic emphasis on "winning hearts and minds." Military strategists' interest in indigenous peoples is thus purely tactical, obsessed with understanding "tribal areas" as a particular kind of battlefield. Save for the odd, committed official, neither the State Department nor the Pentagon has much interest in the intricacies of Miskito, Zapotec, or Wazir life. Institutionally speaking, their concerns lie solely with U.S. national security. The main problem is not "geographical ignorance," as Dobson has repeatedly claimed. It is ignorance of the broader context that drives indigenous mapping that presents the real danger.

The conundrum is not a new one for indigenous mapping. In mapping their lands, First Nations in Canada, the Miskito in Nicaragua, and many others have faced the challenge of creating and protecting a space in which they could maintain a collective way of life *and at the same time* be recognizable to state officials. Phoebe Nahanni of the Dene Nation was right to insist that gaining this recognition was fundamentally a political challenge in which cartographic precision was a useful, though limited tool. What mattered was the Canadian government's *recognition* of their status as a nation, and its commitment to figuring out a way of negotiating the cohabitation of these spaces . . . *claimed by two nations*. That the Sandinistas used the alleged existence of the Polanco map to launch a counterinsurgency campaign against the Miskito underscores precisely the political nature of this challenge. Not surprisingly, the overwhelming response from state officials, judges, and World Bank administrators alike has been an insistence on the technical accuracy of the maps. Giving states and international institutions accurate information—that is, information commensurable with their own ways of understanding the world—allows them to decide on the technical merits of claims before getting into the political situation.

Herlihy and Dobson have repeatedly sided with state officials here, insisting on the technical merit of their work. Providing useful, actionable intelligence is, after all, what Dobson says good applied geographic research should be doing. Herlihy goes one further, insisting that it's the technical accuracy and precision of his maps that defines their utility to the indigenous peoples with whom he's worked. It's a simple equation, one that makes common sense. With more accurate maps comes an improved understanding of the terrain, and with a better understanding of the terrain comes more efficient governance—a governance capable of calculating the right mix of "soft" measures, such as legal recognition and property reform, with the "hard" force of military intervention.

For mappers such as Herlihy, the goal is very simple: "Map everything." The phrase isn't Herlihy's. It actually comes from another Haskell presenter, Google Earth co-founder—*and* University of Kansas alumnus—Brian McClendon. Intended to summarize Google's own mapping ambitions, it succinctly captured the convergence in thinking of the military officials and academic geographers in the auditorium. It was also a goal that conveyed a clear, if unspoken, message to indigenous peoples and critics of the Bowman Expeditions alike: They could either throw their efforts behind mapping everything on terms understandable to Google, the U.S. Army, and the other powers that be, or they could get out of the way. It's a position that makes perfect sense if you concede that geography's scholarly purpose is to avail itself to military needs and that indigenous participation in that task is only a matter of consent and not justice.

Herlihy underscored this last point in his conference presentation. He recapitulated his career in indigenous mapping from his dissertation work with the Emberá and Wounaan in Panama during the 1980s to the present, and he used this experience to substantiate—yet again—his claim to have introduced participatory research mapping methods to Latin America as well as the utility of his maps for making indigenous communities visible to state officials.[16] Herlihy bolstered his case by bringing two of his long-time Miskito collaborators from Honduras. Both men testified to the tremendous importance of the maps that Herlihy had helped them make during the 1990s. Evincing the strength of their belief in Herlihy, one of the Miskito men made a passing reference to Herlihy's role in a new, as yet unannounced but clearly welcome, Bowman Expedition to Honduras.

The choice of Haskell as a venue for the AGS conference was intended to bolster Herlihy's and Dobson's efforts to silence critiques about their insensitivity to indigenous rights. Here's what Dobson had to say in the conference summary he sent to attendees three months later:

> The assembled speakers and audience represented the entire range of viewpoints, and yet not a single rancorous word was heard. The biggest accolade was simply attendees marveling that the conference took place with all parties present and yet without discord. That was due to Native American customs, such as the Eagle Staff and talking circle, which invoke certain courtesies while fully engaging dissenting views.[17]

Of course this wasn't true. There was plenty of discord, if perhaps less visible on stage; and of course there was no one there to represent the people who had necessitated the conference in the first place: the Zapotec of the Oaxacan Sierra.

Dobson's and Herlihy's efforts to push aside the controversy over the México Indígena project and to burnish the image of the Bowman Expeditions seemed to work well for the military officials present. Not that they needed much convincing. The Bowman Expedition to Honduras, mentioned by Herlihy's Miskito collaborator, was funded by the U.S. Army Research Office, the same military entity funding the Haskell conference. This funding offered a partial explanation of the new expedition's name: the Bowman Expedition to the Borderlands Region. When pressed during one of the breaks at the conference, Herlihy confirmed the name and the funding source for the project and declared his willingness to continue working with military money. Any military interest in the project, Herlihy insisted, was a function of the quality and utility of the information—the intelligence—gathered rather than the design and purpose of the mission.[18] It's also an approach entirely in keeping with the Army Research Office's commitment to "science to shape the future of the Army."

That would be all the confirmation of the Honduran project's existence for some time. This go around, there was no Expedition website or summary on Herlihy's webpage charting the progress of their work until after it was finished. This time everything was being done free from the scrutiny of would-be critics, underlining the point that this was a military mission.[19] Still the questions won't go away.

If indigenous lands in Mexico sound like an unpressing security concern, in Central America they pose a more readily identifiable threat. Since the 1990s, the region as a whole has become a primary conduit for trafficking drugs to the United States. Traffickers often make use of infrastructure built up during the 1980s to support U.S.-backed proxy forces in the region. Indeed, officials in the Reagan administration allowed traffickers to use proxy bases in Costa Rica and Honduras to move cocaine as long as they gave a portion of the sales to the Contras.[20] The State Department now estimates that as much as 80% of all cocaine entering the United States passes through Central America, much of it via the Miskito communities of eastern Nicaragua and Honduras.[21]

Less than a year after the AGS conference at Haskell, the New York Times ran a front-page story detailing the Drug Enforcement Administration's (DEA) efforts to apply counterinsurgency tactics developed in Iraq to its drug war efforts in Honduras.[22] Nor was it just any part of Honduras: it was the Mosquitia, the very region recently mapped by Herlihy's Bowman Expedition. Known as "Operation Anvil," the DEA's efforts were styled after similar counternarcotics operations in

Afghanistan, pairing members of the DEA's elite Foreign-Deployed Advisory Support Team (FAST) with Honduran security forces. The story focused on FAST's use of "forward operating bases" to get troops off large bases and spread "across remote, hostile areas." In Iraq, the use of forward operating bases was a signature element of the counterinsurgency campaign engineered by Petraeus. Their application in Honduras, the *Times* reported, "showcases the nation's new way of war: small-footprint missions with limited numbers of troops, partnerships with foreign military and police forces that take the lead in security operations, and narrowly defined goals, whether aimed at insurgents, terrorists or criminal groups that threaten American interests." It was a strategy that Edson would have approved.

The DEA's approach did more than apply lessons from Iraq and Afghanistan. It also revisited terrain previously militarized during in the 1980s. Two of the forward operating bases repurposed infrastructure built with U.S. support during the Contra war; but instead of training and equipping proxy forces, this time around the DEA's actions "combined the legal framework of a police action with the hardware and rhetoric of war."[23] The DEA's actions transformed the Mosquitia into a "battlespace," integrating aerial surveillance with enforcement actions, with lethal consequences, in terrain thoroughly populated by Miskito people. Shortly after midnight on May 11, 2012 a DEA-led team opened fire on a boat traveling on the Río Patuca near the Miskito community of Ahuas, killing four unarmed civilians.[24] The DEA has insisted that the boat opened fire on them first, accusing the victims of collaborating with drug traffickers in spite of numerous accounts to the contrary.

Did the DEA agents have Bowman Expedition maps? If they did, they were of little use when it came to opening fire. If they didn't, could Herlihy's maps have prevented the incident? It seems unlikely. Since the incident occurred, both drug traffickers and antinarcotics forces have continued to kill indigenous residents of the region.[25] These killings occur in spite of the fact that the Honduran government titled more than 2,000 square miles of land to Miskito communities in the area. The titles came after decades of struggle, and relied extensively on maps made by the Miskito communities in the region, with help from the likes of Herlihy, Chapin, and the Caribbean Central America Research Council.[26] There's ample evidence to suggest, however, that neither the maps nor the titles have stopped the dispossession of Miskito communities. If anything, the titles may have been a last-ditch effort by the Honduran government to win some loyalty in a region increasingly under the control of drug traffickers.[27] Either way you put it, the Bowman Expedition to the Borderlands is working in a de facto war zone with funds from the U.S. military.

Whether or not DEA agents weaponized Herlihy's maps is beside the point. What matters is how Herlihy's work and that of other Bowman Expedition leaders have converged with US. military efforts to rethink and expand understanding of national security writ large. Notwithstanding claims to the contrary, neither the AGS, Dobson, nor Herlihy have done a thing to challenge that relationship. Indeed, in June, 2013 Dobson announced that he had received a three-year, $1.8 million grant from the Pentagon's Minerva Research Initiative to expand the Bowman Expedition to the Borderlands Region to include indigenous areas throughout Central America, creating a new expedition that has since been dubbed "CA [Central America] Indígena." Together with the earlier Bowman Expeditions to Mexico and the Antilles, the inclusion of Central America encompasses all of the area previously targeted by the United States under the Mérida Initiative first proposed in 2007.[28] The United States has committed $1.6 billion to the project under the guise of coordinating efforts to combat trafficking, money laundering, and organized crime. All of these are activities that Demarest identifies as proliferating in indigenous areas due to the lack of clearly defined property rights. It should come as no surprise either that the Mérida Initiative is loosely based on Plan Colombia, the U.S.-backed counterinsurgency campaign that brought Demarest to that country in the early 1990s and has since shaped his approach to security. The newly expanded Bowman Expedition to the Borderlands Region scales Demarest's vision for Colombia up to the regional level "to include all Latin American countries bordering the Gulf of Mexico and the Caribbean Sea."[29]

It also brings the Bowman Expeditions to new heights as the primary source of social science intelligence in the region. Herlihy is once again slated to lead the three-year expedition. Initially dispersed by the National Science Foundation before being returned to the Pentagon amidst controversy over military funding for research, the Minerva Initiative's mission is "to build deeper understanding of the social, cultural, and political dynamics that shape regions of strategic interest around the world."[30] Among the stated goals of the project is to create a "land stability index" for Central America focused primarily on indigenous areas.[31]

The new grant brings the total Pentagon funding for the AGS Bowman Expeditions to $3.9 million, $2.1 million of which has come from the Foreign Military Studies Office alone.[32] The Minerva project also includes an option to extend the project for an additional two years and another $1.2 million in funding. It's a princely sum in a time of dramatic budget cuts for other, more conventional sources of government funding, such as the National Science Foundation. The grant also burnishes the value of the México Indígena project as prototype for a advancing a "political-military strategy" focused on mapping property.

*   *   *

Indigenous maps keep on coming. More of them every year. Maps of indigenous territories, of use and occupancy, of communal tenure, of cultural resources. Maps that subtly rework existing geographies, bending them in new ways. Indeed, as we write, Google has responded to its own call to map everything by convening an "Indigenous Mapping Day" in conjunction with the National Congress of American Indians.[33] But the maps still guard their silences as well. U.S. reservations and reserves in Canada are still not labeled on Google Maps, and rank lower than national forests, national parks, and military lands on the service's hierarchy of mapped features.

Not that indigenous peoples always have a choice in the matter. Quite often they face conditions where mapping is imperative. If they do not map themselves, most assuredly someone will map them on terms not to their liking. It's map or be mapped, as Nietschmann liked to say. Nor is this profusion of maps a bad—or good—thing. Maps have played a critical role in recognition of indigenous rights to territory, increasingly defined as *community property*, as recent events in Central America demonstrate. In Canada, the flurry of mapping work in the 1970s set the stage for the creation of an Inuit-governed territory, Nunavut. Inuit people are leading efforts to create two more territories, Nunavik and Nunatsiavut, from the areas mapped during the 1970s in Québec and Labrador. Further outside official political channels, indigenous peoples across Canada have mobilized in opposition to energy development and for better living conditions through the Idle No More protest movement.[34]

All this mapping has allowed indigenous peoples to develop their own, critically informed take on mapping itself, influenced, in part, by the inability to fit the intricate relationships that make up territories as spaces of life into the coordinates of the map. For some, this has led to a renewed engagement with other forms of communicating spatial knowledge through travel, song, and storytelling.[35] For others, it's engaged new forms of communication, often approximated loosely as "art." Rather than being driven by latitude and longitude or legal definitions, these approaches creatively strive to find ways adequate to communicating crucial spatial knowledge, such as the qualities of water as it flows over rocks at the site where the Creator emerged.

Or the histories of violent displacement that define the boundaries of "official" maps.

Others have simply resorted to practice, to doing the kinds of things required to maintain a collective way of life, and all these are sometimes referred to as "maps," a term trying to approximate indigenous efforts to . . . convey detailed spatial information. To our ears calling these maps deadens their complexity and potential, translating those realities—yet again—into the gridded space of states and armies. Of course, indigenous peoples aren't entirely free from that geography, but does that mean they have to accede to its dictates? If the maps are going to be useful to communities, they—and not the U.S. Army—will have to be able to control how they're made, with whom they're shared, and when they're used.

Geographers like ourselves, like Herlihy, are uniquely positioned to comply with that request. It's a request that challenges all involved to resist the urge to flatten differences to meet the needs of states or armies, that requires new ways of not only mapping but of *understanding* geography—of understanding *space*. It's a challenge captured by the Zapatistas' call for "a world where many worlds fit."[36] It's a challenge far too important to be left to geographers. It's one that demands new ways of learning from and working with indigenous peoples. After all these years of being mapped by others, isn't it time they had more of a say?

As Gregorio Urbano, erstwhile trainee of Herlihy in the Sierra Juárez, put it: "When we decide to, we are going to make our own maps however they be, but they're going to be ours and belong to us." Anyone who wants do otherwise, as Urbano put it with the firm but infinitely polite tone we heard time and again in the Rincón, "can go to hell."

A Note on Maps

Given what we've said, you might be wondering if we think there's any hope for "indigenous mapping" whatsoever. Certainly some have accused us of saying that *any* map is compromised. They're not wrong, as we've made clear in the preceding chapters. But it was not our intention to disparage all mapping. Far from it. Our intention was to call attention to the compromises mapping entails, in the hope—however remote—that this new awareness would contribute to efforts to use maps more creatively in support of the political transformations that led everyone—from Phoebe Nahanni and the Dene to the Miskito to the Zapotec communities of the Rincón—to their interest in mapping. The best of these efforts have sought to use mapping not just to show the world in a different way, but to leverage that difference to challenge long-standing issues of racial inequality, political marginalization, and militarization.

How can this be done?

We don't know. What we do know is that it *can't* be done by following the approach of the Bowman Expeditions, or by having GIS-equipped computers in every indigenous community. What has to happen is a thorough rethinking of the world itself from the perspective of indigenous peoples, of African-descendant communities, and of other politically marginalized groups.

The list shouldn't stop there. The task is too important to be left to indigenous peoples and other marginalized groups. It will take nothing short of a dedicated effort to shape a new world from a multitude of smaller ones. These won't nest together the way the old cartographic concept of scale would suggest. Instead their shapes will bump up against each other. As we've made clear, no boundary can solve that problem. Boundaries can only magnify it. But these overlaps point to one of the most fundamental political problems: how to cohabit spaces without resorting to violent exclusions. This doesn't mean allowing settlers—*invaders* by any other name—to make their lives by displacing indigenous peoples. But it can't mean drawing lines designed to automatically exclude people either.

With this complexity in mind, here are some ideas for this conversation. The first and most basic involves control over mapping. As everyone from San Miguel Tiltepec to the Dene Nation has made clear, communities need to have full control over the mapping process if maps are to serve their interests. This doesn't mean there won't be compromises along the way about what to map and how to do it so that others can read and make sense of it. Of course there'll be compromises. What it means is that communities have to have control over when mapping projects are initiated in their name, over what information is mapped by these projects, and over the resulting maps' storage and accessibility. Particulars will vary by community. What *can't* happen, though, is for the process to benefit primarily entities such as the Foreign Military Studies Office and the U.S. Army. That approach compromises what is often most fundamental to indigenous struggles: When space is already plotted, defined, and controlled by the U.S. Army or the Government of Mexico, it concedes community members' efforts to make their own spaces on that map. This is an aspect of the discipline of geography we are eager to leave behind.

Second, there can be no pretext about maps simply documenting something already there, even if it's something as obvious, say, as a sacred site. The approach has to be about finding ways to create spaces conducive to the kinds of social change communities desire. We would go further and propose to map the ways in which power works spatially, to make some communities more vulnerable than others, say, to military violence. There is a long tradition of counter-mapping here that aims to show how proposed mines, nuclear facilities, logging concessions, and military strategies impinge on collective ways of life without resorting to defining that way of life cartographically. Instead, they leave the question of how people might live to social interactions, to *politics*.

Third, there are other efforts from which to learn. Most of the ones discussed in this book provide examples of what not to do and situations to avoid. Reflecting on these examples is a first, if necessary, step toward doing something else. We're the first to admit that sometimes you have to make a map before you can figure out what not to do. We are continually inspired by the immense creativity that people bring to mapping. Artists are getting a lot of this attention right now, but communities can and do design approaches to mapping that are every bit as thought-provoking, and often with a far more deliberate sense of their political relevance.

These projects invariably move beyond mapping, expanding into the variety of ways through which deep, spatial knowledge of a people, place, and time is shared and communicated. A good deal of this work has *nothing to do with maps.* Instead, it veers into stories, songs, and performances, into entire ways of organizing people and securing a collective way of life. *Serrano* scholar Jaime Martínez Luna's notion of *comunalidad* is one example that has inspired us. So too have the community assemblies in the Rincón that collectively maintain their elaborate systems of coordinating access to land and governing. The Dene mapping project is another example.

All of these examples have notable drawbacks and disadvantages. But they do approximate what it might take to create spaces conducive to identifying and maintaining a collective way of life. And really there can't be too many of these sorts of maps. On the contrary: Their proliferation makes it harder for any single entity, be it a mining company or a military, to flatly proclaim definitive knowledge of the human terrain.

So go make your own maps. Please. But our hope has been to draw attention to how those maps can be compromised—made to do things against the desires of the people they represent—so that future efforts might avoid the pitfall of reproducing the very hierarchies of race and economic inequality they were meant to counter.

Notes

A Narrative Table of Contents

1. When describing indigenous peoples in general in the text, we use the lowercase "i." We recognize that this goes against the growing tendency to capitalize "Indigenous," particularly in the United States and Canada. This preference is specific to the ways that term is used to command respect for a political identity that spans across peoples who might otherwise identify themselves Crow or Cree, for example. Outside of the United States, and in Latin America in particular, there is considerably less enthusiasm for the use of a capital "I," particularly among those who regard the term itself as a colonial imposition that ought to be critiqued in its own right as a necessary step in its political mobilization. The keyboard affords us no neutrality on this matter. Our use of the lowercase "i" is informed by our sympathy to these efforts to decolonize the concept of "indigeneity" that parallels our endeavor in this book. See Pablo Mamani, *Geopolíticas Indígenas* (CADES: Centro Andino de Estudios Estratégicos, El Alto, Bolivia, 2005), and Silvia Rivera Cusicanqui, *Ch'ixinakax Utxiwa: Una Reflexión Sobre Prácticas y Discursos Descolonizadores* (Tinta Limón, Buenos Aires, Argentina, 2010).
2. Richard H. Pratt, "The Advantages of Mingling Indians with Whites," in Francis P. Prucha, ed., *Americanizing the American Indian: Writings by "Friends of the Indian," 1890–1900* (Harvard University Press, Cambridge, MA, 1973), pp. 260–271. Pratt's speech was delivered in 1892.

Chapter 1

1. See John K. Chance, *Conquest of the Sierra: Spaniards and Indians in Colonial Oaxaca* (University of Oklahoma Press, Norman, OK, 1989), pp. xiii–xv, et passim, on our general ignorance of Zapotec societies peripheral to that of the Valley of Oaxaca.
2. Roberto González, *Zapotec Science: Farming and Food in the Northern Sierra of Oaxaca* (University of Texas Press, Austin, TX, 2001), p. 1; and Walton

Galinat, "Maize: Gift from America's First Peoples," in Nelson Foster and Linda Cordell, eds., *Chiles to Chocolate: Food the Americas Gave the World* (University of Arizona Press, Tucson, AZ, 1992), pp. 47–60. For a broader overview of community economy in the region, see Tad Mutersbaugh, "Migration, Common Property, and Communal Labor: Cultural Politics and Agency in a Mexican Village," *Political Geography* 21(4), 2002, pp. 473–494.

3. Galinat, op. cit., p. 47.
4. "Perhaps the first Mesoamerican community to achieve urban status," says John Paddock in his "Oaxaca in Ancient Mesoamerica," in John Paddock, ed., *Ancient Oaxaca: Discoveries in Mexican Archeology and History* (Stanford University Press, Stanford, CA, 1966), p. 233.
5. The Zapotec script, used in the Central Valleys of Oaxaca, remains the earliest evidence of writing in the American continent. See Javier Urcid, *Zapotec Hieroglyphic Writing: Studies in Pre-Columbian Art and Archaeology 34*, Dumbarton Oaks Research Library and Collection, Washington, DC, 2001; and his elaboration in *Zapotec Writing: Knowledge, Power, and Memory in Ancient Oaxaca*, Foundation for the Advancement of Mesoamerican Studies, 2005, www.famsi.org/zapotecwriting/index.html (accessed March 6, 2012).
6. See Jaime Martínez Luna, "The Fourth Principle: Comunalidad," in Lois Meyer and Benjamín Maldonado, eds., *New World of Indigenous Resistance: Noam Chomsky and Voices from North, South, and Central America* (City Lights, San Francisco, 2010), pp. 85–99. Martínez Luna is a Zapotec scholar and musician born, raised, and living in the Sierra Juárez.
7. Chance, op. cit., p. 12; and Joyce Marcus, "Aztec Military Campaigns against the Zapotecs: The Documentary Evidence," in Kent V. Flannery and Joyce Marcus, eds., *The Cloud People: Divergent Evolution of the Zapotec and Mixtec Civilizations* (Academic Press, New York, 1983), pp. 314–318.
8. Chance, op. cit., p. 16.
9. Ibid., p. 17.
10. Ibid., p. 29.
11. For an interpretation of the *lienzo*'s narration of Tiltepec's genealogy and its relevance to Sierra politics, see Yanna Yannakakis, *The Art of Being In-between: Native Intermediaries, Indian Identity, and Local Rule in Colonial Oaxaca* (Duke University Press, Durham, NC, 2008).
12. Starr had been curator of the ethnographic collections at the American Museum of Natural History, but most of the time he was in Mexico, he was a professor of anthropology at the then new University of Chicago. See Frederick Starr, *In Indian Mexico: A Narrative of Travel and Labor* (Forbes, Chicago, 1908). On the same topic also see his *Indians of Southern Mexico: An Ethnographic Volume* (printed for the author, Lakeside Press, Chicago, 1899). This is a large album of his ethnographic photographs dedicated to General Porfirio Díaz.
13. Starr, op. cit., pp. 112–113. To get people to sit for casting in plaster was so hard that most of the time Starr ended up casting prisoners. We're quoting here from his stop in Ayutla, a Mixe town in the district of Villa Alta that is

immediately adjacent to the district of Ixtlán where Tiltepec is. These are all in the Sierra Norte.

14. Salvador Aquino-Centeno's anthropology dissertation (University of Arizona, Tucson, AZ, 2009), *Contesting Social Memories and Identities in the Zapotec Sierra of Oaxaca, Mexico*, is a history of mining concessions in the Sierra Norte and community resistance to them during the late 19th and 20th centuries.

15. Herlihy has a history of leaving with the data. See Mac Chapin and Bill Threlkeld, *Indigenous Landscapes: A Study in Ethnocartography* (Center for the Support of Native Lands, Arlington, VA, 2001), especially pp. 82–85.

16. Herlihy's claims to the contrary verge on the ludicrous. In public presentations where he's defended the project, he's argued that the community will use the map to transmit Zapotec place names and knowledge to future generations—despite the fact that the Zapotec have been doing this without his map for thousands of years. As for its utility in managing resources, it *would* be useful in digital form *if* he'd given the communities the relevant GIS software and computers. Instead he left them GPS units.

17. J. B. Harley, "Maps, Knowledge, and Power," in Paul Laxton, ed., *The New Nature of Maps: Essays in the History of Cartography* (Johns Hopkins University Press, Baltimore, MD, 2001), pp. 51–81; this quotation is from p. 75.

18. This comes pretty close to Peter Minuit's fabled "twenty-six dollars and a bottle of booze," from Rodgers and Hart's "Give It Back to the Indians." The twist, of course, is that residents of the Sierra Juárez are hardly the ignorant savages stereotyped in accounts such as the apocryphal tale of Manhattan's sale. As their response upon learning of the U.S. Army's involvement in Herlihy's project attests, they were deeply aware of the importance of controlling their land and access to it.

19. But see David Quist and Ignacio H. Chapela, "Transgenic DNA introgressed into traditional maize landraces in Oaxaca, Mexico," *Nature* 414(29), November 2001, pp. 541–542, for a sobering indication of the extent to which, even in the Sierra Juárez, and far from the road that crosses it from Oaxaca de Juárez to Tuxtepec, transference of transgenic DNA has taken place.

20. We interviewed Ramírez in Oaxaca in February 2012. Ramírez is an Ixtlán native and is well known, and respected, in the area. Among his list of accomplishments, Isaiah Bowman led an AGS expedition to the Andes in 1913 before serving as an advisor to Presidents Woodrow Wilson, at Versailles at the conclusion of World War I, and Franklin Roosevelt, during World War II, on territorial issues. For the decisive account of Bowman's significance, see Neil Smith, *American Empire: Roosevelt's Geographer and the Prelude to Globalization* (University of California Press, Berkeley, CA, 2004).

21. We interviewed González in Oaxaca in February 2012.

22. We interviewed participants in these activities, independently, from all three communities in February 2012.

23. Geoffrey Demarest, *Mapping Colombia: The Correlation between Land Data and Strategy* (Strategic Studies Institute, U.S. Army War College, Carlisle

Barracks, PA, 2003), p. 1. He uses the quoted phrase with respect to Colombian insurgencies and narcotraffickers, or, as he puts it elsewhere, "groups that are variously revolutionary, anarchist, and criminal" (as Geoff Demarest, in *Geoproperty: Foreign Affairs, National Security, and Property Rights*; Frank Cass Publishers, Portland, OR, 1998 p. 179). For more on the military importance of a global cadaster, see Douglas E. Batson, *Registering the Human Terrain: A Valuation of Human Cadastre* (National Defense Intelligence College, Washington, DC, 2008).

24. Herlihy's Curriculum Vitae, at *www2.ku.edu/~geography/Vitas/Herlihy_P.pdf* (accessed March 7, 2012), contains this, for example, under "Consultancies (Selected)":

> 2005–present. Leader of the American Geographical Society Bowman Expedition Prototype for global place-based GIS research. Funded by the Foreign Military Studies Office (DoD), our tri-national collaboration is retooling regional geography and foreign area studies to show how Mexico's gargantuan neoliberal land reform, PROCEDE, converts communal ejido lands from social to private property. Our database aims at producing the "digital human terrains" of indigenous Mexico (see our ESRI award-winning website, *http://web.ku.edu/~mexind*).

The website has since been taken down and a revised version reposted. See the note at the beginning of the Bibliography (p. 237).

25. See Herlihy's Curriculum Vitae on the University of Kansas (KU) geography website, op. cit.

26. We interviewed both Ramírez and González, independently, in February 2012, but we also draw on interviews we conducted during the same period with Jaime Martínez Luna, Salvador Aquino, and others.

27. Although Urbano told us essentially the same story when we interviewed him in February 2012, this version of his story comes from the interview included in Simon Sedillo's 2010 film, *The Demarest Factor*. You can watch this online at *http://elenemigocomun.net/2011/09/demarest-factor-military-mapping-indigenous-communities* (accessed March 8, 2012), but it is sold in Oaxaca as a DVD.

28. Indeed:

> Before starting upon any given journey, I secured letters from the department of Fomento, one of the Executive Departments of the Federal Government. These letters were directed to the governors of the states; they were courteously worded introductions. From the governors, I received letters of a more vigorous character to the jefes of the districts to be visited. From the jefes, I received stringent orders upon the local governments; these orders entered into no detail, but stated that I had come, recommended by the superior authorities, for scientific investigations; that the local authorities should furnish the necessaries of life at just prices, and that they should supply such help as was necessary for my investigations. In addition to the orders from the jefes to the town authorities, I carried a general letter from the governor of the state to officials of every grade within its limits. (Starr, *In Indian Mexico*, op. cit., pp. 1–2)

29. We quote from Herlihy's "MI Team's statement to participants and concerned parties—Includes responses to frequently asked questions," p. 6, as posted on México Indígena's "Response to Accusations of Geopiracy" page at *http://web.ku.edu/~mexind/Herlihy_MexicoIndigenaBowmanEthics.pdf* (accessed March 8, 2012). Whereas here he refers to *"a* training workshop" in his "Project Status Reports," for August 2006 he had written, "UNOSJO (described in detail in our previous monthly reports) *installations have been repeatedly used* for our participatory workshops and project meetings," *web.ku.edu/~mexind/FMSO_WebReport.doc* (downloaded September 9, 2012).

30. Although Perez told us essentially the same story when we interviewed him in February 2012, this version of his story comes, like that of Urbano, from the interview with him in *The Demarest Factor*, op. cit.

31. Herlihy's constant assertion that Tiltepec was okay with his work was one of the things that prompted Tiltepec to issue its 2009 proclamation (e.g., "Information has been circulated in different news media and on the internet, alleging that our community agrees with the results of the investigation"). In his "MI Team's statement," op. cit., Herlihy describes a celebratory assembly in Tiltepec at which he and his team were presenting "the final map (printed in color and protected in plastic)," being ruined by González's accusations of geopiracy. Pretending to know scarcely anything about UNOSJO ("It is a very small, independent, un-audited NGO"), Herlihy describes his and his team's careful response and their endorsement by the community, waving UNOSJO away with: "We do not understand UNOSJO's motives, but we do know that it is by no means a politically neutral organization." That is, by implication, that Herlihy's is.

Chapter 2

1. Hans Koning, *Columbus: His Enterprise, His Mission* (Monthly Review Press, New York, 1976/1991), pp. 83–85.
2. Harvey K. Flad, "Audubon Terrace, the American Geographical Society, and the Sense of Place," *Geographical Review* 95(4), 2004, pp. 519–529.
3. See Richard White's *Railroaded: The Transcontinentals and the Making of Modern America* (Norton, New York, 2011), where he *means* railroaded.
4. John Kirtland (J. K.) Wright, *Geography in the Making: The American Geographical Society, 1851–1951* (American Geographical Society, New York, 1952), p. 25.
5. Karen Morin, *Civic Discipline: Geography in America, 1860–1890* (Ashgate, Farnham, Surrey, UK, 2011), pp. 106–177; this quotation is from pp. 109–110. Her treatment of the AGS's involvement in the Nicaraguan Canal Company is both understated and devastating.
6. The American Geographical Society was anything but alone here (Wright, op. cit., pp. 69, 83). It could be said to be the predominant view in the 19th century. See David Livingstone, *The Geographical Tradition: Episodes in the*

History of a Contested Enterprise (Blackwell, Oxford, UK, 1992), especially here his seventh chapter, "A 'Sternly Practical' Pursuit: Geography, Race and Empire," pp. 216–259.

7. Wright, op. cit., pp. 69–70.

8. For a gendered history of the AGS in the 19th century, see Karen Morin's "Charles P. Daly's Gendered Geography, 1860–1890," *Annals of the Association of American Geographers* 9(4), 2008, pp. 897–919; and her *Civic Discipline*, op. cit. Morin's detailed and valuable analysis is vitiated by a doubtful assumption that the AGS mattered all that much in the 19th century, but far more by the entirely questionable assertion that Charles Daly, New York Common Pleas judge and AGS president for many years, was a significant geographer who's been unjustly neglected. It's unclear that he was a geographer at all, much less a significant one. That he's been *neglected* is wholly debatable. That the AGS was run by "manly men" committed to the promotion of New York's mercantile interests is not, and she quashes any doubt that it was ever anything else.

9. Jeremy Crampton notes that "explorer societies such as the Royal Geographical Society (RGS) and the American Geographical Society (AGS) not only promoted racialized imperial visions, but included racists in positions of authority. Madison Grant, for example, the author of the racist pro-Aryan book *The Passing of the Great Race*, was a member of the AGS Board of Governors for 22 years," and so on in his useful "The Cartographic Calculation of Space: Race Mapping and the Balkans at the Paris Peace Conference of 1919," *Social and Cultural Geography* 7(5), 2006, pp. 731–752; the quotation is from p. 734.

10. For a critical history of the magazine, see Catherine Lutz and Jane Collins, *Reading National Geographic* (University of Chicago Press, Chicago, 1993). This is a book about "the magazine and the Society as a key middlebrow arbiter of taste, wealth and power in America." For a view more tightly focused on the magazine's "imperialist" imagery, see Rothenberg's "Voyeurs of Imperialism: *The National Geographic Magazine* before World War II," in Anne Godlewska and Neil Smith, eds., *Geography and Empire* (Blackwell, Oxford, UK, 1994), pp. 155–72. Howard Abramson's *National Geographic: Behind America's Lens on the World* (Crown Publishing, New York, 1987) may have been the first extended critical take on the magazine. Despite their differences (and Lutz and Collins provide the most nuanced reading), the emphasis in all these is on the *Geographic's* photographs (especially of naked brown women) and their captions, and so, on the magazine's construction of human culture. For a critical take on the magazine's *maps*, see Denis Wood and John Fels, *The Natures of Maps* (University of Chicago Press, Chicago, 2008), especially the third and fourth chapters. For a Society-approved history, see C. D. B. Bryan, *National Geographic Society: 100 Years of Adventure and Discovery*, rev. ed. (Abrams, New York, 1997).

11. For this bizarre story, in which "the National Geographic Society assumed the mantle of scientific authority by asking both men [Peary and Frederick Cook] to submit their records to a committee of the organization's officers for verification," despite the Society's long support of Peary, see Susan Schulten, *The*

Geographical Imagination in America, 1880–1950 (University of Chicago Press, Chicago, 2001), pp. 148–149. The immediate support of then President Roosevelt helped.

12. Ibid., p. 74. Schulten is good on the rise of all the American explorer societies and especially on their relationship to schooling.
13. This is from an "Address" Davis gave to the Section of Geology and Geography of the American Association for the Advancement of Science. We're quoting from a copy printed in 1903 of advanced pages from the *Proceedings of the American Association for the Advancement of Science 53*, 1904, pp. 1–32, this quotation is from p. 5.
14. Ibid., p. 6.
15. Ibid., p. 8.
16. Ibid., p. 9.
17. Albert Perry Brigham, "The Association of American Geographers, 1903–1923," *Annals of the Association of American Geographers* 14(3), 1924, pp. 109–116; the quotation is from p. 109.
18. Davis, "Address," op. cit., pp. 30–31.
19. The Association of American Geographers and National Geographic Society figures come from W. M. Davis, "The Progress of Geography in the United States," *Annals of the Association of American Geographers* 14(4), 1924, pp. 159–215; pp. 178 (for the AAG) and 177 (for the NGS). The AGS figures come from Wright, op. cit., p. 193.
20. Davis, "Progress," op. cit., p. 178.
21. Wright, op. cit., pp. 166–167.
22. Neil Smith, *American Empire: Roosevelt's Geographer and the Prelude to Globalization* (University of California Press, Berkeley, CA, 2004).
23. Wright, op. cit., where he's quoting Gladys Wrigley.
24. For the Inquiry being headquartered in the AGS building, and Bowman's calendar during the period, see Wright, ibid., et passim. For the nature of his role and the immediate geopolitical questions Bowman addressed, see Crampton, "The Cartographic Calculation," op. cit., and Smith, op. cit., especially pp. 142–143.
25. The most recent biography of Maury is Chester Hearn's *Tracks in the Sea: Matthew Fontaine Maury and the Mapping of the Oceans* (McGraw-Hill, New York, 2002).
26. Wright, op. cit., pp. 24–25. Maury completely ignored "the broad realm covered by 'statistics.'" Given this, *and* the society's *immediate* support of Arctic exploration, it is hard to credence Morin's insistence in "Charles P. Daly's Gendered Geography," op. cit., that Daly "shifted the focus of the society away from statistical representation of the world toward the action-packed narrative descriptions of the world provided by embodied explorers in the field." In fact, Wright notes, "Although there is no positive evidence that Grinnell's polar interests had any direct influence on the *founding* of the institution, his election as President in October 1851 suggests this" (pp. 15–16). In its first

year the society helped organize Kane's expedition to the Arctic. Then, at the second sitting, in 1852, a letter from Livingstone was read; and so on. It was an explorer club from the start.
27. Wright, op. cit., p. 33.
28. Ibid., p. 155. Huntington ponied up $262,000 of the $300,000 cost. Other members of the council came up with the balance.
29. Ibid., p. 158. That's according to Wright. Beatrice Gilman Proske, however, claims the doors were sculpted by Berthold Nebel, who also did the doors for the Museum of the American Indian across the Terrace (in her *Archer Milton Huntington*, Hispanic Society of America, New York, 1963, p. 19).
30. Ibid., pp. 158–166.
31. Wright, op. cit., p. 25.
32. Jerome E. Dobson, "Restoring Geography in America," *Ubique* 27(3), 2006, pp. 1–2, quoted from p. 2. At the same time he's capable of saying things like "foreign intelligence is geography," which is the title of an article, "Foreign Intelligence Is Geography," which he published in *Ubique* 25(1), pp. 1–2.
33. It's what scientists do! Among other things, see Bruno Latour, *Pandora's Hope: Essays on the Reality of Science* (Harvard University Press, Cambridge, MA, 1999), especially the second chapter, "Circulating Reference," pp. 24–80. Though with respect to the Inquiry, perhaps Hans-Jörg Rheinberger's, *Toward a History of Epistemic Things: Synthesizing Protein in the Test Tube* (Stanford University Press, Stanford, CA, 1997) would be more appropriate, substituting "nation" for "protein" throughout.
34. See Crampton here, "The Cartographic Calculation," op. cit., p. 744, as well as his essay on "Maps, Race, and Foucault: Eugenics and Territorialization Following World War I," in Jeremy Crampton and Stuart Elden, eds., *Space, Knowledge, and Power: Foucault and Geography* (Ashgate, Aldershot, UK, 2007), pp. 223–244. Our reading here is heavily indebted to his of the Inquiry's role.
35. Ignoring here the manifold problems in what "linguistic predominance" could possibly mean. How many fraught issues are buried there? To say nothing of "racial predominance." Like science could settle questions like these!
36. Crampton, "Maps, Race, and Foucault," op. cit., p. 241.
37. Wright, op. cit., p. 199, again quoting Wrigley.
38. Ibid., pp. 301–302. Wright here is quoting from a recollection Bowman sent him in 1949.
39. This is also from Wright, but pp. 202–203. Where we aren't quoting, we're paraphrasing. Bowman confirms this account as well in his application for a Rockefeller Foundation grant to support the work.
40. Proske, *Archer Milton Huntington*, op. cit.; coins, pp. 6–7; map, pp. 13–14.
41. It's important to keep in mind how local so many of these connections are. When John D. Rockefeller moved from his native Cleveland to New York, he bought his first home at 4 West 54th Street from, who else? Archer Milton Huntington's mom. (It's now part of the Museum of Modern Art's garden.)

On the Rockefeller Foundation in Latin America, see Marcos Cueto, ed., *Missionaries of Science: The Rockefeller Foundation and Latin America* (Indiana University Press, Bloomington and Indianapolis, IN, 1994).

42. On these aspects of Eckert's role in the academization of cartography, see Denis Wood, *Rethinking the Power of Maps* (Guilford Press, New York, 2010), pp. 121–122.
43. Wright, op. cit., p. 309.
44. Ibid., 310.
45. For this, see Wolfgang Scharfe's, "Max Eckert's 'Kartenwissenschaft': The Turning Point in German Cartography," *Imago Mundi 38*, 1986, pp. 61–66; the quotation is from p. 62.
46. Wright, op. cit., p. 304.
47. Wright includes an appendix that lists and categorizes the Society's publications through 1951. The Society continued to publish during the 1960s, but other than a couple of publications in 1971 and 1972, it's published no new books since (it's reprinted a couple) and few new maps. For the past 40 years the Society really has operated as a scholarly institution only marginally, maintaining support of its *Geographical Review* and awarding the McColl Family Fellowship for research in geography since 1999. It also publishes *Focus on Geography* in applied geographical research.
48. Wright, op. cit., p. 79.
49. Ibid., p. 110.
50. Once the benefactors disappeared, the Society was dependent on dues, and members were increasingly attracted to the AAG, leaving the Society with ever fewer funds. See, for example, Wright's discussion about receipts in the post-Bowman period, p. 351.
51. Flad, op. cit., p. 525.
52. Which, however, he published in the AAG's *Annals*: "Geography, Experience, and Imagination: Towards a Geographical Epistemology," *Annals of the Association of American Geographers 51*(3), 1961, pp. 241–260.
53. Warntz's appointment was in Harvard's Graduate School of Design since Harvard had no geography department. Warntz was also part of the research team in the Laboratory for Computer Graphics. See Donald Janelle's "In Memoriam: William Warntz, 1922–1988," *Annals of the Association of American Geographers 87*(4), 1997, pp. 723–731.
54. Personal communication, October 2011. Woldenberg was there, from 1965 to 1967, mostly as a researcher on Warntz's grant from the Office of Naval Research on "Geography and the Properties of Surfaces." Woldenberg notes: "The AGS library was magnificent! Personally, it was responsible for my doctoral dissertation. I was randomly reading a journal, FENNIA, and found a keystone article by Palomaki. This helped me join and explain hierarchies in rivers, blood vessels, cities, trees, etc." Positive recollections of the library are common. See Baruth, below.

55. Michael Woldenberg, "Energy Flow and Spatial Order: Mixed Hexagonal Hierarchies of Central Places," *Geographical Review* 58(4), 1968, pp. 552–574.
56. John Kirtland Wright, *Human Nature in Geography: Fourteen Papers, 1925–1965* (Harvard University Press, Cambridge, MA, 1966).
57. Christopher Baruth, "The American Geographical Society Library: A Treasure Trove for Twenty-First-Century Geographical Scholarship," *Geographical Review* 96(3), 2006, pp. 459–472, quoted from p. 460.
58. Ibid.
59. Shortly afterwards the Museum of the American Indian abandoned Audubon Terrace for the Customs House in Lower Manhattan, though before the decade was out it had been absorbed by the Smithsonian, and in 2004 was carted off to Washington. In 2004 the American Numismatic Society also left Audubon Terrace. For the past five years the Hispanic Society has also been engaged in efforts to relocate.
60. Flad, "Audubon Terrace," op. cit., p. 527.
61. John Krygier and I (Denis Wood) visited the offices in February 2012, where we not only photographed the plaque, saw the globe and the spanking new replica of the globe (which is getting too fragile to travel much), but also signed the guest book. It had been six years since the last guest had signed it. The office doesn't get a lot of visitors.
62. Ibid., p. 528.
63. Finley became editor-in-chief of the *Times* in 1937, resigning for reasons of health a year later. See Wright, op. cit., p. 208 for details.
64. Though they can always give medals. Since Wright completed his history of the first 100 years, the AGS has created four new medals (five weren't enough!): the Osborn Maitland Miller Cartographic Medal (in 1968), the Van Cleef Memorial Medal (in 1970), the Paul P. Vouras Medal (in 1988), and more recently the Alexander & Ilse Melamid Medal (in 2002).

Chapter 3

1. Merritt A. Edson, "The Coco Patrol: Operations of a Marine Patrol Along the Coco River in Nicaragua, 1928–29," Sections I–III, *Marine Corps Gazette*, August 1936, pp. 18–23, 38–48; "Section III," continued from the August issue, *Marine Corps Gazette*, November 1936, pp. 40–41, 60–72; and "Section IV: The Advance to Poteca," *Marine Corps Gazette*, February 1937, pp. 35–43, 57–63. We cite these as Edson 1, Edson 2, and Edson 3. This 22,000-word report on the first half of his 14 months along the Coco is compulsively readable. He never finished his projected record of the patrols, but the rest of the story is told in an Edsonian fashion in Jon T. Hoffman, *Once a Legend: "Red Mike" Edson of the Marine Raiders* (Presidio Press, Novato, CA, 1994), pp. 76–93. The quote here comes from Edson 3, p. 41.
2. Edson 3, op. cit., p. 42.

3. Ibid., p. 63.
4. In his "The Strategy and Tactics of Small Wars, Part 1" (which Edson may be presumed to have read), Samuel Harrington writes of the Marines' occupation of Santo Domingo that "the reasons for this occupation are unimportant for this discussion," and indeed this was the general attitude among Marines: they were just doing what they were told (in the *Marine Corps Gazette*, December 1921, pp. 474–491; the quotation is from p. 478).
5. Edson 1, p. 18.
6. For example, after a firefight where the Marines might have been bested, Edson observed that "the outlaws were well organized, well disposed and well led. They were uniformly dressed in blue denim and looked less like the orthodox bandit than did we with our month's growth of beard, worn out clothing and broken shoes" (Edson 3, op. cit., p. 61).
7. Among other things, Sandino regarded Yankees as "our legitimate enemies by race and language." See David Whisnant, *Rascally Signs and Sacred Places: The Politics of Culture in Nicaragua* (University of North Carolina Press, Chapel Hill, NC, 1995), p. 351.
8. According to Alan MacPherson:

   > Sandinism was so transnational in scope that it became a symbol for those fighting imperialism everywhere. In 1928, Chinese nationalists marched through Beijing with banners of Sandino, and one of their divisions was named after him. Guatemalans sold a cigarette called "Cigarrillo Sandino," and in El Salvador, a liquor called "Nectar Sandino." Marxists and leftists throughout the world supported the Nicaraguan struggle. As one French journalist wrote in support, "Today attention is directed towards General Sandino, whose figure resembles those of the great historical Liberators, and who according to the words of Manuel Ugarte, represents with his heroic troops the popular revenge of the Spanish-speaking countries against the Anglo-Saxon imperialism and against the treacherous local oligarchies, opposed to every movement of liberation."

   In his "Anti-Americanism in Latin America and the Caribbean—'False Promise' or Coming Full Circle," in Alan MacPherson and Ivan Krastev, eds., *The Anti-American Century* (Central European University Press, New York, 2007), pp. 49–76; the quotation is from p. 63.

9. Whisnant, op. cit., p. 371.
10. Ibid., p. 344. Sandino was speaking to Carlton Beals reporting for *The Nation* (about which Whisnant has more to say on p. 354, where he compares Beals with others writing about Sandino). The Beals articles appeared in February and March, 1928, and he expanded on his time with Sandino at length in his book, *Banana Gold* (Lippincott, Philadelphia, 1932), especially pp. 195–301.
11. The last word here is Michael Schroeder's "Horse Thieves to Rebels to Dogs: Political Gang Violence and the State in the Western Segovias, Nicaragua, in the Time of Sandino, 1926–1934" (*Journal of Latin American Studies 28*, 1996, pp. 383–414), which is not only nuanced and detailed but very clear.
12. Ibid., p. 400 for a list of these gangs operating in the Segovias. The quotes come from throughout the paper.

13. Nonetheless, Schroeder makes the point that "Sandino's rebellion was as much a regional civil war as a nationalist, anti-imperialist crusade," ibid., p. 426, and in his doctoral dissertation in history, *To Defend Our Nation's Honor": Toward a Social and Cultural History of the Sandino Rebellion in Nicaragua, 1927–1934* (University of Michigan, Ann Arbor, MI, 1993).

14. The astonishing list, *One Hundred Eighty Landings of the United States Marines, 1800–1934*, was prepared by Captain Harry Ellsworth as a mimeographed pamphlet in 1934. A 1974 reprint, with added material, was published by the History and Museums Division of the U.S. Marine Corps and is available online as a pdf at *www.tecom.usmc.mil/HD/PDF_Files/Pubs/One%20 Hundred%20Eighty%20Landings%20of%20United%20States%20 Marines%201800-1934.pdf* (accessed May 19, 2012). The reasons for the landings are almost as varied as their number is large.

15. On the Banana Wars, see Lester Langley, *The Banana Wars: United States Intervention in the Caribbean, 1898–1934*, rev. ed. (Scholarly Resources, Wilmington, DE, 2002), and *Banana Wars: Power, Production, and History in the Americas*, Steve Striffler and Mark Moberg, eds. (Duke University Press, Durham, NC, 2003).

16. Though it must be acknowledged that the route had been a favorite since the founding of a trans-Nicaraguan canal company during the John Quincy Adams administration. See Langley, ibid., pp. 49–51. See also Christian Brannstrom, "Almost a Canal: Visions of Interocean Communication across Southern Nicaragua," *Ecumene* 2(1), 1995, pp. 65–87.

17. On the early history of León and Granada and its resonance, see Whisnant, op. cit., pp. 30–46.

18. See Karl Offen, "The Sambo and Tawira Miskitu: The Colonial Origins and Geography of Intra-Miskitu Differentiation in Eastern Nicaragua and Honduras," *Ethnohistory* 49(2), 2002, pp. 319–372.

19. See the second chapter of my (Joe Bryan's) dissertation in geography, *Map or Be Mapped: Land, Race, and Property in Eastern Nicaragua* (University of California, Berkeley, CA, 2007), pp. 28–94, for a detailed reading of the nature, arrival, and influence of the Moravian church on the Miskito Coast. For a parallel discussion of the church in Honduras, see Benjamin F. Tillman, *Imprints on Native Lands: The Miskito–Moravian Settlement Landscape in Honduras* (University of Arizona Press, Tucson, AZ, 2011).

20. Karl Offen, "The Geographical Imagination, Resource Economies and Nicaraguan Incorporation of the Mosquitia, 1838–1909," in Christian Brannstrom, ed., *Territories, Commodities and Knowledges: Latin American Environmental History in the Nineteenth and Twentieth Centuries* (Institute for the Study of the Americas, London, 2004, pp. 50–89).

21. Michael Gismondi and Jeremy Mouat, "Merchants, Mining and Concessions on Nicaragua's Mosquito Coast: Reassessing the American Presence, 1893–1912," *Journal of Latin American Studies* 34(4), 2002, pp. 845–879; and Michael Gismondi and Jeremy Mouat, "'La Enojosa Cuestión De Emery': The Emery Claim in Nicaragua and American Foreign Policy, C. 1880–1910," *The Americas* 65(3), 2009, pp. 375–409.

22. The largest UNIA affiliate was, by far, Bluefields. However, there was an affiliate in Puerto Cabezas as well that was active from 1927 to 1933. See Theodore G. Vincent, *Black Power and the Garvey Movement* (Black Classic Press, Baltimore, MD, 2006), pp. xvi–xvii.
23. Frutos Ruiz y Ruiz, *Informe del Doctor Frutos Ruiz y Ruiz, Comisionado del Poder Ejecutivo en la Costa Atlántica de Nicaragua*, Managua, September, 1927; CIDCA–UCA archives, Managua.
24. Edson 1, op. cit., p. 18.
25. Ibid., p. 22. The immediate directive came from Major Harold Utley, but the Marine commander in Nicaragua, Brigadier General Logan Feland, had first issued the guideline. David Brooks quotes Feland's orders in a footnote at the bottom of p. 315 of his "U.S. Marines, Miskitos and the Hunt for Sandino: The Río Coco Patrol in 1928," *Journal of Latin American Studies* 21(2), 1989, pp. 311–342. Further detail is to be found in his dissertation in political science, *Rebellion from Without: Culture and Politics along Nicaragua's Atlantic Coast in the Time of the Sandino Revolt, 1926–1934* (University of Connecticut, Storrs, CT, 1998).
26. Edson expiates at length on the high quality of Miskito boat handing, Edson 3, op. cit., pp. 39–40.
27. Brooks, "U.S. Marines," op. cit., p. 326.
28. Though Edson soon learned that manaca leaves were more effective than banana: "A well built manaca leaf lean-to would shed the heaviest rains for three or four weeks before the leaves had to be renewed; banana leaves had to be replaced every week or ten days," Edson 3, op. cit., p. 38.
29. Edson 1, op. cit., pp. 39–40.
30. Ibid., p. 41. Brooks certainly concurs with this self-judgment, but Charles Hale has reservations. See his *Resistance and Contradiction: Miskitu Indians and the Nicaraguan State, 1894–1987* (Stanford University Press, Stanford, CA, 1994), p. 253, footnote 51. Brooks responds to Hale's arguments ("Unfortunately this class-based analysis of the cooperation Edson received is inadequate") in his later dissertation, op. cit., p. 127, but also more generally, pp. 165–166. In any case, this is a minor issue for Hale's argument, and for our purposes it hardly matters: that Edson *thought* he had the Miskitos' loyalty is all that counts for his future work on the *Small Wars Manual*.
31. It's also salient that the Miskito had no time for Spanish Nicaragua, against which Sandino was in revolt.
32. Edson 3, op. cit., p. 43.
33. See Edson 1, op. cit., p. 43 in particular, though the same thing occurs later as well.
34. Ibid., p. 62.
35. Brooks, "U.S. Marines," op. cit., p. 311.
36. Geoffrey Parker, "Maps and Ministers: The Spanish Habsburgs," in David Buisseret, ed., *Monarchs, Ministers, and Maps: The Emergence of Cartography as a Tool of Government in Early Modern Europe* (University of

Chicago Press, Chicago, 1992), pp. 124–152, with the maps illustrated on pp. 140–141.

37. David Buisseret "Monarchs, Ministers, and Maps in France before the Accession of Louis XIV," in his *Monarchs, Ministers, and Maps*, op. cit., pp. 99–123, with the quote on p. 109. One of Fougeu's maps is reproduced on p. 111.

38. See Martin van Creveld's valuable, and surprisingly amusing, *Supplying War: Logistics from Wallenstein to Patton* (Cambridge University Press, Cambridge, UK, 1977), with the quote from p. 182. Suffice it to say that in so modern a war as that of World War II, German troops lived off the land in their initial push through Belgium, where the demolition of railroads cut their supply lines, to cite just one counterintutive example, out of many, from that war.

39. "Beginning the next day (May 8th) the entire patrol would be completely dependent on the country for food," reads a typical entry (Edson 2, op. cit., p. 40).

40. Long goes on to say, "But the human geography of the same place would reveal much more—tribal boundaries, political ideology, ethnicity and languages." William Matthews says, "Long means to include such elements as birth and death rates, degree of education, access to media, principal market commodities and proximity to health care facilities," all in his Nextgov blog at *www.nextgov.com/health/2011/07/mapping-human-terrain/49401* (accessed May 29, 2012).

41. Edson 1, op. cit., p. 23.

42. The best place to go for insight into these maps is historian Michael Schroeder's increasingly comprehensive web archive, The Sandino Rebellion: Nicaragua 1927–1934, at *www.sandinorebellion.com/index.htm*. The "Christian Brothers' map" likely came from their textbook, *Geografía de Nicaragua: Para Uso de los Grados 3°, 4° y 5° de las Escuelas Primarias*, published in Managua in 1928. The textbook is filled with maps, many of which were used extensively by the U.S. Marines; *www.sandinorebellion.com/PhotoPgs/2maps-ChristianBros1928.html* (accessed May 28, 2014; original from the U.S. National Archives, Record Group 127, Entry 38, Box 20). The Moravians had also published a detailed map of the "Miskito Coast" for inclusion in *The Moravian Missionary Atlas; containing an account of the various countries in which the Missions of the Moravian Church are carried on, and of its missionary operations. New edition with eighteen maps and a Mission field and station index* (Moravian Church and Mission Agency, London, 1908; *http://divdl.library.yale.edu/dl/OneItem.aspx?qc=AdHoc&q=10160*; accessed May 28, 2014). The map of Nicaragua in this was at a scale of 1:3,000,000 and was made by Wagner and Debes of Leipzig. The "Ham Map" was a traced copy of a 1924 document created by Clifford D. Ham. Ham was the Collector-General of Customs in Nicaragua at the time, but earlier he'd been the surveyor of the Port of Manila, so he had mapmaking skills. For maps, go directly to *www.sandinorebellion.com/PhotoPgs/2maps-RioCocoPatrol.html* (accessed September 4, 2012).

43. That this is the map Edson described is confirmed, among other things, by the fact that trying to coordinate with headquarters during the Wanks Patrol, he

informs them, "MUSAWAS SHOWN ON MAP AS TULULUK" (Edson 2, op. cit., p. 40), which is what it is on the map we've illustrated (very bottom of detail, near center). This is one of the discrepancies to which Edson referred.
44. Edson 1, op. cit., p. 22.
45. Edson 2, op. cit., p. 40.
46. Edson 1, op. cit., p. 43.
47. Indeed, Kittle replaced the first guide that had been found for Edson who had proved wholly unsatisfactory. See early pages of Edson 1, op. cit., especially pp. 23 and 38.
48. Edson 3, op. cit., p. 43. Unfortunately, the soldiers leaving this correspondence in their flight immediately informed Sandino of Edson's proximity, leading to an ambush a little later.
49. These were notable for their ineffectiveness. See Hoffman, *Once a Legend*, op. cit., pp. 84–93, for Edson's time in Poteca.
50. For all this, see Hoffman, *Once a Legend*, pp. 97–100. But also see Samuel Harrington, "The Strategy and Tactics of Small Wars, Part 1," op. cit.; and "The Strategy and Tactics of Small Wars (continued)," *Marine Corps Gazette*, March 1922, pp. 84–92.
51. Major Harold H. Utley, "An Introduction to the Tactics and Technique of Small Wars," *Marine Corps Gazette* 15(5), 1931, pp. 50–53. This quote and the one in the following sentence are from page 50.
52. Still in print: C. E. Callwell, *Small Wars: Their Principles and Practice*, 3rd ed. (University of Nebraska Press, Lincoln, NE, 1996), though it's available in a number of editions.
53. W. C. G. Heneker's *Bush Warfare* is not in print at the moment, but an edition of the book, together with other early writings and an introduction by Andrew Godefroy, was published in 2009 by the Directorate of Land Concepts and Designs of the Canadian Department of National Defense, Kingston, Ontario, which is currently available online at *www.army.forces.gc.ca/DLCD-DCSFT/pubs/bushwarfare/BushWarFare.pdf* (accessed June 1, 2012). Heneker had been just about to deploy to West Africa—the Gold Coast, Ashanti country—when Callwell's *Small Wars* appeared. Heneker would apply, challenge, and improve on many of Callwell's ideas to produce what is really a small wars *textbook*.
54. E. H. Ellis, "Bush Brigades," *Marine Corps Gazette*, March 1921, pp. 1–15. In addition to a series of recommendations, this is a defense of the Marines against their detractors. Ellis concludes: "Yes, the Marines are down in jungle-land, and they did kill a man in a war, and a great many people did not know anything about it. This is most unfortunate, but the Marines are only doing their job as ordered by the people of the United States."
55. Harold Utley, "An Introduction to the Tactics and Technique of Small Wars," *Marine Corps Gazette*, May 1931, pp. 50–53; "The Tactics and Technique of Small Wars: Part II–Intelligence," *Marine Corps Gazette*, August 1933, pp. 44–48; "The Tactics and Technique of Small Wars: Part III–Functions of the Personnel (First) Section of the Staff," *Marine Corps Gazette*, November

1933, pp. 43–46. But also see Keith Kopets, "Why Small Wars Theory Still Matters: The Extension of the Principles on Irregular Warfare and Non-Traditional Missions of the *Small Wars Manual* to the Contemporary Battlespace," *Small Wars Journal* 2(3), 2006, pp. 10–11, *http://smallwarsjournal.com/content/journal-way-back-issues#v6* (accessed June 2, 2012).

56. *Small Wars Operations* is extremely rare, and is not online. WorldCat lists four copies, two of them at Quantico. There's apparently a fifth copy in the Edson papers at the Library of Congress, of which Hoffman wrote, in 1994, "Of special note is the copy of *Small Wars Operations* in Box 21, perhaps the only copy left in existence." We have not seen a copy.

57. "Chapter VIII, Operations Orders and Instructions," was never printed. See Richard C. McMonagle, "The Small Wars Manual and Military Operations Other Than War" (master's thesis, U.S. Army and General Staff College, Ft. Leavenworth, 1996), p. 58.

58. The familiar term *guerilla*, itself Spanish for "small war," entered into use during Napoleon's efforts to conquer Spain and Portugal during the Peninsular War of 1807–1814. The term was used to describe the bloody, spontaneous resistance waged by units organized by remnants of the Spanish army. For a decidedly continental (European) take on approaches to "small wars," see Walter Laqueur, "The Origins of Guerrilla Doctrine," *Journal of Contemporary History* 10(3), 1975, pp. 341–382. Laqueur traces the military history of the term back through the Thirty Years' War (1618–1648), the Spanish War of Succession (1701–1714), and the wars of Fredrick the Great (1740–1786). Although we would expand that history considerably to include, at the very least, the numerous Indian wars fought in North America, Lacqueur's account lends credence to an understanding of small wars as a catch-all term used to describe armed conflicts *not* fought exclusively between states.

59. Hoffman, *Once a Legend*, op. cit., pp. 122–123; McMonagle, op. cit., 59–62. The *Small Wars Manual* is available online at *www.au.af.mil/au/awc/awcgate/swm/index.htm* (accessed June 2, 2012), as well as in a variety of print editions. In addition to McMonagle and Hoffman's *Once a Legend*, see the latter's "Small Wars Manual," in *The War of 1898 and U.S. Interventions, 1898–1934: An Encyclopedia*, Benjamin R. Beede, ed. (Garland Publishing, New York, 1994), pp. 511–516; Ronald Schaffer, "The 1940 Small Wars Manual and the 'Lessons of History,'" *Military Affairs* 36, 1972, pp. 46–51; and Keith B. Bickel, *Mars Learning: The Marine Corps Development of Small Wars Doctrine, 1915–1940* (Westview, Boulder, CO, 2001).

60. Ellsworth's *One Hundred Eighty Landings of the United States Marines, 1800–1934*, op. cit., had been prepared as part of the *Operations* project: This all went by the wayside.

61. Hoffman, *Once a Legend*, op. cit., p. 123.

62. We don't provide page numbers for citations to the *Small Wars Manual*. Each chapter is individually numbered, so page numbers are redundant. We could give chapter and section numbers, but they're not keyed to any sort of through pagination. The easier thing to do, assuming you have the digital copy cited above, is to search for the phrase if you want to find it in context.

63. Hoffman, *Once a Legend,* op. cit., p. 123.
64. Max Boot, *The Savage Wars of Peace* (Basic Books, New York, 2002), pp. 284–285.
65. On the CIA's recruitment of ethnic minorities in Laos and Vietnam, see Kenneth J. Conboy and James Morrison, *Shadow War: The CIA's Secret War in Laos* (Paladin Press, Boulder, CO, 1995); Jane Hamilton-Merritt, *Tragic Mountains: The Hmong, the Americans, and the Secret Wars for Laos, 1942–1992* (Indiana University Press, Bloomington, IN, 1993); Gerald C. Hickey,*Window on a War: An Anthropologist in the Vietnam Conflict* (Texas Tech University Press, Lubbock, TX, 2002); and Alfred McCoy, *The Politics of Heroin: CIA Complicity in the Global Drug Trade*, 2nd ed., (Lawrence Hill Books, Chicago, 2003). Former CIA agent Thomas L. Briggs' *Cash on Delivery: CIA Special Operations during the Secret War in Laos* (Rosebank Press, Rockville, MD, 2009) presents a firsthand account of this process.
66. Allen Ford notes: "Physical access to the manual as well as its distribution became difficult. The Marines classified the *SWM* as 'Restricted,' thus immediately creating physical access and distribution friction. . . . Consequently, the *SWM* most likely languished unused and unread in a unit's security vault, inside a three-drawer safe, with access strictly controlled. Physical handling was limited to Publications Clerks and Security Managers," in "The *Small War Manual* and Marine Corps Military Operations Other Than War Doctrine" (master's thesis, U.S. Army Command and General Staff College, Fort Leavenworth, 2003), p. 27.

Chapter 4

1. The date of the original filing with the Supreme Court of British Columbia is variously reported as 1967 and 1968. Our date is given by Alex Rose in his *Spirit Dance at Meziadin: Chief Joseph Gosnell and the Nisga'a Treaty* (Harbor Publishing, Madeira Park, BC, 2000), p. 89. Given the uncertainty, however, we asked Thomas Berger, who filed the case, to confirm it (personal communication, July 4, 2012).
2. Fort Simpson represented an effort on the part of the Hudson's Bay Company to curtail inroads the maritime fur trade had made into its inland fur-trading profits. The maritime fur trade drew the indigenous peoples of Alaska and the Northwest Coast into an international "triangular trade" in which, mainly, sea otter pelts were exchanged in China for silk, tea, and other Chinese trade goods that were subsequently sold in the Eastern United States and Europe. The fort was moved from the mouth of the Nass River to the nearby Tsimpsean Peninsula in 1834. See James Gibson's *Otter Skins, Boston Ships, and China Goods: The Maritime Fur Trade of the Northwest Coast, 1785–1841* (McGill-Queen's University Press, Montréal, QC, Canada, 1992) for background.
3. Mary Hurley's "The Nisga'a Final Agreement" (Law and Government Division, Parliament of Canada, 1999, revised in 2001) provides a succinct overview of the background to the Nisga'a land question. It's available online at

www.parl.gc.ca/Content/LOP/ResearchPublications/prb992-e.htm#(9) txt (accessed June 26, 2012). But see also Alex Rose, op. cit. (chronology on pp. 243–246), and Hamar Foster, Heather Raven, and Jeremy Webber, eds., *Let Right Be Done: Aboriginal Title, the Calder Case, and the Future of Indigenous Rights* (University of British Columbia Press, Vancouver, BC, Canada, 2007).

4. The story has enormous currency. Among others, Nisga'a Chief Joseph Gosnell told it at the gold bar of the House in the legislative chamber in question in 1998 in a widely reproduced speech opening debate on the Nisga'a land treaty. The speech is reprinted in full in Alex Rose, op. cit., pp. 16–24; but see also Paul Tennant, *Aboriginal Peoples and Politics: The Indian Land Question in British Columbia, 1849–1989* (University of British Columbia Press, Vancouver, BC, Canada, 1990), p. 58; and for the perspective of a government negotiator, Tony Penikett's *Reconciliation: First Nations Treaty Making in British Columbia* (Douglas & McIntyre, Vancouver, BC, Canada, 2006), p. 125. You can also find the speech in the official record at *www.leg.bc.ca/hansard/36th3rd/h1202PM.HTM* (accessed July 7, 2012).

5. Unless it was in 1907 or 1909. There are sources for all three dates. The confusion is indicative of the ferment, the numerous efforts, the frustration; and most current sources acknowledge the multiplicity of dates. See Hurley, 2001 op. cit., but also Stephanie Hanna and Hamar Foster's "A Select Chronology," in Foster, Raven, and Webber, op. cit., pp. 231–240.

6. This amounts to the entire Nass watershed, and although it's undoubted that the Nisga'a used it all in one degree or another, some ~17,000 square kilometers of the claim involves overlap land variously claimed by the neighboring Gitanyow, the Gitxsan, the Tahltan, and the Tsimshian. For an introduction to the complexities, see Neil Sterritt's "The Nisga'a Treaty: Competing Claims Ignored!", *BC Studies 120*, 1998–1999, pp. 73–94. For *far more* detail, see Neil Sterritt, Susan Marsden, Robert Galois, Peter Grant, and Richard Overstall, *Tribal Boundaries in the Nass Watershed* (University of British Columbia Press, Vancouver, BC, Canada, 1998)—all 350 pages of it.

7. Hurley, "Background," Part B, "The Nisga'a Land Question," op. cit.; Alex Rose, op. cit., pp. 79–80.

8. That Job was Arthur's father we understand from Hamar Foster's "One Hundred Years of Advocating for Justice: Litigating the Calder Case," online at The Osgoode Society site at *www.osgoodesociety.ca/pdfs/abstracts_hamar-foster.pdf* (accessed July 7, 2012). Foster told us this: "My recollection—which may even be accurate—is that Frank told me that Job and Arthur were father and son" (personal communication, July 1, 2012).

9. This story too is widely reported. We've excerpted from the version in Alex Rose, op. cit., p. 82.

10. The decision has been variously interpreted, but the language is fairly extraordinary:

> There is an aboriginal Indian interest usufructuary in nature which is a burden on the title of the Crown and is inalienable except to the Crown and extinguishable only by

a legislative enactment of the Parliament of Canada. This aboriginal title does not depend on treaty, executive order or legislative enactment but flows from the fact that the owners of the interest have from time immemorial occupied the areas in question and have established a pre-existing right of possession. In the absence of an indication that the sovereign intends to extinguish that right the aboriginal title continues.

Furthermore, the Proclamation of 1763, since it applies to "all the Lands and Territories lying to the Westward of the Sources of the Rivers which fall into the Sea from the West and North West as aforesaid" indicated that the framers of the Proclamation were well aware that there was territory to the west of the sources of the rivers and showed that it was intended to include therein the lands west of the Rocky Mountains. In addition, the recorded activities of the explorers at the time do not support the view that the territory west of the Rockies was *terra incognita*.

Once aboriginal title is established it is presumed to continue until the contrary is proven and when the predecessors of the appellants came under British sovereignty they were entitled to assert their Indian title as a legal right. This right could not therefore be extinguished except by surrender to the Crown or by competent legislative authority and then only by specific legislation. However, there was no surrender by the Nishgas and neither the Colony of British Columbia nor the Province, after Confederation, nor the Parliament of Canada, enacted legislation specifically purporting to extinguish the Indian title. It must be presumed that the British Crown intended to respect native rights and the onus of proving that the Sovereign intended to extinguish the Indian title was on the respondent. The Proclamations and Ordinances relied on to establish an exercise of sovereignty and the assertion of title to lands by the Crown in fee were not relevant to the claim brought by the appellants which did not challenge the fee of the Crown but rather sought a declaration that the appellants possessed a right of occupation against the world except the Crown and that the Crown had not to date lawfully extinguished that right. In any event, the Proclamations and Ordinances relied on, to the extent that they extinguished aboriginal Indian title, were *ultra vires* since the Commission, Letters Patent and Instructions forming an integral part of the Commission, of the colonial Governor did not give any power or authorization to extinguish Indian title. (CALDER v. A.-G. B.C. 91, 1973, 34 D.L.R. (3d) 145) (also reported: [1973] S.C.R. 313, [1973] 4 W.W.R. 1)

11. Hugh Brody refers to the Royal Proclamation of 1763 as "the so-called Magna Carta of Indian rights in British North America," reserving, as it did, "as a hunting territory for Indians all lands west of the Allegheny Mountains" (*Maps and Dreams*, Douglas & McIntyre, Vancouver, BC, 1981), p. 63. As *Calder* made clear, these lands could be alienated only by treaty with the Crown— or later the Canadian and American federal governments—and was the legal basis for the treaties they signed. The Proclamation, *Calder* implied, meant that any group that had not signed a treaty continued to enjoy the aboriginal title that flowed "from the fact that the owners of the interest have from time immemorial occupied the areas in question and have established a pre-existing right of possession." It's this clause that led to the land use and occupancy studies and the map biographies.

12. Comprehensive claims always involve land and "deal with the unfinished business of treaty-making in Canada," as it says on the Land Claims page of the Aboriginal Affairs and Northern Development Canada website. Specific claims may involve land but more often deal with other grievances of First

Nations. The page has links to a wealth of detailed information, including status reports on all treaties and claims under negotiation: *www.aadnc-aandc. gc.ca/eng/1100100030285/1100100030289* (accessed July 8, 2012).

13. The statement is online at *www.aadnc-aandc.gc.ca/eng/1100100010189* (accessed July 8, 2012). The aboriginal response, *Citizens Plus*, is online at *http://ejournals.library.ualberta.ca/index.php/aps/article/view/11690/8926* (accessed September 4, 2012). It's better known as the *Red Paper*. Its sophistication came as a shock to many Canadians. For example, see that of Alex Rose, op. cit, pp. 99–102. For a thorough critique of the "White Paper Multiculturalism," see Dale Turner, *This Is Not a Peace Pipe; Towards a Critical Indigenous Philosophy* (University of Toronto Press, Toronto, ON, Canada, 2006).

14. This is Alex Rose, op. cit., pp. 105–106.

15. For a copy of the 1928 letter, see Frank T'Seleie's statement to the Mackenzie Valley Pipeline Inquiry in Mel Watkins, ed., *Dene Nation: The Colony Within* (University of Toronto Press, Toronto, ON, Canada, 1977), pp. 12–17, with the letter on p. 14. For the broader history of the Dene claims, see the "Native Claims" chapter of Thomas Berger's *Northern Frontier, Northern Homeland: The Report of the Mackenzie Valley Pipeline Inquiry: Vol. One* (Supply and Services Canada, Ottawa, ON, Canada, 1977). The entire inquiry report, a model of its kind, can be downloaded at *https://www.neb-one.gc.ca/ ll-eng/livelink.exe?func=ll&objId=238336&objAction=browse&redirect=3* (accessed August 17, 2012).

16. See Harvey Feit's "Hunting and the Quest for Power: The James Bay Cree and Whitemen in the 20th Century," in R. Bruce Morrison and C. Roderick Wilson, eds., *Native Peoples: The Canadian Experience* (McClelland & Stewart, Toronto, ON, Canada, 1986), pp. 171–207. This is available in second (1995) and third editions (2004). For a detailed account of the conflict between the Cree and Hydro-Québec, see Boyce Richardson's *Strangers Devour the Land: A Chronicle of the Assault upon the Last Coherent Hunting Culture in North America, the Cree Indians of Northern Québec, and Their Vast Primeval Homelands* (Alfred A. Knopf, New York, 1976).

17. Milton Freeman, "Looking Back—and Looking Ahead—35 Years After," *Canadian Geographer* 55(1), 2011, pp. 20–31; the quotation is from p. 21.

18. The Committee on Original Peoples Entitlement was an organization of the Inuvialuit of the Western Arctic and directly responded to oil and minerals exploration in the Mackenzie Delta/Beaufort area. The Inuit Tapirisat represented the Inuvualuit, Nunatsiavut (of Labrador), the Nunavik (of northern Québec), and the Nunavut. Environmentalists were also concerned about these developments. For their relationship with the Inuit, see Peter Burnett, "Environmental Politics and Inuit Self-determination," in Franklyn Griffiths' *Politics of the Northwest Passage* (McGill-Queen's University Press, Montréal, QC, Canada, 1987), pp. 181–199.

19. This was also going on in Alaska, where the same 1968 Prudhoe oil discovery wanted to drive a pipeline across Alaskan aboriginal lands to Valdez. In Alaska this led *within three years* to the signing of the Alaska Native Claims

Settlement Act, whereas in Canada the Mackenzie pipeline has yet to be built 40 years later. See Arthur Lazarus, Jr., and W. Richard West, Jr., "The Alaska Native Claims Settlement Act: A Flawed Victory," *Law and Contemporary Problems* 40(1), 1976, pp. 132–165, for a thoughtful parsing of the act. This was an example Canadian aboriginals did *not* want to follow. Thomas Berger was later contracted by the Inuit Circumpolar Conference to conduct an inquiry into the effects of the Alaska Native Claims Settlement Act, published as *Village Journey: The Report of the Alaska Native Review Commission*, rev. ed. (Hill & Wang, New York, 1995).

20. "The high number of American military personnel in the Eastern Arctic during the Second World War worried Ottawa who felt its sovereignty in the area was threatened by their presence," writes André Légaré in "The Process Leading to a Land Claims Agreement and Its Implementation: The Case of the Nunavut Land Claims Settlement," *Canadian Journal of Native Studies* 16(1), 1996, pp. 139–163; the quotation is from p. 143. Also see Frank Tester and Peter Kulchyski's *Tammamiit (mistakes): Inuit Relocation in the Eastern Arctic, 1939–1963* (University of British Columbia Press, Vancouver, BC, Canada, 1994), especially pp. 14–17, and Winona LaDuke's *All Our Relations: Native Struggles for Land and Life* (South End Press, Cambridge, MA, 1999).

21. This claim was long denied by the Canadian government, which argued that the move was made for other reasons, but with its formal apology to the Inuit in 2010 it would seem to have conceded the point (if not explicitly). For irrefutable acknowledgments by the government of its sovereignty interests, see the *Submission to the Royal Commission on Aboriginal Peoples of Makavik and Inuit Tapirisat of Canada on Behalf of the Inuit Relocated to the High Arctic (Griese Fiord and Resolute Bay) in the 1950s by the Federal Government*, June 28, 1993 (*http://pubs.aina.ucalgary.ca/makivik/CI034.pdf*, accessed July 18, 2012), pp. 5–8. See as well, Tester and Kulchyski, op. cit., pp. 113–118.

22. Milton Freeman, ed., *Inuit Land Use and Occupancy Project Report* (Supply and Services Canada, Ottawa, ON, Canada, 1976). The report consists of three volumes: *Vol. One: Land Use and Occupancy* (Freeman, ed., 1976a); *Vol. Two: Supporting Studies* (Freeman, ed., 1976b); and *Vol. Three: Land Use Atlas* (Freeman, ed., 1976c).

23. Freeman, "Looking Back," op. cit., p. 20.

24. For descriptions of the project methodology, see "Fieldwork Methodology: Rationale and Assessment," in Freeman, ed., 1976b, pp. 47–59; see also Freeman, "Looking Back," pp. 22–26.

25. Freeman, "Looking Back," op. cit., p. 22.

26. Peter Usher, "Environment, Race, and Nation Reconsidered: Reflections on Aboriginal Land Claims in Canada," *Canadian Geographer* 47(4), 2003, pp. 365–382, p. 375.

27. Brody, *Maps and Dreams*, op. cit., p. 147.

28. The general point was made by Peter Herlihy and Gregory Knapp in their "Maps of, by, and for the Peoples of Latin America," *Human Organization* 62(4), Winter 2003, pp. 303–314, with the key remark in the notes on p. 310,

as later by Mac Chapin, Zachary Lamb, and Bill Threlkeld in their "Mapping Indigenous Lands," *Annual Review of Anthropology 34*, 2005, pp. 618–638, where attention is drawn to the work of Boas and his students on p. 621. Alfred Kroeber, Carl Sauer, Julian Steward, and William Denevan were important conduits.

29. On Steward's relevance for indigenous mapping, see Joe Bryan, "Where Would We Be without Them?: Knowledge, Space and Power in Indigenous Politics," *Futures* 41(1), 2009, pp. 24–32.

30. Joseph Sonnenfeld, *Changes in Subsistence among Barrow Eskimo*, Project No. ONR–140, Arctic Institute of North America, 1956. In their 250-page typescript—"Subsistence Mapping: An Evaluation and Methodological Guidelines," Technical Paper Number 125, Division of Subsistence, Alaska Department of Fish and Game, Juneau, AK, 1985—Linda Ellanna, George Sherrod, and Steven Langdon describe the Alaskan and Canadian contexts out of which "map biographies" emerged and to which they contributed, and noted that Sonnenfeld was less clear about his methods than Freeman was. Sonnenfeld writes:

> Certainly, I didn't consider the maps that a few Inupiat helped me develop during that study to represent examples of "indigenous mapping," which was not a concept of mapping that I entertained at the time. I worked with USAF Pilotage Charts which I found to be available at Barrow's Arctic Research Lab (ARL) at the time, most of which were on a scale of 1:500,000, but with remarkably good detail of lakes and rivers in the areas mapped, as well as bathymetric data for coastal portions of northern Alaska's Arctic Slope. I also had an equally detailed USGS map of the "Geology of the Arctic Slope of Alaska," prepared in cooperation with the Office of the Director, Naval Petroleum and Oil Shale Reserves, and developed by Thomas G. Payne and others in 1951 ("Oil and Gas Investigations Map OM 126" [in three sheets]). Most of the Arctic Slope of Alaska north of the Brooks Range is included in Naval Petroleum Reserve No. 4 (Pet 4), which is at least part of the reason for the excellent maps available for much of this area; in fact many Inupiat worked in a variety of roles with the Navy's petroleum exploration program which was based at Barrow at the time, including assisting as guides in Pet 4 field studies. I'm not sure that many didn't learn their map-reading skills at this time, but my experience with Inupiat who helped me to identify changes in the subsistence economy of the Barrow Eskimo (Inupiat), to include an attempt to identify the distribution of subsistence hunting activities central to Inupiat culture, was similar [to the way Brody described collecting map biographies]. (personal communication, September 14, 2011)

Map biographies were in the air.

31. Key here is the work of MIT planner, Kevin Lynch, on urban imagery. See his germinal *Image of the City* (MIT Press, Cambridge, MA, 1960). The psychologist David Stea brought Lynch's ideas to geographers at Clark University, where one of us (Denis Wood) collected sketch maps in his master's and doctoral work, in the former case collecting maps from nearly 300 residents of San Cristobal de las Casas in Chiapas (*Fleeting Glimpses*, Clark University Cartographic Laboratory, Worcester, MA, 1971) and in the latter, 300 maps from 30-some students visiting Europe for the first time (*I Don't Want to, But I Will*, Clark University Cartographic Laboratory, Worcester, MA, 1973).

Jeremy Anderson, Roger Hart, Tom Saarinen, and many others used sketch maps in related ways.

32. Herlihy and Knapp, op. cit., explicitly cite the work in Chiapas of George Collier (*Fields of the Tzotzil: The Ecological Bases of Tradition in Highland Chiapas*, University of Texas Press, Austin, TX, 1975). We would also point to that of Gary Gossen (*Chamulas in the World of the Sun: Time and Space in a Maya Oral Tradition*, Harvard University Press, Cambridge, MA, 1974), who actually asked Chamulas to draw maps. Also see Evon Vogt's *Aerial Photography in Anthropological Field Research* (Harvard University Press, Cambridge, MA, 1974), the result of a conference (in which I [Denis Wood] happened to be a participant). Conklin's monument is the amazing *Ethnographic Atlas of Ifugao* (Yale University Press, New Haven, CT, 1980), capped by its practically obsessive large-scale maps of pond field parcels. It is further worth noting the importance that the University of California at Berkeley played as a center for training and developing this field. Alfred Kroeber, a student of Boas's, made the approach central to the university's anthropology program, an effort later bolstered by the arrival of Carl Sauer in geography. Kroeber and Sauer influenced the research careers of Julian Steward and Harold Conklin, who in turn influenced a raft of future cultural ecologists who either studied or taught at UC Berkeley up through 2000. Bernard Nietschmann, discussed at length in Chapters 5 and 6, might be considered the last in this long line of cultural ecologists. Alternatively, Nietschmann is also counted among the forerunners of "political ecology," the Marxist-inspired subfield that subsumed much of cultural ecology in the 1980s and 1990s.

33. Published as Peter Usher, *The Bankslanders: Economy and Ecology of a Frontier Trapping Community* (Indian Affairs and Northern Development, Ottawa, ON, Canada, 1971).

34. We've stitched together this description of Patsah's mapping from pp. 2–11 of Brody's very beautiful *Maps and Dreams*, op. cit., in which alternating chapters describe his work among the Beaver and their situation in Canada. It was a task he'd been asked to take on because of his work on the Inuit studies in both the Northwest Territories and Labrador. The second edition of *Maps and Dreams* (Douglas & McIntyre, Vancouver, BC, Canada, 1988) has an impassioned new introduction. "A final genocide," Brody insists, "is not going to be achieved."

35. Brody, *Maps and Dreams*, op. cit., p. 177. A more recent take on this point is that of Charles Hale, in his "Activist Research v. Cultural Critique: Indigenous Land Rights and the Contradictions of Politically Engaged Anthropology," *Cultural Anthropology* 21(1), 2006, pp. 96–120): "By *activist research,* I mean a method through which we affirm a political alignment with an organized group of people in struggle and allow dialogue with them to shape each phase of the process, from conception of the research topic to data collection to verification and dissemination of the results" (p. 97). Hale refers to the mapping he and his colleagues do as "participatory ethnomapping." You may recall Hale from our chapter on Edson in Nicaragua (Chapter 3).

36. Freeman, "Looking Back," op. cit., p. 22; but also see Brody, *Maps and Dreams*, op. cit., pp. 174–176. It helped that "the study virtually included

every male head of household, some of their older sons, and a few widows who have for some time supported their families by hunting and trapping" (Freeman, *Inuit Land Use*, op. cit., Vol. Two, p. 48). There was no sampling. No one has contributed more to making the process "scientific" than Terry Tobias. See his *Chief Kerry's Moose: A Guidebook to Land Use and Occupancy Mapping, Research Design and Data Collection*, Union of British Columbia Indian Chiefs and Ecotrust Canada, Vancouver, BC, Canada, 2000. More recently, Tobias authored what is perhaps *the* compendium on use and occupancy mapping in Canada, the lavishly illustrated *Living Proof: The Essential Data-Collection Guide for Indigenous Use-and-Occupancy Map Surveys* (Ecotrust Canada and the Union of British Columbia Indian Chiefs, Vancouver, BC, Canada, 2009). Tobias acknowledges Peter Usher as a mentor.

37. See, for example, Mac Chapin and Bill Threlkeld's *Indigenous Landscapes: A Study in Ethnocartography*, Center for the Support of Native Lands, Arlington, VA, 2001.

38. Freeman, "Looking Back," op. cit., p. 22.

39. Freeman, *Inuit Land Use*, op. cit., Vol. Three, p. xv.

40. Although this is far from "twenty-six dollars and a bottle of booze," what is reinforced in the process is the authority of the British Crown to have made such determinations. For a detailed account of the negotiations, see Légaré, op. cit.

41. The *Inuit Land Use and Occupancy Project* goals were explicit: "We seek to say in what way a certain piece of land was used by the local people. We do not attempt to determine whether that piece of land yielded a certain number of animals in a certain number of years, nor do we attempt a qualitative assessment of perceived 'usefulness' of that piece of land" (Freeman, *Inuit Land Use*, op. cit., Vol. Two, p. 47).

42. Here reference is made to the *Northern Land Use Information Series* maps published jointly by the federal Departments of Indian Affairs and Northern Development and Environment between 1972 and 1984 (Canada Map Office, Ottawa, ON, Canada).

43. Rick Riewe, ed., *Nunavut Atlas* (Canadian Circumpolar Institute and the Tungavik Federation of Nunavut, Edmonton, AB, Canada,1992). Riewe had been part of the original Inuit Land Use and Occupancy Project team.

44. "Totally ignored by the Québec government when the hydro scheme was planned," is how the Cree put it in the post, "Role of the Canadian Courts in Aboriginal Rights," on the site of the Grand Council of the Crees, at *www.gcc.ca/archive/article.php?id=103* (accessed August 13, 2012).

45. Also see Keith Crowe, *A History of the Original People of Northern Canada* (McGill-Queen's University Press, Montréal, QC, Canada, 1991), pp. 219–220.

46. A few months later, after the formation of the Grand Council of Crees, the Grand Council assumed the funding. For details, see Martin Weinstein, *What the Land Provides: An Examination of the Fort George Subsistence Economy and the Possible Consequences on It of the James Bay Hydroelectric Project*

(Grand Council of the Crees of Québec, Montréal, QC, Canada, 1976), pp. iii–v.
47. Again, see Feit's "Hunting and the Quest for Power," op. cit.; and Richardson's *Strangers Devour the Land*, op. cit. Feit's piece has been excerpted and is available online at *http://arcticcircle.uconn.edu/HistoryCulture/Cree/Feit1* (accessed August 13, 2012).
48. Weinstein, op. cit., throughout, with the interviewing and mapping methods detailed on pp. 87–93. For the 1971–1972 harvests, see Richard Salisbury, *Not by Bread Alone* (Indians of Québec Association, Montréal, QC, Canada, 1972). Nathan Elberg, Jaqueline Hyman, Kenneth Hyman, and Richard F. Salisbury, *Not by Bread Alone: The Use of Subsistence Resources among James Bay Cree* (Program in the Anthropology of Development, Department of Sociology and Anthropology, McGill University, Montréal, QC, Canada, 1975) is a revised edition. The Hymans carried out the study of the size and composition of the 1971–1972 subsistence economy.
49. This is online at *www.autochtones.gouv.qc.ca/grands_dossiers/index_en.htm* (accessed August 14, 2012), which is a portal to a collection of Québec records dealing with aboriginal affairs. For this text, click on "Complementary Agreement No. 13."
50. Microfilm copies of much of the evidence (on 28 reels) is available as *Mackenzie Valley Pipeline Inquiry* (Mackenzie Valley Pipeline Inquiry, Toronto, ON, Canada, 1977). For a collection of some of the evidence presented by the Dene at the community hearings, see Martin O'Malley's extremely valuable *Past and Future Land: An Account of the Berger Inquiry into the Mackenzie Valley Pipeline* (P. Martin Associates, Toronto, ON, Canada, 1976). For Dene testimony and descriptions of Berger traveling from Dene community to Dene community, the inquiry spending the night on the floor of a gym or church, see Patrick Scott, *Stories Told: Stories and Images of the Berger Inquiry* (Edzo Institute, Yellowknife, NT, Canada, 2007). For northern development, see Berger, *Northern Frontier, Northern Homeland*, op. cit., and for the Dene Nation, see Watkins, op. cit. For the claim about *Northern Frontier's* popularity, see "The Online Guide to Thomas R. Berger," at *http://web.uvic.ca/~mharbell/a1/workshop2/mvp_inquiry.html* on the University of Victoria website (accessed August 14, 2012).
51. Berger, *Northern Frontier, Northern Homeland*, op. cit., p. vii.
52. Scott, op. cit., p. 76
53. Ibid.
54. Scott Rushford says: "Developers, such as the pipeline applicants, frequently allege that the Dene no longer rely substantially on country food derived from traditional use of the land," in his "Country Food," in Watkins, op. cit., pp. 32–46. His study of its continuing importance was conducted in 1974–1975 for the National Indian Brotherhood of the Northwest Territories for submission to the Berger Inquiry. We quote from p. 32. Watkins himself pushes this further when he writes, "To cast the Dene simply as wage-earners, as the applicants do, is then not only to cast them in a role they may not want, but to deny them their role as the land-owners who should be entitled to appropriate

the rents from projects which they choose to let proceed on their land," in his "From Underdevelopment to Development," pp. 84–99 in Watkins; the quotation is from p. 90.
55. Berger, *Northern Frontier, Northern Homeland*, op. cit., p. xix.
56. Scott, op. cit., p. 74. For an example of one of the maps, see Phoebe Nahanni, "The Mapping Project," in Watkins, op. cit., pp. 21–27; the map on p. 25.
57. Nahanni, op. cit., p. 23.
58. Our reading here is indebted in part to Joanne Rappaport's interpretation of "translation strategies" used by indigenous Nasa in Colombia, as described in *Intercultural Utopias: Public Intellectuals, Cultural Experimentation, and Ethnic Pluralism in Colombia* (Duke University Press, Durham, NC, 2005), pp. 96–97. A similar analysis is suggested by Matthew Sparke in his article "A Map That Roared and an Original Atlas: Canada, Cartography, and the Narration of Nation," *Annals of the Association of American Geographers* 88(3), 1998, pp. 463–495, as well as in Hugh Brody's *Maps and Dreams*, op. cit.
59. In her *The People of Denendeh: Ethnohistory of the Indians of Canada's Northwest* (University of Iowa Press, Iowa City, IA, 2000), anthropologist June Helm, who had long lived among the Dene in Rae, recalls:

> Within a year [of August 1973], Beryl Gillespie and I were working for the Brotherhood in land claims research. I developed a field project in which young members of the Brotherhood, as they interviewed Dene of the Northwest Territories, plotted on maps the routes and resources (e.g., good furbearer locations) that men and their families used across the land. The project's aim was to establish the facts of native land use and deployment as part of the Brotherhood's effort to maximize Indian control over land use and development through negotiations and/or judicial process. (p. 92)

Nahanni and Helm knew each other, and doubtless there were exchanges, but be that as it may, Nahanni makes no mention of Helm in her 1977 paper. Nor does the Dene Mapping Project, which used Nahanni's research in putting together their *Dogrib and Chipewyan Land Use in the Dene/Inuit Overlap Region* (Dene Mapping Project, Edmonton, AB, Canada, 1985); nor do Michael Asch, Thomas Andrews, and Shirleen Smith in their "The Dene Mapping Project on Land Use and Occupancy: An Introduction" (in Philip Spaulding, ed., *Anthropology in Praxis*, University of Calgary Press, Calgary, AB, Canada, 1986). Asch, Andrews, and Smith *were* the Dene Mapping Project. Nor, finally, does G. G. Tychon in his presentation, "The Dene Mapping Project: Past and Present," at the 7th Annual Symposium on Geographic Information Systems in Forestry, Environment and Natural Resources Management, Vancouver, BC, Canada, 1993, where Tychon's interest was in moving the Asch, Andrews, and Smith mainframe digitization of Nahanni's maps to a microprocessor environment for use in ongoing land claims negotiations.
60. Scott, op. cit., p. 74.
61. Nahanni, op. cit., p. 27.
62. From the "Dene Declaration," in Watkins, op. cit., pp. 3–4.
63. George Manuel, *The Fourth World: An Indian Reality* (Free Press, New York, 1974).

64. Watkins, *Dene Nation: The Colony Within,* op. cit., p. ix.
65. Berger, 1977, op. cit., p. xviii.
66. Ibid., pp. xxii–xxiii. Peter Puxley notes that the problem penetrates to the words *land claims*: "The term suggests that the federal government in fact holds all the land by right and that the Dene wish to claim some of it. Not only does this misrepresent the Dene concept of land, but it totally ignores the elements of political right so vital to the decolonization of the Dene," in his "The Colonial Experience," in Watkins, op. cit., pp. 103–119; the quotation is from p. 116.
67. For example, in 1975 the Department of Indian Affairs and Northern Development agreed to carry out a project for the Labrador Inuit Association based on that of the Inuit Land Use and Occupancy Project, which Carol Brice-Bennett and her colleagues describe in *Our Footprints Are Everywhere: Inuit Land Use and Occupancy in Labrador* (Labrador Inuit Association, Nain, Labrador, NL, Canada, 1977).
68. Matthew Sparke, op. cit.
69. On the controversy surrounding treaty negotiations, see Taiaiake Alfred's *Peace, Power, Righteousness: An Indigenous Manifesto,* 2nd ed. (Oxford University Press, Oxford, UK, 2009).

Chapter 5

1. Bernard Q. Nietschmann, "Defending the Miskito Reefs with Maps and GPS: Mapping with Sail, Scuba and Satellite," *Cultural Survival Quarterly* 18(4), 1995, pp. 34–37; the quotation is from p. 36.
2. Nietschmann, "Defending," op. cit., p. 36.
3. Rudolph C. Rÿser, "Peaceful Warriors Passing Through," *Fourth World Eye,* December 2000, http://cwis.org/publications/FWE/archive-2000-2003/peaceful-warriors-passing-through (accessed August 9, 2013).
4. Bernard Q. Nietschmann, *Between Land and Water: The Subsistence Ecology of the Miskito Indians, Eastern Nicaragua* (Seminar Press, New York, 1973), p. 4.
5. The novel *Waikna: Adventures on the Mosquito Shore* was written by Ephraim George Squier, an American journalist and amateur archeologist, and published under the pseudonym of Samuel Bard. Squier likely did complete much of the journey described in the novel in his role as the American chargé d'affaires to Central America between 1849 and 1853. Toward the end of his appointment, Squier traveled the length of the Mosquito Shore, then a British protectorate, to survey potential routes for a transcontinental railroad. American gold miners heading to California had already established such a route through Nicaragua, traveling up the Río San Juan. Squier's novel enjoyed considerable popularity in the lead-up to U.S. efforts to negotiate a treaty that would divest Britain of its commercial and political influence on the Mosquito

Shore. The novel is still read for its close descriptions of Miskito life in the mid-19th century.

6. Nietschmann writes:

   > I was tired of writing things down in a notebook—things I hadn't done, places I hadn't seen, and facts I really didn't understand. It wasn't as if the data weren't good; they were; that is, they conformed to my preexisting theories. The trouble was that most of the information was secondhand, told to me in the comfort of the village by people who related the content but not the structure or the meaning of the information. My research methodology consisted of being a slipstream scribe; in the academic literature it is usually called "participant observation."

   Bernard Q. Nietschmann, *Caribbean Edge: The Coming of Modern Times to Isolated People and Wildlife* (Bobbs-Merrill, New York, 1979), p. 11.

7. Nietschmann, *Between Land and Water*, op. cit., p. 24.
8. Nietschmann, *Between Land and Water*, op. cit., p. 25.
9. Jenkins Molieri calculates that timber exports from eastern Nicaragua totaled 602.2 million boards between 1945 and 1964; Jorge Jenkins Molieri, *El Desafío Indígena en Nicaragua: El Caso de los Mískitos* (Editorial Vanguardia, Managua, Nicaragua, 1986), p. 21.
10. NIPCO (The Nicaraguan Long Leaf Pine Company) purchased Standard Fruit Company's concession in 1946 with the approval of the Somoza regime.
11. The names given for this region on Nicaraguan maps (the Miskito Coast, the Atlantic Coast, and, more recently, the Caribbean Coast) illustrate persistent nationalist anxieties about the region's location within Nicaraguan territory and separated from it historically, socially, economically, and politically. They also underscore the region's ties to the rest of the Anglophone Caribbean, in particular Jamaica, Grand Cayman, and New Orleans, as well as the "black Atlantic" that extends beyond to include the routes traversed by the slave trade.
12. Mary W. Helms, *Asang: Adaptations to Culture Contact in a Miskito Community* (University of Florida Press, Gainesville, FL, 1971), p. 221.
13. William Dampier, *A New Voyage Round the World* (J. Knapton, London, 1697–1703); Olaudah Equiano, *The Interesting Narrative of the Life of Olaudah Equiano, or Gustavus Vassa, the African* (Penguin Classics, New York, 2003/1814). See also Nietschmann, *Between Land and Water*, op. cit., and Peter Linebaugh and Marcus Rediker, *The Many-Headed Hydra: Sailors, Slaves, Commoners, and the Hidden History of the Revolutionary Atlantic* (Beacon Press, New York, 2001) p. 265.
14. Nietschmann, *Caribbean Edge*, op. cit., p. 188.
15. Nietschmann, *Caribbean Edge*, op. cit. Nietschmann's phrase borrows from an antidote that repeated variously about an exchange between a philosopher (William James in Nietschmann's account, Bertrand Russell in others) and an old woman or Hindu. In the exchange, the latter refutes the former's claim that the world orbits the sun by claiming that the earth is really the back of a turtle that stands on the back of another turtle who stands on another, such that

"it's turtles all the way down." See also, Bernard Q. Nietschmann, "When the Turtle Collapses, the World Ends," *Natural History* 83(6), 1974, pp. 34–43.

16. Bernard Q. Nietschmann, "Turtles and the Revolution," *Natural History* 89(2), 1980, pp. 8, 12.
17. Nietschmann, *Caribbean Edge*, op. cit., p. 213.
18. *The Situation of the Indian in South America: Contributions to the Study of Inter-Ethnic Conflict in the Non-Andean Regions of South America*, Walter Dostal, ed., World Council of Churches, Geneva, Switzerland, 1972.
19. But as Joseph Hill's sonic recharting of the Caribbean in the song "Pirate Days" reminds, "Arawaks was here first." Barbados controversially has a population that claims Arawak descent. On the politics of indigenous "extinction" in the Caribbean, see Maximilian Forte's essay "The Dual Absences of Extinction and Marginality—What Difference Does an Indigenous Presence Make?" that introduces his edited volume, *Indigenous Resurgence in the Contemporary Caribbean*, Peter Lang Publishing, New York, 2006. Those claims aside, it mattered tremendously that the Barbadian state officials saw no relevance in the indigenous question and thus agreed to host the meeting.
20. "The Declaration of Barbados: For the Liberation of the Indians," *Current Anthropology* 14(3), 1973, pp. 267–270. The text of the Declaration is also available at *www.nativeweb.org/papers/statements/state/barbados1.php*, accessed March 27, 2012.
21. Dostal, op. cit. The Spanish language version was simultaneously published, with Georg Grünberg and Pedro Agostinho da Silva as authors, *La Situación del Indígena en América del Sur: Aportes al Estudio de la Fricción Interétnica en los Indios No-andinos: Edición Bi-lingüe Español-portugués*, Biblioteca Ciéntifica, Montevideo, Uruguay, 1972.
22. Stefano Varese, *Witness to Sovereignty: Essays on the Indian Movement in Latin America*, International Work Group for Indigenous Affairs, Copenhagen, Denmark, 2006, pp. 36–37.
23. Alison Brysk, *From Tribal Village to Global Village: Indian Rights and International Relations in Latin America* (Stanford University Press, Stanford, CA, 2000), p. 64.
24. *Irredeemable America: The Indians' Estate and Land Claims*, Imre Sutton, ed., Olympic Marketing, Minnetonka, MN, 1986; and Ward Churchill, "Charades, Anyone? The Indian Claims Commission in Context," *American Indian Culture and Research Journal* 24(1), 2000, pp. 43–68.
25. ALPROMISU is an acronym for "Alliance for Progress of the Miskito and Sumo." As the name suggests, the organization engaged U.S. President John F. Kennedy's call for a new "Alliance for Progress" in the Americas that used economic development to counter poverty and fight Communism. Nicaragua briefly aligned itself with that project in the 1960s under the leadership of Luis Somoza Debayle. Unlike his father, Anastasio Somoza García, and younger brother, Anastasio Somoza Debayle, Luis was not trained as a military man. Whereas his younger brother studied at West Point, Luis enrolled at Louisiana State University. After his father's assassination in 1958, Luis became president

and presided over a brief period of reform that included national presidential elections in 1963. After leaving office, however, Luis and the Somoza family continued to control national politics. When Luis died of a heart attack in 1968, Anastasio, the youngest son, assumed the presidency. Anastasio was head of the U.S.-trained National Guard and used his position as president to accelerate economic development projects that added to his family's wealth while cracking down on internal dissent.

26. Eric R. Meringer, "The Local Politics of Indigenous Self-Representation: Intraethnic Political Division among Nicaragua's Miskito People during the Sandinista Era," *Oral History Review* 37(1), 2010, pp. 1–17.
27. "MI*skito,* SU*mo,* RA*ma, Sandinista Asla TAkanka.*"
28. Theodore Macdonald, "Nicaragua: National Development and Atlantic Coast Indians," *Cultural Survival Quarterly* 5(3), 1981, pp. 9–11; *www.culturalsurvival.org/publications/cultural-survival-quarterly/nicaragua/nicaragua-national-development-and-atlantic-coast* (accessed January 27, 2011); Bernard Q. Nietschmann, *The Unknown War: The Miskito Nation, Nicaragua and the United States* (Freedom House, New York, 1989), p. 30.
29. Fray Gregorio Smutko, *La Presencia Capuchina Entre Los Miskitos 1915–1995* (Universidad de las Regiónes Autónomas de la Costa Caribe Nicaragüense y La Vice Provincia de los Capuchinos de América Central y Panama, San Jose, Costa Rica, 1996); see also Susan Hawley, "Protestantism and Indigenous Mobilisation: The Moravian Church among the Miskitu Indians of Nicaragua," *Journal of Latin American Studies* 29(1), 1997, pp. 111–129.
30. MISURASATA, "Plan of action 1981," in *National Revolution and Indigenous Identity: The Conflict between the Sandinistas and Miskito Indians on Nicaragua's Atlantic Coast,* Klaudine Ohland and Robin Schneider, eds., Document 47, International Work Group for Indigenous Affairs (IWGIA), Copehagen, Denmark, November, 1983.
31. The Carter Administration initiated this arrangement under the name of "Operation Charly," delegating the task of training anti-Communist forces to the military architects of Argentina's notorious "Dirty War." See Martha Honey, *Hostile Acts: U.S. Policy in Costa Rica in the 1980s* (University Press of Florida, Gainesville, FL, 1994); María Seoane, "Los Secretos de la Guerra Sucia Continental de la Dictadura," *Clarín Especiales: TIENE,* March 24, 2006; and Timothy Brown, *The Real Contra War: Highlander Peasant Resistance in Nicaragua,* University of Oklahoma, Norman, OK, 2001, p. 76.
32. The Miskito mobilization bears a strong resemblance to the experience of ethnic groups in the highlands of Vietnam, Laos, and Cambodia who were aligned with the CIA during the Vietnam War. For comparison, see Hickey, op. cit.
33. Nietschmann, "Turtles and the Revolution," op. cit., p. 8.
34. Harold C. Conklin, Pugguwon Lupaih, and Miklos Pinther, *Ethnographic Atlas of Ifugao: A Study of Environment, Culture, and Society in Northern Luzon* (Yale University Press, New Haven, CT, 1980). See Michael R. Dove, "*Ethnographic Atlas of Ifugao*: Implications for Theories of Agricultural

Evolution in Southeast Asia," *Current Anthropology* 24(4), 1983, pp. 516–519, for a concise summary of the *Atlas*'s scholarly contributions. The American Geographical Society is listed as a corporate author for the *Atlas* for its supervising the production of the final maps. However, as a sign of the decline, the AGS disbanded the cartography unit prior to the *Atlas*'s publication while the organization continued its financial slide.

35. Nietschmann, *Caribbean Edge*, op. cit., p. xii.
36. Bernard Q. Nietschmann, "Bruno Gabriel: A Miskito Nationalist and Revolutionary," *Fourth World Journal* 2(3), 1990, p. 164–184.
37. Bernard Q. Nietschmann, "The Unreported War against the Sandinistas," *Policy Review* 29, 1984, pp. 32–39.
38. On Miskito participation in the war, see Gilles Bataillon, "Trabajo del Antropólogo y Trabajo de los Testigos, la Mosquitia 1982–2007," *Estudios Sociológicos* 26(3), 2008, pp. 509–555; Joe Bryan, *Map or Be Mapped: Land, Race, and Property in Eastern Nicaragua* (University of California, Berkeley, CA, 2007); Charles Hale, *Resistance and Contradiction: Miskitu Indians and the Nicaraguan State, 1894–1987* (Stanford University Press, Stanford, CA, 1994); and Reynaldo Reyes, *Rafaga: The Life Story of a Nicaraguan Miskito Comandante* (University of Oklahoma Press, Norman, OK, 1992).
39. Robert Owen to Oliver North, "Costa Rica Trip," May 20, 1985; from the National Security Archives at *http://gateway.proquest.com/openurl?url_ver=Z39.88-2004&res_dat=xri:dnsa&rft_dat=xri:dnsa:article:CIC01153* (accessed August 9, 2013). It is worth noting that the Costa Rican-based Contra forces, including MISURASATA, were far less strident in their anti-Communist rheotoric than the Honduran-based forces, a critical difference that is often obscured by generic references to the "Contras."
40. Comandante Coyote/Santiago Benjamin, "Resena Historica de la Lucha Indígena en Nicaragua: Los Hechos más Relevantes de la Lucha Indígena de 1973 a 1989, Tanto en Nicaragua como en el Exilio (Honduras y Costa Rica)," Compilado por Gilles Bataillon, *TRACE (Traveux et Recherches dans les Ameriques du Centre): Relatos De La Vida 41*, 2002, pp. 50–64.
41. Charles R. Hale, "Contested Notions of Land Rights in Miskitu History," in Jonathan Boyarin, ed., *Remapping Memory: The Politics of Timespace* (University of Minnesota, Minneapolis, MN, 1994); Nietschmann, *The Unknown War*, op. cit.
42. YATAMA, "La Nueva Alternativa—A Proposal to the National Endowment For Democracy," 1988. Availble from the Chief George Manuel Library, Center for World Indigenous Studies, *http://cwis.org/GML/?post=494* (accessed August 9, 2013).
43. Nietschmann, *The Unknown War*, p. 37.
44. See, for example, the report by Robert Matthews, "Sowing Dragon's Teeth," *NACLA Report on the Americas*, 20(4), July/August 1986, pp. 14–40, and Gary Webb, *Dark Alliance: The CIA, the Contras, and the Crack Cocaine Explosion* (Seven Stories Press, New York, 1999).

45. *Human Rights in Nicaragua under the Sandinistas: From Revolution to Repression*, U.S. Department of State, Washington, DC, 1986. See also "Public Diplomacy Action Plan: Support for the White House Educational Campaign," Confidential, Project Proposal, March 12, 1985, *http://gateway.proquest.com/openurl?url_ver=Z39.88-2004&res_dat=xri:dnsa&rft_dat=xri:dnsa:article:CNI02394* (accessed August 9, 2013]).
46. Nietschmann, *The Unknown War*, op. cit., p. 2.
47. Ibid., p. 52.
48. Nietschmann's Curriculum Vitae lists two appearances on *The 700 Club* that aired on February 7, 1984 and July 27, 1984; *http://oldweb.geog.berkeley.edu/PeopleHistory/Nietschmann/NietschmannBio.html* (accessed August 6, 2013). The interviews were also reported by Paul Rauber, "The Nietschmann File," *The Express*, (Berkeley, CA) 6(46), August 31, 1984, pp. 1, 16–23, and later summarized in Sara Diamond, *Spiritual Warfare: The Politics of the Christian Right* (Black Rose Books, Montréal, QC, Canada, 1990), pp. 16–17.
49. Bernard Q. Nietschmann, "The Third World War," *Cultural Survival Quarterly* 11(3), 1987, pp. 1–16; see also Bernard Q. Nietschmann, "The Fourth World: Nations versus States," in George Demko and William B. Wood, eds., *Reordering the World: Geopolitical Perspectives on the Twenty-First Century* (Westview, Boulder, CO, 1994).
50. Nietschmann, "The Third World War," op. cit., p. 15.
51. Ibid., p. 15.
52. Ibid., p. 15.

Chapter 6

1. Indigenous Alliance of the Americas on 500 Years of Resistance, Declaration of Quito, July 1990, *www.nativeweb.org/papers/statements/quincentennial/quito.php* (accessed May 20, 2013).
2. For an overview of mapping projects in the early 1990s, see Peter Poole's 1995 report prepared for the Biodiversity Support Program on "Indigenous Peoples, Mapping and Biodiversity Conservation: An Analysis of Current Activities and Opportunities for Applying Geomatics Technologies," *http://rmportal.net/library/content/frame/indigenous-people.pdf/view* (accessed October 24, 2012). Founded after the 1992 U.N. Earth Summit in Rio de Janeiro, the Biodiversity Support Program was a USAID-funded project implemented by a trio of conservation NGOs: the World Wildlife Fund, The Nature Conservancy, and the World Resources Institute. Several of these projects are described in detail in a special issue of *Cultural Survival Quarterly*, titled *Geomatics: Who Needs It?*, 18(4), 1994, edited by Peter Poole.
3. Kahnawake Mohawk scholar Taiaiake Alfred has described the ongoing Canadian treaty process as "a path to assimiliation" (*Peace, Power, Righteousness: An Indigenous Manifesto*, 2nd ed., Oxford University Press, Oxford, UK, 2009), pp. 144–155. See also Joel Wainwright and Joe Bryan, "Cartography,

Territory, Property: Postcolonial Reflections on Indigenous Counter-Mapping in Nicaragua and Belize," *Cultural Geographies* 16(2), 2009, pp. 153–178.

4. For an early, generative critique of the spread of mapping in the 1990s, see Nancy Peluso's article "Whose Woods Are These?: Counter-Mapping Forest Territories in Kalimantan, Indonesia," *Antipode* 27(4), 1995, pp. 383–406. Our discussion in this chapter of how indigenous mapping transforms the spaces in which people live and struggle over addresses one of her key questions raised in the article, namely "how new notions of territory [produced through mapping] reflect older ones" (p. 393). For a related commentary, see Jefferson Fox, Krisnawati Suryanata, Peter Hershock, and Albertus Hadi Pramano, "Mapping Power: Ironic Effects of Spatial Information Technology," in Fox et al., eds., *Mapping Communities: Ethics, Values, Practice*, (East–West Center, Honolulu, HI, 2005, pp. 1–10); and Dorothy Hodgson and Richard Schroeder, "Dilemmas of Counter-Mapping Community Resources in Tanzania," *Development and Change* 33(1), 2002, pp. 79–100.

5. On the transformation from territory to property, see Charles R. Hale, "Neoliberal Multiculturalism: The Remaking of Cultural Rights and Racial Dominance in Central America," *Political and Legal Anthropology Review* 28(1), 2005, pp. 10–28; Eric Olund, "From Savage Space to Governable Space: The Extension of United States Judicial Sovereignty over Indian Country in the Nineteenth Century," *Cultural Geographies* 9(2), 2002, pp. 129–157; and Wainwright and Bryan, op. cit.

6. U.S. Army Command and General Staff College, *Field Circular 100-20: Low Intensity Conflict*, Fort Leavenworth, KS, July 16, 1986, p. 3-1.

7. USAID, *Country Development Strategy Statement: USAID/Nicaragua 1991–1996*, Washington, DC, June 14, 1991, pp. 62–63, as quoted on p. 75 of William Robinson, *Transnational Conflicts: Central America, Social Change and Globalization*, Verso, London, 2003. The five-year period covered by the report coincided with the presidential term of Violeta Barrios de Chamorro, whose UNO (National Opposition Union) coalition defeated the Sandinistas in the 1990 national elections. Whereas UNO's victory came about in part as a result of serious divisions within Nicaraguan society, the coalition's scope and rapid rise to power would have been unthinkable without massive financial support from the U.S. government, delivered via the National Endowment for Democracy and as part of a US$49.75 million "non-lethal" Contra aid package approved in 1989 by the U.S. Congress.

8. The phrase "slow-motion counterrevolution" comes from Robinson, op. cit., p. 76.

9. Basso's account of that work, *Wisdom Sits in Places: Landscape and Language among the Western Apache* (University of New Mexico Press, Albuquerque, NM, 1996), remains a classic in the field of indigenous mapping. In contrast to the Berkeley school of mapping forged by cultural ecologists, Basso's approach emphasized culture as the medium through which meanings are produced, landscapes organized, and environments percieved. His approach draws from his training at Harvard training in culutral anthropology with Clyde Kluckhohn as well as other faculty involved with the Harvard Chiapas Project.

10. William Davidson and Melanie Counce, "Mapping the Distribution of Indians in Central America," *Cultural Survival Quarterly* 13(3), 1989, pp. 37–40. Our account draws on numerous personal conversations with Chapin carried out by me (Joe Bryan).
11. Bernard Q. Nietschmann, *The Unknown War: The Miskito Nation, Nicaragua and the United States* (Freedom House, New York, 1989), p. 51 (Map 7). Nietschmann used the map to illustrate the geopolitical significance of indigenous nations as a "firebreak" to the spread of Cuban and Soviet influence in Central America.
12. Herlihy's dissertation advisor at Louisiana State University, Bill Davidson, was the lead author of the article that Chapin had reviewed on "Mapping the Distribution of Indians in Central America," cited in Note 10. For his dissertation, Herlihy mapped Emberá and Wounaan lands in Panama. Following the receipt of his PhD in 1986, Herlihy launched a similar project in the Honduran Mosquitia with Tawahka communities living in a remote area near the Nicaraguan border.
13. Bernard Q. Nietschmann, "The Third World War," *Cultural Survival Quarterly* 11(3), 1987, pp. 1–16.
14. Bernard Q. Nietschmann, "The Fourth World: Nations versus States," in George Demko and William B. Wood, eds., *Reordering the World: Geopolitical Perspectives on the Twenty-First Century* (Westview, Boulder, CO, 1994). The quotation is from p. 229.
15. Nietschmann reiterated this point many times, but one of the clearest formulations appears on page 239 in "The Fourth World," ibid. Nietschmann gave another iteration in his essay on "Geographical Security: The Co-Existence of Biological and Cultural Diversity," in the *Briefing Book on International Security: The Environmental Dimension*, published by Tufts University, Boston, 1992, pp. 96–98. This argument is widely made by indigenous environmentalists in the present, and is a common feature of indigenous political mobilization worldwide. See Jerry Mander and Victoria Tauli-Corpuz, eds., *Paradigm Wars: Indigenous Peoples' Resistance to Globalization* (Sierra Club Books, San Francisco, 2006).
16. "The Coexistence of Indigenous Peoples and the Natural Environment in Central America," Special Map Supplement to *Research & Exploration*, Spring, 1992.
17. "The Coexistence of Indigenous Peoples and Natural Environments in Central America," op. cit.
18. On the cultural currency of rainforests, see Susanna B. Hecht and Alexander Cockburn, *The Fate of the Forest: Developers, Destroyers, and Defenders of the Amazon*, Updated Edition (University of Chicago Press, Chicago, 2010; first edition published 1990); and Candace Slater, ed., *In Search of the Rain Forest* (Duke University Press, Durham, NC, 2004).
19. International Labor Organization Convention No. 169 on the Rights of Indigenous and Tribal Peoples, Article 13.2, *www.ilo.org/dyn/normlex/en/f?p=1000:12100:0::NO::P12100_ILO_CODE:C169* (accessed August 9, 2013).

20. The Miskito Coast Protected Area was initially supported by the World Wildlife Foundation and the Caribbean Conservation Corporation (CCC), with funding from USAID. The project fell apart when Nietschmann and the Miskito communities involved in the project accused the CCC of mishandling USAID funds and refusing to support community efforts to develop their own approaches to conservation. Nietschmann accused the CCC of resorting to "colonialist conservation" that limited community participation to following the rules set down by scientists. Nietschmann's argument is laid out in Bernard Q. Nietschmann, "Defending the Miskito Reefs with Maps and GPS: Mapping with Sail, Scuba and Satellite," *Cultural Survival Quarterly* 18(4), 1995, pp. 34–37; and again in Bernard Q. Nietschmann, "Protecting Indigenous Coral Reefs and Sea Territories, Miskito Coast, RAAN, Nicaragua," in Stan Stevens, ed., *Conservation through Cultural Survival*, Island Press, Washington, DC, 1997, pp. 193–224.

21. I (Denis Wood) argue extensively for reading maps as propositions in *Rethinking the Power of Maps* (Guilford Press, New York, 2010).

22. See Nietschmann, "Protecting Indigenous Coral Reefs," op. cit. For an elaboration of this approach, see Nietschmann's initial argument for the Miskito Coast Protected Area and the subsequent debate it sparked in reply to his "Field Notes: Miskito Coast Protected Area," *Research and Exploration* 7(2), 1991, pp. 232–237; Katrina Brandon, "Integrating Conservation and Development," *Research and Exploration* 7(3), 1991, pp. 371–373, followed by Nietschmann's reply, "Conservation by Self-Determination," *Research and Exploration* 7(3), 1991, pp. 372–373.

23. This approach is closely linked to notions of the "ecological native," a concept whose polyvalent meanings are incisively discussed in Astrid Ulloa's *The Ecological Native: Indigenous Peoples' Movements and Eco-Governmentality in Colombia* (Routledge, New York, 2005); and Beth Conklin and Laura Graham's "The Shifting Middle Ground: Amazonian Indians and Eco-politics," *American Anthropologist* 97(4), 1995, pp. 711–737.

24. For critical appraisals of conservationists' take, see Mac Chapin, "A Challenge to Conservationists," *World Watch Magazine* 17(6), 2004, pp. 17–31; and Hecht and Cockburn, *The Fate of the Forest*, op. cit.

25. See the Preface by Anthony George Coates and Archie Carr, III in *Central America: A Natural and Cultural History* (Yale University Press, New Haven, CT, 1997), p. xii. Coates was the deputy director of the Smithsonian Tropical Research Institute in Panama. Carr was the son of famed sea turtle expert, Archie Carr, and the director of the Wildlife Conservation Society's Central America program. Carr initially worked with Nietschmann to secure USAID funds for the Miskito Coast Protected Area, though the relationship fell apart when CCC insisted on a science-based approach over the community-oriented one advocated by Nietschmann; see Nietschmann, "Protecting Indigenous Coral Reefs," op. cit., 1997.

26. Herlihy's recommendations, published as "Central American Indian Peoples and Lands Today," appear in Coates and Carr, op. cit., pp. 215–240. Nietschmann's searing critique of this omission is elaborated in his "Defending

the Miskito Reefs," op. cit., as in his "Protecting Indigenous Coral Reefs," op. cit.

27. Jorge Illueca, "The Paseo Pantera Agenda for Regional Conservation," in Coates and Carr, *Central America*, op. cit., pp. 247–257.

28. Joe Bryan, "Walking the Line: Participatory Mapping, Indigenous Rights, and Neoliberalism," *Geoforum* 42(1), 2011, pp. 40–50; Mac Chapin and Bill Threlkeld, *Indigenous Landscapes: A Study in Ethnocartography* (Center for the Support of Native Lands, Arlington, VA, 2001); Isabel Pérez Chiriboga, *Espíritus De Vida y Muerte; Los Miskitu Hondureños En Época De Guerra* (Editorial Guaymuras, Tegucigalpa, Honduras, 2002); and Peter Herlihy, "Indigenous Peoples and Biosphere Reserve Conservation in the Mosquitia Rainforest Corridor, Honduras," in Stan Steves, ed., *Conservation through Cultural Survival*, Island Press, Washington, DC, 1997, pp. 99–130.

29. Nietschman, "Protecting Indigenous Coral Reefs," op. cit.; see also Bill Weinberg, "The Battle for the Miskito Coast: Piracy and Ecology on Nicaragua's Wild Frontier," *Native Americas: Akwe:kon's Journal of Indigenous Issues*, Fall 1995, pp. 22–33.

30. Tim Johnson, "Hitting Rock Bottom: Addiction Grips Caribbean Coast," *The Miami Herald*, April 2, 1995, A1.

31. For a discussion of USAID's role in postwar conservation in the Petén, see Juanita Sundberg, "NGO Landscapes in the Maya Biosphere Reserve, Guatemala," *Geographical Review* 88(3), 1998, pp. 388–412; and Megan Ybarra, "Taming the Jungle, Saving the Maya Forest: Sedimented Counterinsurgency Practices in Contemporary Guatemalan Conservation," *Journal of Peasant Studies* 39(2), 2012, pp. 479–502.

32. See Anthony Stocks, "Mapping Dreams in Nicaragua's Bosawas Reserve," *Human Organization* 62(4), 2003, pp. 344–356.

33. Ibid., p. 351.

34. On the World Bank, indigenous rights, and Bosawas, see, David Kaimowitz, Angelica Faune, and Rene Mendoza, "Your Biosphere Is My Backyard: The Story of Bosawas in Nicaragua," *CIFOR Working Paper No. 23*, Bogor, Indonesia, April 2003.

35. For an incisive and highly accessible take on this process, see Hecht and Cockburn, *The Fate of the Forest*, op. cit.

36. See, for example, Shelton H. Davis and Alaka Wali, "Indigenous Territories and Tropical Forest Management in Latin America," *Working Paper Series 1100*, Environmental Assessments and Programs, The World Bank, Washington, DC, 1993. Davis, in particular, drew on the ILO's Misión Andina project of the 1950s and 1960s as a historical example of how engaging indigenous communities, rather than forcibly assimilating them, could improve development outcomes. Intriguingly, the ILO's role in the project led the organization to draft its first policy on "indigenous and tribal populations" (Convention 107), which was later reformed as the basis for Convention 169. For an account of this process and its rationale in the ILO, see Luis Rodríguez-Piñero,

Indigenous Peoples, Postcolonialism, and International Law: The ILO Regime, 1919–1989, Oxford University Press, Oxford, UK, 2005.

37. Davis and Wali, op. cit., p. 18, italics added.
38. See Mac Chapin, "Defending Kuna Yala: PEMASKY, the Study Project for the Management of the Wildlands of Kuna Yala, Panama—A Case Study for Shifting the Power: Decentralization and Biodiversity Conservation," Biodiversity Support Program, U.S. Agency for International Development, Washington, DC, 2000. The Kunas's control over land and resources stems in part from their rebellion against Panama in 1925. In a move partially aimed at maintaining regional stability vital to the newly established Panama Canal, the United States encouraged Panamanian officials to negotiate with the Kuna, ultimately resulting in the creation of a semiautonomous department known as the Comarca Kuna Yala. For a detailed account of this history, see James Howe, *A People Who Would Not Kneel: Panama, the United States, and the San Blas Kuna*, Smithsonian Institution Press, Washington, DC, 1998.
39. Peter Poole, *Developing a Partnership of Indigenous Peoples, Conservationists, and Land Use Planners in Latin America*, World Bank Publications, Washington, DC, 1989, p. i.
40. Ibid., p. i.
41. Ibid., p. 2.
42. Chapin and Threlkeld, *Indigenous Landscapes*, op. cit.
43. Alaka Wali and Shelton H. Davis, *Protecting Amerindian Lands: A Review of World Bank Experience with Indigenous Land Regularization Programs in Lowland South America*, Latin America and the Caribbean Regional Office, Technical Department, Regional Studies Program, The World Bank, Washington, DC, 1992.
44. See Kiran Asher, *Black and Green: Afro-Colombians, Development, and Nature in the Pacific Lowlands* (Duke University Press, Durham, NC, 2009); and Bettina Ng'weno, *Turf Wars: Territory and Citizenship in the Contemporary State* (Stanford University Press, Stanford, CA, 2007).
45. The Quintín Lame Armed Movement initially formed for the purpose of defending indigenous communities in the Cauca Valley from attacks by guerilla and paramilitary forces. Led by indigenous Nasa communities, the movement came to include blacks and mestizos as well as members of other indigenous groups in its ranks. Its "intercultural" approach and relative political success provide an interesting point of comparison with the Miskito armed insurgency in Nicaragua. For more on the Quintín Lame movement, see Joanne Rappaport, *Intercultural Utopias: Public Intellectuals, Cultural Experimentation, and Ethnic Pluralism in Colombia* (Duke University Press, Durham, NC, 2005); and David Gow, *Countering Development: Indigenous Modernity and the Moral Imagination* (Duke University Press, Durham, NC, 2008).
46. Joanne Rappaport, *Intercultural Utopias*, op. cit.
47. David Gow and Joanne Rappaport, "The Indigenous Public Voice: The Multiple Idioms of Modernity in Native Cauca," and Jean E. Jackson, "Contested

Discourses of Authority in Colombian National Indigenous Politics: The 1996 Summer Takeovers," in Kay B. Warren and Jean E. Jackson, eds., *Indigenous Movements, Self-Representation, and the State in Latin America* (University of Texas Press, Austin, TX, 2002), pp. 47–80 and pp. 81–122, respectively.

48. Davis and Wali, "Indigenous Territories," op. cit.
49. See Karl Offen, "The Territorial Turn: Making Black Territories in Pacific Colombia," *Journal of Latin American Geography* 2(1), 2003, pp. 43-73; and Kiran Asher, op. cit.
50. On the Bolivian reform, see Penelope Anthias and Sarah A. Radcliffe, "The Ethno-Environmental Fix and Its Limits: Indigenous Land Titling and the Production of Not-Quite-Neoliberal Natures in Bolivia," *Geoforum*, 2013, in press, and Bret Gustafson, "Manipulating Cartographies: Plurinationalism, Autonomy, and Indigenous Resurgence in Bolivia," *Anthropological Quarterly 82*(4), 2009, pp. 985–1016.
51. We elaborate this point in the next chapter. For an overview of the World Bank's approach, see Klaus Deininger, *Land Policies for Growth and Poverty Reduction*, The World Bank and Oxford University Press, Washington, DC, 2003.
52. Peter Poole, "Indigenous Peoples, Mapping, and Biodiversity Conservation," op. cit.
53. Shelton H. Davis and Alaka Wali, "Indigenous Land Tenure and Tropical Forest Management in Latin America," *Ambio*, 1994, pp. 485–490. The quote is on page 490.
54. Nikolas Rose defines "governable space" as a means of modeling, in thought, an understanding of space through the application of governmental logics such as political economy, security, and sovereignty. As "a little machine for producing conviction in others," the map is used to guide material practices of bringing its view of the world into existence in *Powers of Freedom: Reframing Political Thought*, Cambridge University Press, Cambridge, UK, 1999, pp. 31–40. I (Denis Wood) echo this point in my argument that maps bring the world, as we know it, into existence in my *Rethinking the Power of Maps* (Guilford Press, New York, 2010). For specific discussions of how this operates through practices of mapping, see Timothy Mitchell, *Rule of Experts*, University of California Press, Berkeley, CA, 2002, Chapter 3; and Joe Bryan, "Walking the Line," op. cit.
55. Thomas Berger, *Northern Frontier, Northern Homeland: The Report of the Mackenzie Valley Pipeline Inquiry: Vol. One* (Supply and Services Canada, Ottawa, ON, Canada, 1977); and Thomas Berger, *Village Journey: The Report of the Alaska Native Review Commission*, rev. ed., Hill & Wang, New York, 1995).
56. Winona LaDuke with Sean Aaron Cruz, *The Militarization of Indian Country* (Michigan State University Press, East Lansing, MI, 2013).
57. On the World Bank's role in the Meso-American Biological Corridor, see Liza Grandia, "Between Bolivar and Bureaucracy: The Mesoamerican Biological Corridor," *Conservation and Society* 5(4), 2007, pp. 478–503.

58. The team was led by Galio Gurdián, Edmund Gordon, and Charles Hale. Gurdián was the Sandinista official who contracted Philippe Bourgeois and Georg Grünberg to produce the study on indigenous rights we talked about in Chapter 5. After Grünberg and Bourgeois were expelled by the Sandinistas, Gordon and Hale were the two remaining U.S. grad students left working under Gurdián. Gordon and Hale have written ethnographic accounts of their work with Creole and Miskito communities, respectively: Edmund Gordon, *Disparate Diasporas: Identity and Politics in an African Nicaraguan Community*, University of Texas Press, Austin, TX, 1998; and Charles R. Hale, *Resistance and Contradiction: Miskitu Indians and the Nicaraguan State, 1894–1987*, Stanford University Press, Stanford, CA, 1994.

59. Edmund Gordon, Galio Gurdián, and Charles Hale, "Rights, Resources, and the Social Memory of Struggle: Reflections on a Study of Indigenous and Black Community Land Rights on Nicaragua's Atlantic Coast," *Human Organization* 62(4), 2003, pp. 369–381; the quotation is from p. 370.

60. Roque Roldán Ortega, *Legalidad y Derechos Étnicos en la Costa Atlántica de Nicaragua*, Programa de Apoyo Institucional a los Consejos Regionales y las Administraciones Regionales de la Costa Atlántica RAAN-ASDI-RAAS, I/M Editores, Bogotá, Colombia, 2000).

61. The World Wildlife Fund's initial interest in the case was less concerned with indigenous rights than it was with sustainable forestry practices essential to making conservation work at the scale envisioned by the Paseo Pantera project. It was the lawyers who eventually made the case into an indigenous rights one, bringing to life Nietschmann's conservation through cultural survival approach. It was no accident. The lead lawyer on the case, S. James Anaya, knew Nietschmann from their shared support for the Miskito in the 1980s and continued to work together on land claims during the 1990s, as discussed later in this chapter. S. Todd Crider and S. James Anaya, "Indigenous Peoples, the Environment, and Commercial Forestry in Developing Countries: The Case of Awas Tingni, Nicaragua," *Human Rights Quarterly*, 18(2), 1996, pp. 345–367.

62. For an overview of the case and its history, see S. James Anaya and Claudio Grossman, "The Case of Awas Tingni v. Nicaragua: A New Step in the International Law of Indigenous Peoples," *Arizona Journal of International and Comparative Law* 19(1), 2002, pp. 1–15.

63. S. James Anaya and Ted Macdonald, "Demarcating Indigenous Territories in Nicaragua: The Case of Awas Tingni," *Cultural Survival Quarterly* 19(3), 1995, pp. 69–73.

64. For overviews of this process, see Peter Usher, Frank Tough, and Robert Galois, "Reclaiming the Land: Aboriginal Title, Treaty Rights and Land Claims in Canada," *Applied Geography* 12(2), 1992, pp. 109–132; and Matthew Sparke, "A Map That Roared and an Original Atlas: Canada, Cartography, and the Narration of Nation," *Annals of the Association of American Geographers* 88(3), 1998, pp. 463–495. In the United States, lawyers had also experimented with this approach in a much more limited fashion; see Imre Sutton, "Preface to Indian Country: Geography and Law," *American Indian Culture and Research Journal* 15(2), 1991, pp. 3–36.

65. Inter-American Court of Human Rights, "The Case of the Mayangna (Sumo) Awas Tingni Community v. Nicaragua, Judgement of August 31, 2001," *Arizona Journal of International Law* 19(1), 2002, p. 441.

66. The problem of "overlaps" was well-known among communities in Honduras, having first been cartographically identified as such in a mapping project led by the Center for Native Land's Mac Chapin and Peter Herlihy in 1992. In that project, Chapin and Herlihy, along with Miskito political leaders, agreed not to map the overlaps but rather to emphasize Miskito communities' collective claim to the entire Mosquitia region. See Chapin and Threlkeld, *Indigenous Landscapes*, op. cit.; and Herlihy, "Indigenous Peoples and Biosphere Reserve Conservation," op. cit. Community efforts to resolve this problem to the satisfaction of state officials, for the purposes of titling while maintaining social networks through which access is negotiated, is described in Bryan, "Walking the Line," op. cit. Conversely, Sharlene Mollet has chronicled the role that mapping continues to play in conflicts between Miskito and (black) Garifuna communities in her "Racial Narratives: Miskito and Colono Land Struggles in the Honduran Mosquitia," *Cultural Geographies* 18(1), 2011, pp. 43–62.

67. On this legal strategy, see Wainwright and Bryan, "Cartography, Territory, Property," op. cit.

68. Geographer Michael Watts used this phrase to eulogize Nietschmann, his colleague and PhD advisor, at a funeral held at UC Berkeley following Nietschmann's death in 2000.

69. Wainwright and Bryan, "Cartography, Territory, Property," op. cit. See also Joel Wainwright, *Decolonizing Development: Colonial Power and the Maya*, Blackwell, Malden, MA, 2008.

70. Chapin was well aware of these shortcomings. In 2002, he oversaw production of a revised version of his 1992 map. *National Geographic* published the revised map with the new title "Indigenous Peoples and Natural Ecosystems in Central America and Southern Mexico." The text was thoroughly revised for the new edition to place emphasis on campaigns by indigenous peoples and Afro-Caribbean groups to "protect their ancestral homelands, natural resources, and distinctive cultures." The revised map further expanded on the geographic area previously depicted, including Chiapas and the Yucatan Peninsula in Mexico, along with "sea territories" such as the Miskito Cays. That revision was matched by an expanded concept of indigenous peoples that included Afro-Caribbean groups such as the Creole in Nicaragua, reflecting the growing role that those groups now play in land rights struggles in the region. The map also includes indigenous areas in the western part of Central America that had previously been shown as devoid of native peoples. This change further demonstrates how the mainstreaming of indigenous rights has opened the door for claims for groups previously disregarded as insufficiently "indigenous" to warrant recognition. Herlihy did not contribute to the revision.

71. Wainwright, *Decolonizing Development*, op. cit., p. 268.

Chapter 7

1. Geoffrey Demarest, *Property and Peace: Insurgency, Strategy and the Statute of Frauds* (Foreign Military Studies Office, Fort Leavenworth, KS, 2008), p. ii. The manuscript was largely completed in 2007 and has been republished by the Defense Intelligence Agency and others.

2. Demarest summarizes the "political-philosophical drift" of his 2011 tome, *Winning Insurgent War: Back to Basics* (Fort Leavenworth, KS, Foreign Military Studies Office, 2011), as "Libertarian Pragmatism, or perhaps Pragmatic Libertarianism." As he later puts it in a separate essay, "the libertarian part favors (in the context of a thoughtful social contract) weighting toward individuals and voluntary associations against concentrated power. The pragmatic part recognizes moments for abandoning theory altogether and just going right to a punch in the face," in Geoffrey Demarest, "Winning Insurgent War and Pragmatic Libertarianism," *Small Wars Journal* (2011), http://smallwarsjournal.com/jrnl/art/winning-insurgent-war-and-pragmatic-libertarianism (accessed April 9, 2013).

3. Robert Ardrey, *The Territorial Imperative: A Personal Inquiry into the Animal Origins of Property and Nations* (Athenaeum, New York, 1966), p. 3.

4. Demarest, *Winning Insurgent War*, op. cit., p. 294.

5. Geoff Demarest, *Geoproperty: Foreign Affairs, National Security, and Property Rights* (Frank Cass, London, 1998), p. ix.

6. C. Reinhold Noyes, *The Institution of Property: A Study of the Development, Substance and Arrangement of the System of Property in Modern Anglo-American Law* (Longmans, Green, New York, 1936); and John P. Powelson, *The Story of Land: A World History of Land Tenure and Agrarian Reform* (Lincoln Institute of Land Policy, Cambridge, MA, 1988).

7. This version actually comes from Demarest's 2002 proposal, "Feasibility of Creating a Comprehensive Real Property Database for Colombia" (Foreign Military Studies Office, Fort Leavenworth, KS, 2002), p. 12.

8. Powelson, op. cit., p. 308.

9. Demarest, *Geoproperty* throughout, op. cit.; this quotation is from p. x.

10. John P. Powelson and Richard Stock, *The Peasant Betrayed: Agriculture and Land Reform in the Third World*, 2nd ed. (Cato Institute, Washington, DC, 1990). In support of its assertion, the book includes case studies from the Philippines, Mexico, Tanzania, Iran, Egypt, Bolivia, Somalia, Algeria, "India/Kerala," Pakistan, South Korea, Taiwan, Peru, Indonesia, El Salvador, and Nicarauga. Much of the discussion reads like an effort to empirically prove economist Friedrich von Hayek's argument in his 1944 book *The Road to Serfdom*, which is now regarded as a fundamental reference for neoliberal economists.

11. Demarest, "Feasibility," op. cit., p. 8.

12. Geoffrey Demarest, *Mapping Colombia: The Correlation between Land Data and Strategy* (Strategic Studies Institute, Carlisle Barracks, PA, 2003). Part of

the U.S. Army War College, the Strategic Studies Institute is the Army's center for strategic and national security research.

13. Hernando de Soto, *The Other Path: The Invisible Revolution in the Third World* (Harper & Row, New York, 1989); and Hernando de Soto, *The Mystery of Capital: Why Capitalism Triumphs in the West and Fails Everywhere Else* (Basic Books, New York, 2000). The influence of both books can be glimpsed by the promotional blurbs that adorn their covers and fly pages: Richard Nixon, Ronald Reagan, George H. W. Bush, and Bill Clinton for *The Other Path*, and Margaret Thatcher, Milton Friedman, William F. Buckley, and Francis Fukuyama for *The Mystery of Capital*.

14. De Soto, *The Other Path*, op. cit., p. 247.

15. De Soto, *The Mystery of Capital*, op. cit., p. 148.

16. De Soto, *The Other Path*, op. cit., p. 3.

17. In this respect, de Soto repeats arguments made nearly two centuries earlier by Thomas Robert Malthus. *The Other Path* is studded with lines such as this one: "The infiltration of violence and criminality into everyday life has been accompanied by increasing poverty and deprivation" (p. 5). Mike Davis points out some of the parallels between Malthus and de Soto in his book *Planet of Slums* (Verso, London, 2006).

18. For an incisive critique of de Soto's work, see Timothy Mitchell, "The Work of Economics: How a Discipline Makes Its World," *European Journal of Sociology* 46(2), 2006, pp. 297–320.

19. Davis, *Planet of Slums*, op. cit., p. 79.

20. De Soto, *Mystery of Capital*, op. cit., p. 8. In his footnote citing the source, Demarest says, "Mr. De Soto is one of the leading proponents of property formalization as a cornerstone of equitable development," in *Property and Peace*, op. cit., p. 2.

21. As well read as Demarest is, we're inclined to think he's seen this quote from Jean-Jacques Rousseau's *Essay on the Origins of Inequality* (Penguin Books, New York, 1984, p. 42):

> The first man who, having enclosed a piece of ground, bethought himself of saying *This is mine*, and found people simple enough to believe him, was the real founder of civil society. From how many crimes, wars and murders, from how many horrors and misfortunes might not any one have saved mankind, by pulling up the stakes, or filling up the ditch, and crying to his fellows, "Beware of listening to this impostor; you are undone if you once forget that the fruits of the earth belong to us all, and the earth itself to nobody."

22. Demarest, *Geoproperty*, op. cit., p. 243.

23. Gerard Gato, "Insurgencies, Terrorist Groups and Indigenous Movements: An Annotated Bibliography," Foreign Military Studies Office, Fort Leavenworth, KS, July–August 1999, *http://fmso.leavenworth.army.mil/documents/insurgbib.htm* (accessed May 20, 2014).

24. Demarest, *Property and Peace*, op. cit., pp. xiii–xiv.
25. Ibid., p. v.
26. Ibid., p. x.
27. Ibid., p. 487.
28. Ambler H. Moss, Jr., "Preface," in Demarest's *Mapping Colombia*, op. cit., p. vi. Moss served as U.S. Ambassador to Panama under Presidents Jimmy Carter and Ronald Reagan, later directing the North–South Center at the University of Miami from 1984–2004.
29. Demarest, "Feasability," op. cit., p. 33.
30. Nicholas Schlosser opens his 2010 article on the continuing relevance of Edson's manual with this episode. See his "The Marine Corps' *Small Wars Manual*: An Old Solution to a New Challenge," *Fortitudine* 35(1), 2010, pp. 4–9.
31. Ibid., p. 9.
32. *FM 3-24/MCWP 3.33.5 Counterinsurgency* (Department of the Army, U.S. Government Printing Office, Washington, DC, 2006). The volume was later published by the University of Chicago Press as *The U.S. Army/Marine Corps Counterinsurgency Field Manual* (Chicago, 2007), with a new Foreword by John Nagl and an Introduction by Sarah Sewell. The box is on p. 5.25.
33. Jon-Paul N. Maddaloni, "An Analysis of the FARC in Colombia: Breaking the Frame of FM 3-24" (School of Advanced Military Studies, U.S. Army Command and General Staff College, Fort Leavenworth, KS, 2009), p. 46.
34. United States Army, *FM 3-24 MCWP 3-33.5: Insurgencies and Countering Insurgencies*, Department of the Army, Washington, DC, 2014. The new edition was released after this book was written.
35. Seth Robson, "Military to unveil new counterinsurgency field manual," *Stars and Stripes*, January 28, 2013, at *http://www.stripes.com/news/military-to-unveil-new-counterinsurgency-field-manual-1.205579* (accessed February 2, 2013).
36. Marshall Sahlins, "Preface," in Network of Concerned Anthropologists Steering Committee, *"The Counter-Counterinsurgency Manual: or, Notes on Demilitarizing American Society* (Prickly Paradigm Press, Chicago, 2009), pp. ii–iii. Ironically, Prickly Paradigm is distributed by the University of Chicago Press.
37. Robert Scales, "Statement of Major General Robert Scales, USA (ret.) Testifying before the House Armed Services Committee on July 15, 2004," p. 2, *www.au.af.mil/au/awc/awcgate/congress/04-07-15scales.pdf* (accessed August 16, 2013).
38. Montgomery McFate and Andrea Jackson, "An Organizational Solution for DOD's Cultural Knowledge Needs," *Military Review*, July–August, 2005, pp. 18–21.
39. For details, see Jacob Kipp, Lester Grau, Karl Prinslow, and Don Smith's "The Human Terrain System: A CORDS for the 21st Century," *Military Review*,

September–October, 2006, pp. 8–15. Kipp directs the Foreign Military Studies Office. Smith put the HTS together. CORDS was a Vietnam-era predecessor to the HTS that "specifically matched focused intelligence collection with direct action and integrated synchronized activities aimed at winning the 'hearts and minds' of the South Vietnamese. CORDS was premised on a belief that the war would be ultimately won or lost not on the battlefield, but in the struggle for the loyalty of the people," as their paper puts it. In the end, HTS had a hard time getting social scientists to participate, and was unable to develop the reachback cells, expert networks, or a working toolkit. Its attempt to develop a mapping tool, MAP-HT, was generally a failure. For more on the MAP-HT system, see *http://zeroanthropology.net/2010/06/02/hts-map-ht-failure-people-not-being-paid-map-ht-cost-overrruns* (accessed August 16, 2013).

40. Lt. Gen. David H. Petraeus, "Learning Counterinsurgency: Observations from Soldiering in Iraq," *Military Review*, January–February 2006, pp. 45–55; the quotation is from p. 52.

41. For Bhatia, go to *http://en.wikipedia.org/wiki/Michael_Bhatia*; for Loyd, see *http://zeroanthropology.net/2009/05/07/whitewashing-a-us-war-crime-in-afghanistan-the-trial-of-don-ayala-human-terrain-mercenary* (accessed February 8, 2012).

42. Or as it says on Montgomery McFate's website: "Formerly, she was the Senior Social Scientist for the U.S. Army's Human Terrain System, where she helped build the program from a 'good idea' with no money attached to a program with over five hundred employees, 27 teams deployed in Iraq and Afghanistan, and $151 million dollar a year budget"; see *http://montgomerymcfate.com/index.html* (accessed Aug. 16, 2013).

43. For the pay, see "A Gun in One Hand, a Pen in the Other," *Newsweek*, April 12, 2008, at *www.thedailybeast.com/newsweek/2008/04/12/a-gun-in-one-hand-a-pen-in-the-other.html* (accessed February 8, 2012). For the number of anthropologists, see David Price, *Weaponizing Anthropology: Social Science in Service of the Militarized State* (CounterPunch and AK Press, Petrolia, CA, 2011), p. 4.

44. Roberto González, *American Counterinsurgency: Human Science and the Human Terrain* (Prickly Paradigm Press, Chicago, 2009); Marshall Sahlins's "Preface," in Network of Concerned Anthropologists, op. cit.

45. Roberto González, *Militarizing Culture: Essays on the Warfare State* (Left Coast Press, Walnut Creek, CA, 2011); David Price, op. cit.

46. Ben Constable, "All Our Eggs in a Broken Basket: How the Human Terrain System is Undermining Sustainable Military Cultural Competence," *Military Review*, March–April 2009, pp. 57–64.

47. John Stanton, *General David Petraeus' Favorite Mushroom: Inside the U.S. Army's Human Terrain System* (Wiseman Publishing, no place of publication, 2009). Stanton published most of this book in a series of blogs he originally posted at *http://zeroanthropology.net* and *http://www.cryptome.org* (accessed February 8, 2012).

48. This was first published by Cryptome, a blog/archive run/hosted by John Young and Deborah Natsios. It's concerned with freedom of speech, cryptography, spying, and related subjects. Go to *http://cryptome.info/0001/hts-nasty.htm* for the original posting of Stanton's story (accessed February 8, 2012).
49. Jerome Dobson, "Fort Leavenworth Hosts AGS Council," *Ubique* 27(3), December 2006, pp. 1–3. *Ubique* is an AGS newsletter.
50. Prepared for the Foreign Military Studies Office, the PowerPoint presentation was previously available in a "brief" and long format, as well as a version in Spanish from the México Indígena website. The Spanish version does not include the Petraeus slide. All the files, along with the project website (*http://web.ku.edu/~mexind*), have since been removed from the Internet. The authors' copies of the presentations were accessed February 12, 2009.
51. Price, *Weaponizing Anthropology*, op. cit., pp. 39–40. Robin Winks's book is *Cloak and Gown Scholars in the Secret War, 1939–1961*, 2nd ed. (William Morrow, New York, 1996).
52. Geoffrey Demarest, "Urban Land Use by Illegal Armed Groups in Medellin," *Small Wars Journal*, October 17, 2011, pp. 1–10 (*http://smallwarsjournal.com/jrnl/art/urban-land-use-by-illegal-armed-groups-in-medellin* (accessed February 8, 2012). This quotation and the next one are from p. 10.
53. Paul Richter and Greg Miller, "Colombia army chief linked to outlaw militias," *Los Angeles Times*, March 25, 2007, Home Edition, A1.
54. Demarest, "Urban Land Use," op. cit., p. 4.
55. Demarest, *Mapping Colombia*, op. cit., pp. 23–24.

Chapter 8

1. For a complementary account of the México Indígena controversy, see Joel Wainwright, *Geopiracy: Oaxaca, Militant Empiricism, and Geographical Thought* (New York, Palgrave Pivot, 2012). Wainwright's book usefully highlights the relevance of the controversy to geographers, raising pressing theoretical and ethical concerns from the controversy. Much as we concur with him on many of his points, our attention here is more directly concerned with the implications for indigenous peoples of the intertwining of mapping, counterinsurgency, and geography.
2. "Posición de San Miguel Tiltepec a México Indígena," March 17, 2009 (authors' translation): "La comunidad no hizo la solicitud de la investigación, sino que fueron los investigadores quienes convencieron a la comunidad de que se realizara; por lo tanto, la investigación realizada no tuvo como origen una necesidad de la comunidad, sino que fueron los investigadores del proyecto México Indígena los que diseñaron el método de investigación para recabar el tipo de información que realmente les interesaba."
3. Autoridades Municipales y Comisariados de Bienes Comunales de las Comunidades de San Juan Tepanzacoalco, Santa María Zoogochi, Santa Cruz

Yagavila, Santiago Teotlaxco y San Juan Yagila, "Declaración Xidza Sobre Geopirateria," July 24, 2011 (authors' translation): "No estamos de acuerdo con la forma en que se realizaron los estudios geográficos en las comunidades de San Juan Yagila y San Miguel Tiltepec por parte del equipo del Proyecto México Indígena, entre los años de 2006 y 2008, porque no se informó a estas comunidades del origen de los recursos que se utilizaron para la realización de esta investigación, ocultando expresamente la participación del Ejército de los Estados Unidos, violando de esta manera el derecho al consentimiento libre previo e informado que las comunidades indígenas tenemos reconocido en la Declaración de las Naciones Unidas sobre los Derechos de los Pueblos Indígenas; asimismo respaldamos a ambas comunidades en los problemas que puedan tener en lo posterior a razí de las investigaciones realizadas."

4. Peter H. Herlihy, "Self-Appointed Gatekeepers Attack the American Geographical Society's First Bowman Expedition," *Political Geography* 29(8), 2010, p. 418.

5. Melquiades K. Cruz, "A Living Space: The Relationship between Land and Property in the Community," *Political Geography* 29(8), 2010, pp. 420–421. Launched in 2007, the Mérida Initiative is a security cooperation agreement between the United States, Mexico, and the countries of Central America targeting drug trafficking, money laundering, and organized crime. All three are activities that Demarest and others in the military claim proliferate in indigenous areas lacking clear property rights and other state infrastructure. The United States has committed $1.6 billion to support the project between 2007 and 2010.

6. See *www2.ku.edu/~geography/peoplepages/Herlihy_P.shtml* (Accessed June 7, 2013). Considered against the backdrop of all the others doing the same, many of them discussed in previous chapters, this claim is remarkable. Relatedly, Herlihy has stated that much of his method was inspired by reading my (Denis Wood) 1973 doctoral dissertation, especially the discussion of map types, pp. 52–59: *I Don't Want to, But I Will* (Department of Geography, Clark University, Worcester, MA), *www.deniswood.net/lp_idwtbiw.htm* (accessed August 8, 2013).

7. Jerome Dobson, "New Exploration Initiative Launched as AGS Celebrates Its Sesquicentennial," *Ubique*, 20(3), December 2000, *www.amergeog.org/ubique_dec00.htm* (accessed June 10, 2013).

8. Curriculum Vitae for Jerome Dobson, March 10, 2010, *www2.ku.edu/~geography/Vitas/Dobson_J.pdf* (accessed August 12, 2013).

9. Dobson, Curriculum Vitae, op. cit.

10. Geoffrey Demarest, "Feasibility of Creating a Comprehensive Real Property Database for Colombia" (Foreign Military Studies Office, Fort Leavenworth, KS, 2002); see also Geoffrey B. Demarest and Lester W. Grau, "Maginot Line or Fort Apache?: Using Forts to Shape the Counterinsurgency Battlefield," *Military Review*, November–December, 2005, pp. 35–40.

11. The Pat Roberts Intelligence Scholars Program ("PRISP") was a source of heated debate among anthropologists who denounced the program as an

ROTC for the CIA and other intelligence agencies. Moos joined the Anthropology Department in 1960, teaching courses on terrorism and intelligence for 49 years at the University of Kansas. He also worked extensively with the U.S. military and intelligence agencies over the course of his career and is one of the more outspoken advocates of anthropology's relevance for intelligence and military purposes. See Hugh Gusterson, "Spies in Our Midst," *Anthropology News* 46(6), 2005; and Felix Moos, "Some Thoughts on Anthropological Ethics and Today's Conflicts," *Anthropology News* 46(6), 2005. An overview of the debate and its context appears in David Glenn, "Cloak and Classroom," *Chronicle of Higher Education,* March 25, 2005, available online at http:// chronicle.com/article/CloakClassroom/16948 (accessed October 15, 2014).

12. "KU, Fort Leavenworth establish faculty, student exchange program," *The University of Kansas Oread* 29(1), August 20, 2004, www.oread.ku.edu/ Oread04/Aug20/leavenworth.html (accessed June 7, 2013).

13. Jerome Dobson, "AGS Conducts Fieldwork in México," *Ubique* 26(1) 2006, pp. 1–3.

14. For an overview, see Willem Assies, "Land Tenure and Tenure Regimes in Mexico: An Overview." *Journal of Agrarian Change* 8(1), 2007, pp. 33–63.

15. Raymond Craib, *Cartographic México: A History of State Fixations and Fugitive Landscapes,* Duke University Press, Durham, NC, 2004.

16. For an assessment of PROCEDE's intended benefits, see Klaus Deininger and Fabrizio Bresciani, "Mexico's 'Second Agrarian Reform': Implementation and Impact," Report for the World Bank and the Food and Agriculture Organization (FAO), January 1, 2009. For more critical assessments, see Kristen Appendini, "Land Regularization and Conflict Resolution: The Case of Mexico," prepared for the Food and Agriculture Organization, December, 2001; Ana de Ita, "Land Concentration in México after PROCEDE," in Peter Rosset, Raj Patel, and Michael Courville, eds., *Promised Land: Competing Visions of Agrarian Reform,* Food First Books, Oakland, CA, 2006; Eric Perramond, "The Rise, Fall, and Reconfiguration of the Mexican Ejido," *Geographical Review* 98(3), 2008, pp. 356–371; Lynn Stephen, *Zapata Lives!: Histories and Cultural Politics in Southern Mexico,* University of California Press, Berkeley, CA, 2001; and Peter R. Wilshusen, "The Receiving End of Reform: Everyday Responses to Neoliberalisation in Southeastern Mexico," *Antipode* 42(3), 2010, pp. 767–799. Herlihy's team acknowledges the link between PROCEDE and free trade initiatives in their publications and on their website, including their reproduction of the image that appears in Figure 8.1, in Derek Smith, Peter Herlihy, John Kelly, and Aida Ramos Viera, "The Certification and Privatization of Indigenous Lands in México," *Journal of Latin American Geography* 8(2), 2009, p. 195.

17. Ibid., p. 198. In the same article, Smith et al. describe PROCEDE as having "laudable" goals that are even "revolutionary." The México Indígena team further addresses the signifance of PROCEDE in John H. Kelly, Peter H. Herlihy, Derek A. Smith, Aida Ramos Viera, Andrew M. Hilburn, and Gerardo A. Hernández Cendejas, "Indigenous Territoriality at the End of the Social Property Era in Mexico," *Journal of Latin American Geography,* 9(3), 2010,

pp. 161–181. Joel Wainwright (*Geopiracy*, op. cit., p. 53) remarks that the two articles are "schizophrenic" in their take on PROCEDE.

18. These figures come from Herlihy's Curriculum Vitae at *www2.ku.edu/~geography/Vitas/Herlihy_P.pdf* (accessed August 12, 2013). Dobson's Vitae lists a total of $862,792 from the Foreign Military Studies Office for the México Indígena project. That sum includes the $766,528 listed for the "Mexican OpenSource Geographic Information Systems (GIS) Project" between 2005 and 2008, plus $96,264 for a "Geographic Analysis of Land Tenure in Mexico" awarded in 2005.

19. Members of the México Indígena team calculate that 91.2% of all *ejidos* and *comunidades agrarias* participated in PROCEDE, surveying 85.7% of all communal lands in Mexico, ibid., p. 181. For further analysis, see Ana de Ita, "Land Concentration in México after PROCEDE," op. cit.; and Eric Perramond, op. cit.

20. Herlihy and Knapp, "Maps of, by, and for the Peoples of Latin America," *Human Organization* 62(4), Winter 2003, pp. 303–314.

21. Smith et al., op. cit.

22. Dan Jaffee, *Brewing Justice: Fair Trade Coffee, Sustainability, and Survival* (University of California Press, Berkeley, CA, 2007), pp. 62–63. Herlihy and other members of the México Indígena team acknowledge this fact in one of their publications, illustrated with a similar photo of the Ixtlán sign included here (Figure 8.2). See Smith et al., op. cit., p. 194.

23. For historical information on the Sierra and an account of UNOSJO's founding, see Roberto González, *Zapotec Science: Farming and Food in the Northern Sierra of Oaxaca* (University of Texas Press, Austin, TX, 2001), pp. 230–231. On the history of forestry in the Sierra, see David B. Bray, "The Struggle for the Forest: Conservation and Development in the Sierra Juárez," *Grassroots Development* 15(3), 1991, pp. 13–25.

24. On biopiracy, see Aldo Gonzalez, "Territory, Autonomy, and Defending Maize," *Seedling*, January 2005, pp. 14–17; and Kathleen McAfee, "Corn Culture and Dangerous DNA: Real and Imagined Consequences of Maize Transgene Flow in Oaxaca," *Journal of Latin American Geography* 2(1), 2003, pp. 18–42.

25. Jaime Martínez Luna, "The Fourth Principle: Comunalidad," in Lois Meyer and Benjamín Maldonado, eds., *New World of Indigenous Resistance: Noam Chomsky and Voices from North, South, and Central America* (City Lights, San Francisco, 2010), pp. 85–99.

26. México Indígena Project Status Report for January 2007. On UNOSJO's relationship to the EZLN (Zapatista Army for National Liberation), see Rosalva Aída Hernández Castillo, "The Indigenous Movement in Mexico: Between Electoral Politics and Local Resistance," *Latin American Perspectives* 33(2), 2006, pp. 115–131; and Kunle Owolabi, "¿La Legalización de los 'Usos y Costumbres' ha Contribuido a la Permanencia del Gobierno Priísta en Oaxaca?: Análisis de las Elecciones para Diputados y Gobernadores, De 1992 a 2001," *Foro Internacional* 177, 2004, pp. 474–508.

27. Personal communication with Gregorio Urbano, February 8, 2012, Santa Cruz Yagavila. Urbano makes a similar claim in Simón Sedillo's film *The Demarest Factor*.
28. México Indígena Project Status Report for August 2006.
29. It's not clear when Teotlaxco dropped out, or even if it was ever really "in" the project. Herlihy's reports frequently mention community work that turns out to involve little more than compiling data from government sources.
30. México Indígena Project Status Report for July 2007.
31. México Indígena Project Status Report for June–December 2007.
32. The México Indígena team's work in Tiltepec is described in Scott Brady, "Participatory Mapping Empowers Patrimony," *Américas* 61(2), 2009, pp. 38–43. Brady's article begins in much the same place that we do, with a brief synopsis of Tiltepec's historical struggle to defend its land from the Spanish conquistadores. The México Indígena team, with which Brady was involved, did not come in search of gold, Brady writes: Instead, "they sought cooperation." Brady's article makes no mention of the other communities' rejection of the project, nor the Foreign Military Studies Office's role in the Bowman Expeditions.
33. Jerome Dobson, "The First Bowman Conference," *Ubique*, 28(2), September 2008, pp. 1, 3, 6.
34. AFRICOM is reportedly the first army command to use HTS teams outside of a combat zone in "Phase Zero" operations aimed at preventing conflicts. A similar effort was launched in Mexico in 2012. See Ben Iannotta, "U.S. Seeks to Calm Terrain Team Controversy," *Defense News*, May 3, 2012 (www.defensenews.com/article/20120503/C4ISR01/305030005/U-S-Seeks-Calm-Terrain-Team-Controversy; accessed August 8, 2013); and Nathan Hodge, "Help Wanted: 'Human Terrain' Teams for Africa," *Wired Magazine*, Danger Room blog, January 12, 2009, www.wired.com/dangerroom/2009/01/help-wanted-hum (accessed August 8, 2013).
35. Dobson, "The First Bowman Conference," op. cit.
36. Geoff Demarest, *Geoproperty: Foreign Affairs, National Security, and Property Rights* (Frank Cass Publishers, Portland, OR, 1998), p. 243.
37. México Indígena Project Status Report for June–December 2007.
38. Jerome Dobson, "Let the Indigenous People of Oaxaca Speak for Themselves," February 5, 2009, www.amergeog.org/newsrelease/dobson-oaxaca09.pdf (accessed June 6, 2013).
39. Ibid.
40. Ibid., p. 10.
41. American Geographical Society, "The American Geographical Society's Bowman Expeditions seek to improve geographic understanding at home and abroad: Spotlight on México Indígena," 2009, www.amergeog.org/newsrelease/bowmanPR-en.pdf (accessed June 6, 2013).
42. Autoridad Municipal y Comisariado de Bienes Comunes de San Miguel Tiltepec, "Posición de San Miguel Titlepec con Respeto al Proyecto México

Indígena," March 17, 2009, *http://www.elenemigocomun.net/es/2009/03/posicion-san-miguel-tiltepec-mexico-indigena/* (accessed October 15, 2014), authors' translation.

43. Peter Herlihy, "AGS Bowman Expedition México Indígena Ethics," 2009, *http://academic.evergreen.edu/g/grossmaz/Herlihy_MéxicoEthics.pdf* (accessed June 6, 2013). The document was initially posted to the México Indígena project website at *http://web.ku.edu/~mexind*. Sometime in early 2013, the website and all its contents were removed from the University of Kansas's servers, as noted in the headnote to these references.

44. Jacob Kipp, Lester Grau, Karl Prinslow, and Don Smith's "The Human Terrain System: A CORDS for the 21st Century," *Military Review*, September–October, 2006, pp. 8–15. Prinslow is Demarest's "boss" at the Foreign Military Studies Office, and his involvement in the México Indígena project is briefly mentioned in the project status report for September 2006:

> The team also attended the Bowman Antilles Expedition kickoff meeting on September 29, 2006 hosted by Dr. Geoff Demarest and FMSO Director Karl Prinslow at the Foreign Military Studies Office inside Ft. Leavenworth, KS. Guests included Mary Lynne Bird, AGS, Jerry Dobson, AGS & KU, Peter Herlihy, KU, John Harrington, KSU, Shawn Hutchinson, KSU, John Kelly, Joe Scarpaci, VA Tech, Andrew Sluyter, LSU, Mike Smith, Radiance, Aida Ramos Viera, UASLP/KU. A variety of project issues were discussed with representatives from the FMSO, NGA, and Marine Core Intelligence Agency (MCIA) representatives. The *México Indígena* research team has maintained regular contact, including periodic meetings, with representatives of the FMSO, especially with Dr. Demarest's and also with Captain Reanier's Geospatial Information Team; meetings have occurred in our KU Geography offices and at Ft. Leavenworth.

45. Mike Belt, "Troops, profs explore 'cultural agility': Discussion to focus on saving lives through social sciences," *Lawrence Journal–World*, November 9, 2007, online edition (accessed June 7, 2013). For an analysis of the University of Kansas's role in the HTS program, see *http://zeroanthropology.net/2010/05/20/imperial-instruction-the-human-terrain-systems-academic-trainers-part-1* (accessed June 7, 2013).

46. See Jeremy Crampton, Susan Roberts, and Ate Poorthuis, "The New Political Economy of Geographical Intelligence," *Annals of the Association of American Geographers*, 104(1), pp. 196–214.

47. The Bowman Expedition controversy sparked a flurry of letters and commentaries among geographers. Joel Wainwright and I (Joe Bryan) wrote a letter to the AAG, endorsed by more than 50 of their colleagues, calling for an investigation into UNOSJO's allegations of professional misconduct. See Joe Bryan and Joel Wainwright, Letter to the AAG requesting an inquiry into a violation of ethical norms of geography, March 16, 2009, *http://academic.evergreen.edu/g/grossmaz/HerlihyLetterSign.pdf* (accessed August 28, 2013); with a follow-up letter dated April 8, 2009, *http://academic.evergreen.edu/g/grossmaz/Bryan%20%26%20Wainwright%208%20April%202009.pdf* (accessed August 28, 2013). The Indigenous Peoples' Specialty Group of the AAG later sent a letter of support to Tiltepec on April 14, 2009, *http://*

academic.evergreen.edu/g/grossmaz/IPSGLetterSanMiguelTiltepec.pdf (accessed August 28, 2013), and another letter to the AAG recommending changes to the organization's Statement of Professional Ethics, *http://academic.evergreen.edu/g/grossmaz/IPSGletterAAGboard.pdf* (accessed August 28, 2013). Though a representative of the Specialty Group did participate in the revision of the AAG's ethics statement, the new version did not address military funding. This point was raised again in my commentary (Joe Bryan, "Force Multipliers: Geography, Militarism, and the Bowman Expeditions," *Political Geography*, 29[8], 2010, pp. 414–416). Former AAG President John Agnew summarily dismissed the entire controversy, including Tiltepec's allegations, in his reply, "Ethics or Militarism?: The Role of the AAG in What Was Originally a Dispute over Informed Consent," *Political Geography*, 29(8), 2010, pp. 422–423. For a summary of this debate, see Joel Wainwright's book, *Geopiracy*, op. cit.

Coda

1. For more on Pratt's philosophy and experiences, see his posthumously published autobiography, *Battlefield and Classroom: Four Decades with the American Indian, 1867–1904* (University of Oklahoma Press, Norman, OK, 2004/1964). With regard to Pratt's views on culture and race, see Lee D. Baker, *Anthropology and the Racial Politics of Culture* (Duke University Press, Durham, NC, 2010), especially Chapter Two, "Fabricating the Authentic and the Politics of the Real."
2. Pratt, op. cit., p. 306.
3. On Indian boarding schools as battlefields, see David W. Adams, *Education for Extinction: American Indians and the Boarding-School Experience, 1875–1928* (University Press of Kansas, Lawrence, KS, 1995); and Brenda J. Child, *Boarding School Seasons: American Indian Families, 1900–1940* (University of Nebraska Press, Lincoln, NB, 1998).
4. For accounts of life at Haskell, see Haskell graduate and faculty member Theresa Milk's *Haskell Institute: 19th Century Stories of Sacrifice and Survival, with Haskell Cemetery Walking Tour* (Mammoth Publications, Lawrence, KS, 2007); and Child, op. cit., pp. 66–68. Both Milk and Child discuss the cemetery at some length. For a further discussion of the cemetery, see the article written by Yvonne E. Miller and published on May 28, 1973 on the front page of the *Lawrence Journal–World*, as part of the paper's coverage of Memorial Day celebrations around town. Miller's story, "Haskell Cemetery Once Was Site of Impressive Services," focused on school officials' efforts in the 1910s and 1920s. She revised and republished the article seven years later under the title "Tribes That Slumber" in *True West* 27(5), 1980, pp. 38–41. An anonymous friend graciously visited the cemetery in 2013 to cross check our count of the acutal tombstones.
5. Milk, op. cit., records 12 unmarked graves in Haskell Cemetery. The unmarked graves in the wetlands have been widely noted, particularly with respect to the

proposed rerouting of Kansas Highway 10, the "South Lawrence Trafficway," through the wetland. For an overview, see the webpage maintained by the Wetlands Preservation Organization, a Haskell Indian Nations University student group, *www.haskell.edu/student_life/wpo.html* (accessed August 27, 2013).

6. "Mission" is a generic term used to refer to the vastly diverse range of groups indigenous to coastal California. Many of these groups were settled by the Spanish on missions, hence the appellation. After Prieto, only three more people were buried in the cemetery.

7. "The Army's premier extramural basic research agency in the engineering, physical, information, and life sciences," *www.arl.army.mil/www/default.cfm?page=29* (accessed July 18, 2013).

8. Authors' notes, World Human Geography Conference, September 15, 2011. This approach has been advocated by others in the Pentagon and State Department. See Secretary of State Hillary Clinton's "Opening Remarks on the President's FY 2009 War Supplemental Request" testimony before the Senate Appropriations Committee, Washington, DC, April 30, 2009, *http://www.state.gov/secretary/rm/2009a/04/122463.htm* (accessed August 8, 2013).

9. Peter Herlihy, "Self-appointed gatekeepers," op cit., p. 418.

10. On this point, see also my (Joe Bryan) article, "Force Multipliers: Geography, Militarism, and the Bowman Expeditions," *Political Geography,* 29(8), 2010, pp. 414–416; and Joel Wainwright's book, *Geopiracy: Oaxaca, Militant Empiricism, and Geographical Thought*, New York, Palgrave, 2012. Our focus on Herlihy and Dobson in this text is aimed squarely at calling attention to the larger problem of militarism, geography, and indigenous peoples. As Joel Wainwright notes:

    We could heap scorn on them but it would mean nothing. Knowingly or not, they are the intellectual bearers of the interests of a particular social group: enlightened, liberal managers of the contemporary U.S. capitalist empire. Their work may not be intellectually interesting but it advances definite interests. They are (to use Gramsci's term) organic intellectuals of a social group that seeks to militarize geographical thought. This is why objecting to the Oaxaca controversy as the result of practical-ethical errors by individual geographers is necessary but wholly inadequate. They are not so much errors as effects. (*Geopiracy,* op. cit., p. 88)

    We wholeheartedly agree.

11. This argument is essential to understanding how maps *work*. For an expanded discussion, see my (Denis Wood) book *Rethinking the Power of Maps* (Guilford, New York, 2010), which reprises arguments I first published in *The Power of Maps* (Guilford Press, New York, 1992). For similar formulations, see Jeremy Crampton's *Mapping: A Critical Introduction to Cartography and GIS* (Blackwell Publishing, Malden, MA, 2009); the volume edited by Martin Dodge, Rob Kitchin, and Chris Perkins, *Rethinking Maps: New Frontiers in Cartographic Theory* (Routledge, London, 2009); and Chapter 3 ("The Character of Calculability") in Timothy Mitchell's *Rule of Experts* (University of California Press, Berkeley, CA, 2002).

12. Matt Erickson, "KU Geographers Win Defense Grant to Study Central American Communities," *Lawrence Journal–World*, June 19, 2013, *www2.ljworld.com/news/2013/jun/19/ku-geographers-win-18-million-department-defense-g* (accessed July 25, 2013).
13. The myth of "The White Man's Indian Law" is stridently critiqued by legal scholar Robert A. Williams in his book *Linking Arms Together: American Indian Treaty Visions of Law and Peace, 1600–1800* (Oxford University Press, New York, 1997).
14. See geographer Zoltán Grossman's article "The Global War on Tribes," which appeared in the June 2010 issue of *Z Magazine*, at *www.zcommunications.org/the-global-war-on-tribes-by-zoltan-grossman* (accessed July 18, 2013). A shortened version of the article was first published in *CounterPunch* on April 13, 2010 at *www.counterpunch.org/2010/04/13/the-global-war-on-tribes* (accessed July 18, 2013). For a detailed case study of this approach in the Philippine–American War (1899–1904), see Richard Drinnon's book *Facing West: The Metaphysics of Indian-Hating and Empire-Building* (University of Oklahoma Press, Norman, OK, 1997), and Alfred McCoy's book *Policing America's Empire: The United States, the Philippines, and the Rise of the Surveillance State* (University of Wisconsin Press, Madison, WI, 2009).
15. Bernard Q. Nietschmann, "The Third World War," *Cultural Survival Quarterly* 11(3), 1987, pp. 1–16
16. The first claim is repeated on Herlihy's faculty website, *www2.ku.edu/~geography/peoplepages/Herlihy_P.shtml* (accessed July 24, 2013): "I developed the first participatory research mapping (PRM) methodology in Latin America in 1992, since pioneering its research and applied use in geography, other disciplines and development work, particularly conservations work in Central America." The second claim is elaborated in his 2003 article with Gregory Knapp, "Maps of, by, and for the Peoples of Latin America," *Human Organization* 62(4), Winter 2003, pp. 303–314.
17. Dobson, via e-mail, December 30, 2011. Dobson painted a different picture in a series of e-mails circulated among members of the conference steering committee sent September 18–20, 2011. In particular, he took issue with the presentation that one of us—Joe Bryan—gave at the conference that rehashed the critique of the military's role in the México Indígena project and the Bowman Expeditions more generally. Dobson characterized Bryan's presentation as an example of irresponsible criticism. In a follow-up e-mail, Dobson painted Bryan as part of "a new breed of 'scholars' who want to be critics but not peers." Dobson further professed his "responsibility to help them mend their ways."
18. Authors' notes from the 2011 World Human Geography Conference, but see also Peter Herlihy, "Self-Appointed Gatekeepers Attack the American Geographical Society's First Bowman Expedition," *Political Geography* 29(8), 2010, p. 418.
19. As this book was finalized, the public status of the Bowman Expeditions changed. The AGS website now includes a list of past and present Expeditions (*https://www.amergeog.org/research-special-projects/bowman-expeditions*,

accessed October 15, 2014). The U.S. Army or the Department of Defense have funded every one of the Expeditions listed. For specific information on the work in Honduras, see the pages for the Bowman Expedition to the Borderlands (*https://www.amergeog.org/research-special-projects/bowman-expeditions/bowman-expedition-to-the-borderlands*, accessed October 15, 2014) and the newly formed "CA [Central America] Indígena Bowman Expedition to Honduras" (*https://www.amergeog.org/research-special-projects/bowman-expeditions/ca-indigena-bowman-expedition-to-central-america*, accessed October 15, 2014). Other aspects of the Honduras program can be found on the website maintained by a group of Tawahka communities in Honduras. The site includes eight maps produced by the Bowman Expedition to the Borderlands showing "admininstrative divisions," "settlements and roads," climate, ecosystems, slope, topography, land cover and use, and zoning for conservation; see *http://krausirpi.wordpress.com/mapas_plan_manejo* (accessed August 28, 2013).

20. See the findings of a U.S. Senate investigation led by Senator John Kerry, "Drugs, Law Enforcement and Foreign Policy: A Report Prepared by the Subcommittee on Terrorism, Narcotics and International Operations of the Committee on Foreign Relations, United States Senate" (U.S. Government Printing Office, Washington, DC, 1989). The findings are further elaborated in Gary Webb's book *Dark Alliance: The CIA, the Contras, and the Crack Cocaine Explosion* (Seven Stories Press, New York, 1999).

21. U.S. Department of State 2013 International Narcotics Control Strategy Report, *www.state.gov/j/inl/rls/nrcrpt/2013/vol1/204050.htm#Honduras* (accessed July 24, 2013).

22. Thom Shanker, "Lessons of Iraq Help U.S. Fight a Drug War in Honduras," *New York Times*, May 5, 2012, p. 1, *www.nytimes.com/2012/05/06/world/americas/us-turns-its-focus-on-drug-smuggling-in-honduras.html?_r=2&pagewanted=all&* (accessed July 24, 2013).

23. Mattathias Schwartz, "A Mission Gone Wrong: Why Are We Still Fighting the Drug War?," *The New Yorker*, January 6, 2014, pp. 44–55.

24. See Schwartz, op. cit.; and Damien Cave, Charlie Savage, and Thom Shanker, "A New Front Line in the U.S. Drug War," *New York Times*, May 31, 2012, *www.nytimes.com/2012/06/01/world/americas/honduran-drug-raid-deaths-wont-alter-us-policy.html?pagewanted=2&src=recg&pagewanted=all* (accessed July 24, 2013).

25. Alexander Main and Annie Bird. "Still Waiting for Justice: An Assessment of the Honduran Public Ministry's Investigation of the May 11, 2012 Killings in Ahuas, Honduras," Center for Economic and Policy Research and Rights Action, Washington, DC, April 2013; and Kendra McSweeney and Zoe Pearson, "Prying Native People from Native Lands: Narco Business in Honduras," *NACLA Report on the Americas*, 46(4), 2013, pp. 7–12.

26. See Chapin and Threlkeld, *Indigenous Landscapes*, op. cit.; Peter Herlihy, "Indigenous Peoples and Biosphere Reserve Conservation in the Mosquitia Rain Forest Corridor, Honduras," in Stan Steves, ed., *Conservation through*

Cultural Survival, 1997, pp. 99–130; Bryan, "Walking the Line: Participatory Mapping, Indigenous Rights, and Neoliberalism," *Geoforum 42*(1), 2011, pp. 40–50.

27. Kendra McSweeney and Zoe Pearson, op cit., pp. 7–12.
28. Jerry Dobson defines the "Borderlands Region" as "all countries surrounding the Gulf of Mexico and the Caribbean Sea." Jerome E. Dobson, "The Why, What, and Where of Bowman Expeditions," *FOCUS on Geography*, 55(4), 2012, pp. 117–118.
29. See *www.amergeog.org/research-special-projects/bowman-expeditions/bowman-expedition-to-the-borderlands* (accessed May 20, 2014).
30. See *http://minerva.dtic.mil* (accessed July 24, 2013). The Social Science Research Council invited a group of scholars to comment on the program when it was first announced in 2009. Their essays argue the pros and cons of the program, mirroring concerns raised about the Bowman Expeditions and the HTS. The essays are posted at *http://essays.ssrc.org/minerva* (accessed July 24, 2013).
31. Jerome Dobson (PI), "Abstract: The Human Geography of Resilience and Change: Land Rights and Stability in Central American Indigenous Societies." This document is available from the Minerva Research Initiative website, *http://minerva.dtic.mil/doc/Dobson_LandRights_FY13.pdf* (accessed July 25, 2013). For more information on the expedition, see *www.amergeog.org/research-special-projects/bowman-expeditions/ca-indigena-bowman-expedition-to-central-america* (accessed May 20, 2014).
32. The figure of $2.2 millon ($2,219,678) comes from Dobson's Curriculum Vitae, op. cit. The sum includes a 2009 Defense Department grant for $115,411 for "Study of Transportation in Kazakhstan" that Dobson lists elsewhere as a Bowman Expedition (*http://kars.ku.edu/research/bowman-expedition-study-transportation-kazakhstan/* and *http://kbs.ku.edu/research/bowman-expedition-study-transportation-kazakhstan*; both accessed August 12, 2013).
33. The press release includes these lines: "Many tribal communities in the United States lack accurate mapping information pertaining to roads, buildings, and information on services available to tribal members and the general public." It's the same idea Herlihy advanced in the Sierra Juárez. And like Herlihy, Google is enlisting tribes to participate in mapping all this and more through the use of their "Map Maker" tool, producing along the way information—data—that will presumably be housed on Google's own servers. Given Google's relationship to the National Security Administration, one has to wonder where these data will end up and what they will be used for. See *www.ncai.org/news/articles/2013/08/02/ncai-google-partner-for-indigenous-mapping-day-on-august-9th* (accessed August 8, 2013).
34. On the spatial extent and locations of the protests, see Toronto-based journalist Tim Groves's report on Google Maps at *http://goo.gl/maps/3WdK4* (accessed May 20, 2014).

35. *A:shiwi A:wan Ulohnanne: The Zuni World*, Jim Enote and Jennifer McLerran, eds. (A:shiwi A:wan Museum and Heritage Center, Zuni, NM, 2011); Jay Johnson, Renee Pualani Louis, and Albertus Hadi Pramono, "Facing the Future: Encouraging Critical Cartographic Literacies in Indigenous Communities," *ACME: An International E-Journal for Critical Geographies*, 4(1), 2006, pp. 80–98; and David Turnbull, "Maps, Narratives, and Trails: Performativity, Hodology, and Distributed Knowledges in Complex Adaptive Systems—an Approach to Emergent Mapping," *Geographical Research*, 45(2), 2007, pp. 140–149.

36. For an elaboration on this phrase, see Subcomandante Marcos, "7 Piezas sueltas del rompecabezas mundial (El neoliberalismo como rompecabezas: la inútil mundial que fragmenta y destruye naciones)," June 1997, *http://palabra.ezln.org.mx/comunicados/1997/1997_06_b.htm* (accessed July 26, 2013). On the geopolitics of Zapatismo, see Alvaro Reyes and Mara Kaufman, "Sovereignty, Indigeneity, Territory: Zapatista Autonomy and the New Practices of Decolonization," *South Atlantic Quarterly* 110(2), 2011, pp. 505–525. In a similar vein Joel Wainright notes the fundamental question this challenge poses for geographers: "How could we take the opportunity presented by these texts from Oaxaca [the declarations from Rincón communities] to rethink the production of geographical knowledge?" See Joel Wainwright, *Geopiracy*, op cit., p. 86.

Bibliography

Many of our sources on the México Indígena project come from the project's own "ESRI-Award winning" website (*http://web.ku.edu/~mexind*). At the outset of the Expedition, the site was repeatedly invoked as evidence of the project's commitment to "open-source research." Most of the references to Tiltepec were taken down in March and April of 2009 at the community's request. We do not reference any of that information here. The rest of the website was removed entirely from the Internet in early 2013. We archived as much of the site as possible as part of our ongoing research, and have copies of all documents cited. An archived copy of the México Indígena site can also be accessed with the "Wayback Machine" at *http://web.archive.org*. More information on the AGS's Bowman Expeditions can be found at *www.amergeog.org/research-special-projects/bowman-expeditions*.

Zoltán Grossman, Professor of Geography and Native American Studies at Evergreen State College in Olympia, Washington, also maintains an excellent site with most of the documents relevant to the Méxcio Indígena project, including the community declarations (*http://academic.evergreen.edu/g/grossmaz/bowman.html*). The community declarations, along with Simón Sedillo's film *The Demarest Factor,* can be found at *http://elenemigocomun.net*.

As with all Internet sources, the pages referenced are continually evolving. We have archived copies of the pages cited here; different versions may have different content.

Abramson, Howard, *National Geographic: Behind America's Lens on the World*, Crown Publishing, New York, 1987.

Adams, David W., *Education for Extinction: American Indians and the Boarding-School Experience, 1875–1928*, University Press of Kansas, Lawrence, KS, 1995.

Agnew, John, "Ethics or Militarism?: The Role of the AAG in What Was Originally a Dispute over Informed Consent," *Political Geography* 29(8), 2010, pp. 422–423.

Alfred, Taiaiake, *Peace, Power, Righteousness: An Indigenous Manifesto*, 2nd ed., Oxford University Press, Oxford, UK, 2009.

"American Geographical Society's Bowman Expeditions Seek to Improve Geographic Understanding at Home and Abroad: Spotlight on México Indígena," 2009, *www.amergeog.org/newsrelease/bowmanPR-en.pdf* (accessed June 6, 2013).

Anaya, S. James, and Claudio Grossman, "The Case of *Awas Tingni v. Nicaragua*: A New Step in the International Law of Indigenous Peoples," *Arizona Journal of International and Comparative Law* 19(1), 2002, pp. 1–15.

Anaya, S. James, and Ted Macdonald, "Demarcating Indigenous Territories in Nicaragua: The Case of Awas Tingni," *Cultural Survival Quarterly* 19(3), 1995, pp. 69–73.

Anthias, Penelope, and Sarah A. Radcliffe, "The Ethno-Environmental Fix and Its Limits: Indigenous Land Titling and the Production of Not-Quite-Neoliberal Natures in Bolivia," *Geoforum* 43(2), 2013, pp. 240–249.

Appendini, Kristen, "Land Regularization and Conflict Resolution: The Case of Mexico," Land Tenure Service, Rural Development Division, Food and Agriculture Organization, December 2001.

Aquino-Centeno, Salvador, *Contesting Social Memories and Identities in the Zapotec Sierra of Oaxaca, Mexico*, PhD dissertation, University of Arizona, Tucson, 2009.

Ardrey, Robert, *The Territorial Imperative: A Personal Inquiry into the Animal Origins of Property and Nations*, Athenaeum, New York, 1966.

Asch, Michael, Thomas Andrews, and Shirleen Smith, "The Dene Mapping Project on Land Use and Occupancy: An Introduction," in Philip Spaulding, ed., *Anthropology in Praxis*, University of Calgary Press, Calgary, AB, Canada, 1986, pp. 36–43.

Asher, Kiran, *Black and Green: Afro-Colombians, Development, and Nature in the Pacific Lowlands*, Duke University Press, Durham, NC, 2009.

Assies, Willem, "Land Tenure and Tenure Regimes in Mexico: An Overview," *Journal of Agrarian Change* 8(1), 2007, pp. 33–63.

Baker, Lee D., *Anthropology and the Racial Politics of Culture*, Duke University Press, Durham, NC, 2010.

Bard, Samuel A. [pseudonym for Ephraim George Squier], *Waikna: Adventures on the Mosquito Shore*, University of Florida Press, Gaineville, FL, 1965/1855.

Baruth, Christopher, "The American Geographical Society Library: A Treasure Trove for Twenty-First-Century Geographical Scholarship," *Geographical Review* 96(3), 2006, pp. 459–472.

Basso, Keith, *Wisdom Sits in Places: Landscape and Language among the Western Apache*, University of New Mexico Press, Albuquerque, NM, 1996.

Bataillon, Gilles, "Trabajo del Antropólogo y Trabajo de los Testigos, la Mosquitia 1982–2007," *Estudios Sociológicos* 26(3), 2008, pp. 509–555.

Batson, Douglas E., *Registering the Human Terrain: A Valuation of Human Cadastre*, National Defense Intelligence College, Washington, DC, 2008.

Beals, Carlton, "With Sandino in Nicaragua," *The Nation*, February 22, 1928, pp. 204–205; February 29, 1928, pp. 232–233; March 7, 1928, pp. 260–261; March 14, 1928, pp. 288–289; March 21, 1928, pp. 304–315; March 28, 1928, pp. 340–341.

Beals, Carlton, *Banana Gold*, Lippincott, Philadelphia, 1932.

Belt, Mike, "Troops, Profs Explore 'Cultural Agility': Discussion to Focus on Saving Lives through Social Sciences," *Lawrence Journal–World*, November 9, 2007, *www2.ljworld.com/news/2007/nov/09/troops_profs_explore_cultural_agility/?more_like_this* (accessed June 7, 2013).

Berger, Thomas, *Northern Frontier, Northern Homeland: The Report of the Mackenzie Valley Pipeline Inquiry*, Supply and Services Canada, Ottawa, ON, Canada, 1977.

Berger, Thomas, *Village Journey: The Report of the Alaska Native Review Commission*, rev. ed., Hill & Wang, New York, 1995.

Bickel, Keith B., *Mars Learning: The Marine Corps Development of Small Wars Doctrine, 1915–1940*, Westview, Boulder, CO, 2001.

Boot, Max, *The Savage Wars of Peace*, Basic Books, New York, 2002.

Bourgois, P., and Grünberg, J. (1980, May 1). *La mosquitia en la revolución: Informe de una investigación rural en la Costa Atlántica Norte*. Managua, Nicaragua: Instituto Nicaragüense de Reforma Agraria, Departamento de Planificación.

Brady, Scott, "Participatory Mapping Empowers Patrimony," *Américas* 61(2), 2009, pp. 38–43.

Brandon, Katrina, "Integrating Conservation and Development," *Research and Exploration* 7(3), 1991, pp. 371–373.

Brannstrom, Christian, "Almost a Canal: Visions of Interocean Communication across Southern Nicaragua," *Ecumene* 2(1), 1995, pp. 65–87.

Bray, David B., "The Struggle for the Forest: Conservation and Development in the Sierra Juárez," *Grassroots Development* 15(3), 1991, pp. 13–25.

Brice-Bennett, Carol, et al., *Our Footprints Are Everywhere: Inuit Land Use and Occupancy in Labrador*, Labrador Inuit Association, Nain, Labrador, 1977.

Briggs, Thomas L., *Cash on Delivery: CIA Special Operations During the Secret War in Laos*, Rosebank Press, Rockville, MD, 2009.

Brigham, Albert Perry, "The Association of American Geographers, 1903–1923," *Annals of the Association of American Geographers* 14(3), 1924, pp. 109–116.

Brody, Hugh, *Maps and Dreams*, Douglas & McIntyre, Vancouver, BC, Canada, 1981.

Brody, Hugh, *Maps and Dreams*, 2nd ed., Douglas & McIntyre, Vancouver, BC, Canada, 1988.

Brooks, David, "U.S. Marines, Miskitos and the Hunt for Sandino: The Rio Coco Patrol in 1928," *Journal of Latin American Studies* 21(2), 1989, pp. 311–342.

Brooks, David, *Rebellion from Without: Culture and Politics along Nicaragua's Atlantic Coast in the Time of the Sandino Revolt, 1926–1934*, PhD dissertation, University of Connecticut, Storrs, CT, 1998.

Brown, Timothy, *The Real Contra War: Highlander Peasant Resistance in Nicaragua*, University of Oklahoma, Norman, OK, 2001.

Bryan, C. D. B., *National Geographic Society: 100 Years of Adventure and Discovery*, rev. ed., Abrams, New York, 1997.

Bryan, Joe, *Map or Be Mapped: Land, Race, and Property in Eastern Nicaragua*, PhD dissertation, University of California, Berkeley, CA, 2007.

Bryan, Joe, "Where Would We Be without Them?: Knowledge, Space and Power in Indigenous Politics," *Futures* 41(1), 2009, pp. 24–32.

Bryan, Joe, "Force Multipliers: Geography, Militarism, and the Bowman Expeditions," *Political Geography,* 29(8), 2010, pp. 414–416.

Bryan, Joe, "Walking the Line: Participatory Mapping, Indigenous Rights, and Neoliberalism," *Geoforum* 42(1), 2011, pp. 40–50.

Bryan, Joe, and Joel Wainwright, letter to the AAG, March 16, 2009, *http://academic.evergreen.edu/g/grossmaz/HerlihyLetterSign.pdf* (accessed August 28, 2013).

Bryan, Joe, and Joel Wainwright, letter to the AAG, April 8, 2009, *http://academic.evergreen.edu/g/grossmaz/Bryan%20%26%20Wainwright%20 8%20April%202009.pdf* (accessed August 28, 2013).

Brysk, Alison, *From Tribal Village to Global Village: Indian Rights and International Relations in Latin America,* Stanford University Press, Stanford, CA, 2000.

Buisseret, David, "Monarchs, Ministers, and Maps in France before the Accession of Louis XIV," in David Buisseret, ed., *Monarchs, Ministers, and Maps,* University of Chicago Press, Chicago, 1992.

Burnett, Peter, "Environmental Politics and Inuit Self-Determination," in Franklyn Griffiths, ed., *Politics of the Northwest Passage,* McGill-Queen's University Press, Montréal, QC, Canada, 1987.

Calder et al. v. Attorney-General of British Columbia, [1973] S.C.R. 313, *http://scc-csc.lexum.com/scc-csc/scc-csc/en/item/5113/index.do* (accessed June 12, 2012).

Callwell, C. E., *Small Wars: Their Principles and Practice,* 3rd ed., University of Nebraska Press, Lincoln, NE, 1996.

Cave, Damien, Charlie Savage, and Thom Shanker, "A New Front Line in the U.S. Drug War," *The New York Times,* May 31, 2012, *www.nytimes. com/2012/06/01/world/americas/honduran-drug-raid-deaths-wont-alter-us-policy.html?pagewanted=2&src=recg&pagewanted=all* (accessed July 24, 2013).

Chance, John K., *Conquest of the Sierra: Spaniards and Indians in Colonial Oaxaca,* University of Oklahoma Press, Norman, 1989.

Chapin, Mac, "Defending Kuna Yala: PEMASKY, the Study Project for the Management of the Wildlands of Kuna Yala, Panama. A Case Study for Shifting the Power: Decentralization and Biodiversity Conservation," Biodiversity Support Program, U.S. Agency for International Development, Washington, DC, 2000.

Chapin, Mac, "A Challenge to Conservationists," *World Watch Magazine* 17(6), 2004, pp. 17–31.

Chapin, Mac, and Bill Threlkeld, *Indigenous Landscapes: A Study in Ethnocartography,* Center for the Support of Native Lands, Arlington, VA, 2001.

Chapin, Mac, Zachary Lamb, and Bill Threlkeld, "Mapping Indigenous Lands," *Annual Review of Anthropology* 34, 2005, pp. 618–638.

Child, Brenda J., *Boarding School Seasons: American Indian Families, 1900–1940,* University of Nebraska Press, Lincoln, NE, 1998.

Christian Brothers, *The Moravian Missionary Atlas; containing an account of the various countries in which the Missions of the Moravian Church are carried on, and of its missionary operations* (new edition with 18 maps and a mission field and station index), Moravian Church and Mission Agency, London, 1908.

Churchill, Ward, "Charades, Anyone?: The Indian Claims Commission in Context," *American Indian Culture and Research Journal* 24(1), 2000, pp. 43–68.

Clinton, Hillary, "Opening Remarks on the President's FY 2009 War Supplemental Request," testimony before the Senate Appropriations Committee, Washington, DC, April 30, 2009.

Coates, Anthony George, and Archie Carr III, "Preface," in Anthony George Coates, ed., *Central America: A Natural and Cultural History*, pp. xi–xiv, Yale University Press, New Haven, CT, 1997.

Collier, George, *Fields of the Tzotzil: The Ecological Bases of Tradition in Highland Chiapas*, University of Texas Press, Austin, TX, 1975.

Comandante Coyote/Santiago Benjamin, "Resena Historica de la Lucha Indígena en Nicaragua: Los Hechos más Relevantes de la Lucha Indígena de 1973 a 1989, Tanto en Nicaragua como en el Exilio (Honduras y Costa Rica)," compilado por Gilles Bataillon, *TRACE (Traveux Et Recherches Dans Les Ameriques Du Centre): Relatos De La Vida 41*, 2002, pp. 50–64.

Conboy, Kenneth J., and James Morrison. *Shadow War: The CIA's Secret War in Laos*, Paladin Press, Boulder, CO, 1995.

Conklin, Beth, and Laura Graham, "The Shifting Middle Ground: Amazonian Indians and Eco-Politics," *American Anthropologist* 97(4), 1995, pp. 711–737.

Conklin, Harold C., Pugguwon Lupaih, and Miklos Pinther, *Ethnographic Atlas of Ifugao: A Study of Environment, Culture, and Society in Northern Luzon*, Yale University Press, New Haven, CT, 1980.

Constable, Ben, "All Our Eggs in a Broken Basket: How the Human Terrain System Is Undermining Sustainable Military Cultural Competence," *Military Review*, March–April 2009, pp. 57–64.

Craib, Raymond, *Cartographic México: A History of State Fixations and Fugitive Landscapes*, Duke University Press, Durham, NC, 2004.

Crampton, Jeremy, "The Cartographic Calculation of Space: Race Mapping and the Balkans at the Paris Peace Conference of 1919," *Social and Cultural Geography* 7(5), 2006, pp. 731–752.

Crampton, Jeremy, "Maps, Race, and Foucault: Eugenics and Territorialization Following World War I," in Jeremy Crampton and Stuart Elden, eds., *Space, Knowledge, and Power: Foucault and Geography*, Ashgate, Aldershot, UK, 2007, pp. 223–244.

Crampton, Jeremy, *Mapping: A Critical Introduction to Cartography and GIS*, Blackwell Publishing, Malden, MA, 2009.

Crampton, Jeremy, Susan Roberts, and Ate Poorthuis, "The New Political Economy of Geographical Intelligence," *Annals of the Association of American Geographers, 104*(1), 2014, pp. 196–214.

Crider, S. Todd, and S. James Anaya, "Indigenous Peoples, the Environment, and Commercial Forestry in Developing Countries: The Case of Awas Tingni, Nicaragua," *Human Rights Quarterly*, 18(2), 1996, pp. 345–367.

Crowe, Keith, *A History of the Original People of Northern Canada*, McGill-Queen's University Press, Montréal, QC, Canada, 1991.

Cruz, Melquiades K., "A Living Space: The Relationship between Land and Property in the Community," *Political Geography* 29(8), 2010, pp. 420–421.

Cueto, Marcos, ed., *Missionaries of Science: The Rockefeller Foundation and Latin America*, Indiana University Press, Bloomington and Indianapolis, IN, 1994.

Dampier, William, *A New Voyage Round the World*, 1500 Books, New York, 2007/ 1697.

Davidson, William and Melanie Counce, "Mapping the Distribution of Indians in Central America," *Cultural Survival Quarterly* 13(3), 1989, pp. 37–40.

Davis, Mike, *Planet of Slums*, Verso, London, 2006.

Davis, Shelton H., and Alaka Wali, "Indigenous Territories and Tropical Forest Management in Latin America," Working Paper Series 1100, Environmental Assessments and Programs, World Bank, Washington, DC, 1993.

Davis, Shelton H., and Alaka Wali, "Indigenous Land Tenure and Tropical Forest Management in Latin America," *Ambio* 23(8), 1994, pp. 485–490.

Davis, William Morris, "Address," *Proceedings of the American Association for the Advancement of Science* 53, 1904, pp. 1–32.

Davis, William Morris, "The Progress of Geography in the United States," *Annals of the Association of American Geographers* 14(4), 1924, pp. 159–215.

"The Declaration of Barbados: For the Liberation of the Indians," *Current Anthropology* 14(3), 1973, pp. 267–270, www.nativeweb.org/papers/statements/state/barbados1.php (accessed March 27, 2012).

Deininger, Klaus, *Land Policies for Growth and Poverty Reduction*, World Bank and Oxford University Press, Washington, DC, 2003.

Deininger, Klaus, and Fabrizio Bresciani, "Mexico's 'Second Agrarian Reform': Implementation and Impact," Report for the World Bank and the Food and Agriculture Organization (FAO), January 1, 2009.

Demarest, Geoff, *Geoproperty: Foreign Affairs, National Security, and Property Rights*, Frank Cass, London, 1998.

Demarest, Geoffrey, *Feasibility of Creating a Comprehensive Real Property Database for Columbia*, Foreign Military Studies Office, Fort Leavenworth, KS, 2002.

Demarest, Geoffrey, *Mapping Colombia: The Correlation between Land Data and Strategy*, Strategic Studies Institute, U.S. Army War College, Carlisle Barracks PA, 2003.

Demarest, Geoffrey, and Lester W. Grau, "Maginot Line or Fort Apache?: Using Forts to Shape the Counterinsurgency Battlefield," *Military Review*, November–December, 2005, pp. 35–40.

Demarest, Geoffrey, *Property and Peace: Insurgency, Strategy and the Statute of Frauds*, Foreign Military Studies Office, Fort Leavenworth, KS, 2008.

Demarest, Geoffrey, "Urban Land Use by Illegal Armed Groups in Medellin," *Small Wars Journal*, October 2011, pp. 1–10, http://smallwarsjournal.com/jrnl/art/urban-land-use-by-illegal-armed-groups-in-medellin (accessed February 8, 2012).

Demarest, Geoffrey, *Winning Insurgent War: Back to Basics*, Foreign Military Studies Office, Fort Leavenworth, KS, 2011.

Demarest, Geoffrey, "Winning Insurgent War and Pragmatic Libertarianism," *Small Wars Journal*, December 2011, http://smallwarsjournal.com/jrnl/art/

winning-insurgent-war-and-pragmatic-libertarianism (accessed April 9, 2013).

Dene Mapping Project, *Dogrib and Chipewyan Land Use in the Dene/Inuit Overlap Region*, Dene Mapping Project, Edmonton, AB, Canada, 1985.

Diamond, Sara, *Spiritual Warfare: The Politics of the Christian Right*, Black Rose Books, Montréal, QC, Canada, 1990.

Dobson, Jerome, "New Exploration Initiative Launched as AGS Celebrates Its Sesquicentennial," *Ubique*, 20(3), 2000, online at *www.amergeog.org/publications/about-ubique/past-issues/2-uncategorised/67-volume-xx-number-3-december-2000* (accessed June 5, 2013).

Dobson, Jerome, "Foreign Intelligence *Is* Geography," *Ubique* 25(1), 2005, pp. 1–2.

Dobson, Jerome, "AGS Conducts Fieldwork in México," *Ubique* 26(1), 2006, pp. 1–3.

Dobson, Jerome, "Fort Leavenworth Hosts AGS Council," *Ubique* 27(3), 2006, pp. 1–3.

Dobson, Jerome, "Restoring Geography in America," *Ubique* 27(3), 2006, pp. 1–2.

Dobson, Jerome, "The First Bowman Conference," *Ubique*, 28(2), 2008, pp. 1, 3, 6.

Dobson, Jerome, "Let the Indigenous People of Oaxaca Speak for Themselves," *Ubique* 29(1), 2009, pp. 1–2, 4, 7–9, *www.amergeog.org/newsrelease/dobson-oaxaca09.pdf* (accessed June 6, 2013).

Dobson, Jerome, "The Why, What, and Where of Bowman Expeditions," *FOCUS on Geography*, 55(4), 2012, pp. 117–118.

Dobson, Jerome (PI), "Abstract: The Human Geography of Resilience and Change: Land Rights and Stability in Central American Indigenous Societies," 2013, *http://minerva.dtic.mil/doc/Dobson_LandRights_FY13.pdf* (accessed July 25, 2013).

Dodge, Martin, Rob Kitchin, and Chris Perkins, *Rethinking Maps: New Frontiers in Cartographic Theory*, Routledge, London, 2009.

Dostal, Walter, ed., *The Situation of the Indian in South America: Contributions to the Study of Inter-Ethnic Conflict in the Non-Andean Regions of South America*, World Council of Churches, Geneva, Switzerland, 1972.

Dove, Michael R., "*Ethnographic Atlas of Ifugao*: Implications for Theories of Agricultural Evolution in Southeast Asia," *Current Anthropology* 24(4), 1983, pp. 516–519.

Drinnon, Richard, *Facing West: The Metaphysics of Indian-Hating and Empire-Building*, University of Oklahoma Press, Norman, OK, 1997.

Edson, Merritt A., "The Coco Patrol: Operations of a Marine Patrol along the Coco River in Nicaragua, 1928–29, Sections I–III," *Marine Corps Gazette*, August 1936, pp. 18–23, 38–48.

Edson, Merritt A., "The Coco Patrol: Operations of a Marine Patrol along the Coco River in Nicaragua, 1928–29, Section III, continued," *Marine Corps Gazette*, November 1936, pp. 40–41, 60–72.

Edson, Merritt A., "The Coco Patrol: Operations of a Marine Patrol along the Coco River in Nicaragua, 1928–29, Section IV: The Advance to Poteca," *Marine Corps Gazette*, February 1937, pp. 35–43, 57–63.

Elberg, Nathan, Jaqueline Hyman, Kenneth Hyman, and Richard. F. Salisbury, *Not by Bread Alone: The Use of Subsistence Resources among James Bay*

Cree, Program in the Anthropology of Development, Department of Sociology and Anthropology, McGill University, Montréal, QC, Canada, 1975.

Ellanna, Linda, George Sherrod, and Steven Langdon, "Subsistence Mapping: An Evaluation and Methodological Guidelines," Technical Paper Number 125, Division of Subsistence, Alaska Department of Fish and Game, Juneau, AK, 1985.

Ellis, Earl Hancock, "Bush Brigades," *Marine Corps Gazette*, March 1921, pp. 1–15.

Ellsworth, Harry, *One Hundred Eighty Landings of the United States Marines, 1800–1934*, pamphlet, History and Museums Division, U.S. Marine Corps, 1974/ 1934, *www.tecom.usmc.mil/HD/PDF_Files/Pubs/One%20 Hundred%20Eighty%20Landings%20of%20United%20States%20 Marines%201800-1934.pdf* (accessed May 19, 2012).

Enote, Jim, and Jennifer McLerran, eds., *A:shiwi A:wan Ulohnanne: The Zuni World*, A:shiwi A:wan Museum and Heritage Center, Zuni, NM, 2011.

Equiano, Olaudah, *The Interesting Narrative of the Life of Olaudah Equiano, or Gustavus Vassa, the African*, Penguin, New York, 2003/1814.

Erickson, Matt, "KU Geographers Win Defense Grant to Study Central American Communities," *Lawrence Journal–World*, June 19, 2013, *www2.ljworld. com/news/2013/jun/19/ku-geographers-win-18-million-department-defense-g* (accessed July 25, 2013).

Feit, Harvey, "Hunting and the Quest for Power: The James Bay Cree and Whitemen in the 20th Century," in R. Bruce Morrison and C. Roderick Wilson, eds., *Native Peoples: The Canadian Experience*, McClelland & Stewart, Toronto, ON, Canada, 1986, pp. 171–207.

Flad, Harvey K., "Audubon Terrace, the American Geographical Society, and the Sense of Place," *Geographical Review* 95(4), 2004, pp. 519–529.

Ford, Allen, "The *Small War Manual* and Marine Corps Military Operations Other Than War Doctrine," master's thesis, U.S. Army Command and General Staff College, Fort Leavenworth, KS, 2003.

Forte, Maximilian, "The Dual Absences of Extinction and Marginality: What Difference Does an Indigenous Presence Make?", in Maximilian Forte, ed., *Indigenous Resurgence in the Contemporary Caribbean*, Peter Lang Publishing, New York, 2006, pp. 1–17.

Foster, Hamar, "One Hundred Years of Advocating for Justice: Litigating the Calder Case," The Osgoode Society, *www.osgoodesociety.ca/pdfs/abstracts_ hamarfoster.pdf* (accessed July 7, 2012).

Foster, Hamar, Heather Raven, and Jeremy Webber, eds., *Let Right Be Done: Aboriginal Title, the Calder Case, and the Future of Indigenous Rights*, University of British Columbia Press, Vancouver, BC, Canada, 2007.

Fox, Jefferson, Krisnawati Suryanata, Peter Hershock, and Albertus Hadi Pramano, "Mapping Power: Ironic Effects of Spatial Information Technology," in Jefferson Fox et al., eds., *Mapping Communities: Ethics, Values, Practice*, East–West Center, Honolulu, HI, 2005, pp. 1–10.

Fray, Gregorio Smutko, *La Presencia Capuchina Entre Los Miskitos 1915–1995*, Universidad de las Regiónes Autónomas de la Costa Caribe Nicaragüense y La Vice Provincia de los Capuchinos de América Central y Panamá, San José, Costa Rica, 1996.

Freeman, Milton, ed., *Inuit Land Use and Occupancy Project Report, 3 vols.*, Supply and Services Canada, Ottawa, ON, Canada, 1976.

Freeman, Milton, ed., *Inuit Land Use and Occupancy Project Report: Vol. One. Land Use and Occupancy*, Supply and Services Canada, Ottawa, ON, Canada, 1976a.

Freeman, Milton, ed., *Inuit Land Use and Occupancy Project Report: Vol. Two. Supporting Studies*, Supply and Services Canada, Ottawa, ON, Canada, 1976b.

Freeman, Milton, ed., *Inuit Land Use and Occupancy Project Report: Vol. Three. Land Use Atlas*, Supply and Services Canada, Ottawa, ON, Canada, 1976c.

Freeman, Milton, "Looking Back—and Looking Ahead—35 Years After," *Canadian Geographer* 55(1), 2011, pp. 20–31.

Galinat, Walton, "Maize: Gift from America's First Peoples," in Nelson Foster and Linda Cordell, eds., *Chiles to Chocolate: Food the Americas Gave the World*, University of Arizona Press, Tucson, AZ, 1992, pp. 47–60.

Gato, Gerard, "Insurgencies, Terrorist Groups and Indigenous Movements: An Annotated Bibliography," Foreign Military Studies Office, Fort Leavenworth, KS, July–August 1999, *http://fmso.leavenworth.army.mil/documents/insurgbib.htm* (accessed May 20, 2014).

Gibson, James, *Otter Skins, Boston Ships, and China Goods: The Maritime Fur Trade of the Northwest Coast, 1785–1841*, McGill-Queen's University Press, Montréal, QC, Canada, 1992.

Gismondi, Michael, and Jeremy Mouat, "Merchants, Mining and Concessions on Nicaragua's Mosquito Coast: Reassessing the American Presence, 1893–1912," *Journal of Latin American Studies* 34(4), 2002, pp. 845–879.

Gismondi, Michael, and Jeremy Mouat, "'La Enojosa Cuestión de Emery': The Emery Claim in Nicaragua and American Foreign Policy, c. 1880–1910," *The Americas* 65(3), 2009, pp. 375–409.

Glenn, David, "Cloak and Classroom," *Chronicle of Higher Education*, March 25, 2005, A14–A17.

Gonzalez, Aldo, "Territory, Autonomy, and Defending Maize," *Seedling*, January 2005, pp. 14–17.

González, Roberto, *Zapotec Science: Farming and Food in the Northern Sierra of Oaxaca*, University of Texas Press, Austin, TX, 2001.

González, Roberto, *American Counterinsurgency: Human Science and the Human Terrain*, Prickly Paradigm Press, Chicago, 2009.

González, Roberto, *Militarizing Culture: Essays on the Warfare State*, Left Coast Press, Walnut Creek, CA, 2011.

Gordon, Edmund, *Disparate Diasporas: Identity and Politics in an African Nicaraguan Community*, University of Texas Press, Austin, TX, 1998.

Gordon, Edmund, Galio Gurdián, and Charles Hale, "Rights, Resources, and the Social Memory of Struggle: Reflections on a Study of Indigenous and Black Community Land Rights on Nicaragua's Atlantic Coast," *Human Organization* 62(4), 2003, pp. 369–381.

Gossen, Gary, *Chamulas in the World of the Sun: Time and Space in a Maya Oral Tradition*, Harvard University Press, Cambridge, MA, 1974.

Gow, David, *Countering Development: Indigenous Modernity and the Moral Imagination*, Duke University Press, Durham, NC, 2008.

Gow, David, and Joanne Rappaport, "The Indigenous Public Voice: The Multiple Idioms of Modernity in Native Cauca," in Kay B. Warren and Jean E. Jackson, eds., *Indigenous Movements, Self-Representation, and the State in Latin America*, University of Texas Press, Austin, TX, 2002, pp. 47–80.

Grandia, Liza, "Between Bolivar and Bureaucracy: The Mesoamerican Biological Corridor," *Conservation and Society* 5(4), 2007, pp. 478–503.

Grossman, Zoltán, "The Global War on Tribes," *CounterPunch*, April 13, 2010, *www.counterpunch.org/2010/04/13/the-global-war-on-tribes* (accessed July 18, 2013).

Grossman, Zoltán, "The Global War on Tribes," *Z Magazine*, June 2010, *www.zcommunications.org/the-global-war-on-tribes-by-zoltan-grossman* (accessed July 18, 2013).

Grünberg, Georg, and Pedro Agostinho da Silva, *La Situación del Indígena en América del Sur: Aportes al Estudio de la Fricción Inter-Étnica en los Indios No-Andinos: Edición Bi-Lingüe Español-Portugués*, Biblioteca Ciéntifica, Montevideo, Uruguay, 1972.

"A Gun in One Hand, a Pen in the Other," *Newsweek*, April 12, 2008, at *www.thedailybeast.com/newsweek/2008/04/12/a-gun-in-one-hand-a-pen-in-the-other.html* (accessed February 8, 2012)

Gustafson, Bret, "Manipulating Cartographies: Plurinationalism, Autonomy, and Indigenous Resurgence in Bolivia," *Anthropological Quarterly* 82(4), 2009, pp. 985–1016.

Gusterson, Hugh, "Spies in Our Midst," *Anthropology News* 46(6), 2005, pp. 39–40.

Hale, Charles R., "'Wan Tasbaia Dukiara': Contested Notions of Land Rights in Miskitu History," in Jonathan Boyarin, ed., *Remapping Memory: The Politics of Timespace*, University of Minnesota, Minneapolis, 1994, pp. 67–98.

Hale, Charles R., *Resistance and Contradiction: Miskitu Indians and the Nicaraguan State, 1894–1987*, Stanford University Press, Stanford, CA, 1994.

Hale, Charles R., "Neoliberal Multiculturalism: The Remaking of Cultural Rights and Racial Dominance in Central America," *Political and Legal Anthropology Review* 28(1), 2005, pp. 10–28.

Hale, Charles, R., "Activist Research v. Cultural Critique: Indigenous Land Rights and the Contradictions of Politically Engaged Anthropology," *Cultural Anthropology* 21(1), 2006, pp. 96–120.

Hamilton-Merritt, Jane. *Tragic Mountains: The Hmong, the Americans, and the Secret Wars for Laos, 1942–1992*, Indiana University Press, Bloomington, IN, 1993.

Hanna, Stephanie, and Hamar Foster, "A Select Chronology," in Hamar Foster, Heather Raven, and Jeremy Webber, eds., *Let Right Be Done: Aboriginal Title, the Calder Case, and the Future of Indigenous Rights*, University of British Columbia Press, Vancouver, BC, Canada, 2007, pp. 231–240.

Harley, J. B., "Maps, Knowledge, and Power," in Paul Laxton, ed., *The New Nature of Maps: Essays in the History of Cartography*, Johns Hopkins University Press, Baltimore, MD, 2001, pp. 51–81.

Harrington, Samuel, "The Strategy and Tactics of Small Wars, Part 1," *Marine Corps Gazette* 6(12), December 1921, pp. 474–491.

Harrington, Samuel, "The Strategy and Tactics of Small Wars (continued)," *Marine Corps Gazette* 7(3), March 1922, pp. 84–92.

Hawley, Susan, "Protestantism and Indigenous Mobilisation: The Moravian Church among the Miskitu Indians of Nicaragua," *Journal of Latin American Studies* 29(1), 1997, pp. 111–129.

Hearn, Chester, *Tracks in the Sea: Matthew Fontaine Maury and the Mapping of the Oceans*, McGraw-Hill, New York, 2002.

Hecht, Susanna B., and Alexander Cockburn, *The Fate of the Forest: Developers, Destroyers, and Defenders of the Amazon*, updated edition, University of Chicago, Chicago, 2010/1990.

Helm, June, *The People of Denendeh: Ethnohistory of the Indians of Canada's Northwest*, University of Iowa Press, Iowa City, IA, 2000.

Helms, Mary W., *Asang: Adaptations to Culture Contact in a Miskito Community*, University of Florida Press, Gainesville, FL, 1971.

Heneker, W. C. G., *Bush Warfare*, Directorate of Land Concepts and Designs, Canadian Department of National Defense, Kingston, ON, Canada, 2009/1907, *www.army.forces.gc.ca/DLCD-DCSFT/pubs/bushwarfare/BushWarFare.pdf* (accessed June 1, 2012).

Herlihy, Peter, "Indigenous Peoples and Biosphere Reserve Conservation in the Mosquitia Rainforest Corridor, Honduras," in Stan Steves, ed., *Conservation through Cultural Survival*, Island Press, Washington, DC, 1997, pp. 99–130.

Herlihy, Peter, "Central American Indian Peoples and Lands Today," in Anthony Coates, ed., *Central America: A Natural and Cultural History*, Yale University Press, New Haven, CT, 1999, pp. 215–240.

Herlihy, Peter, "AGS Bowman Expedition México Indígena Ethics," 2009, *http://academic.evergreen.edu/g/grossmaz/Herlihy_MéxicoEthics.pdf* (accessed June 6, 2013).

Herlihy, Peter, "Self-appointed Gatekeepers Attack the American Geographical Society's First Bowman Expedition," *Political Geography* 29(8), 2010, p. 418.

Herlihy, Peter, and Gregory Knapp, "Maps of, by, and for the Peoples of Latin America," *Human Organization* 62(4), 2003, pp. 303–314.

Hernández Castillo, Rosalva Aída, "The Indigenous Movement in Mexico: Between Electoral Politics and Local Resistance," *Latin American Perspectives* 33(2), 2006, pp. 115–131.

Hickey, Gerald C., *Window on a War: An Anthropologist in the Vietnam Conflict*, Texas Tech University Press, Lubbock, TX, 2002.

Hodge, Nathan, "Help Wanted: 'Human Terrain' Teams for Africa," *Wired Magazine*, Danger Room blog, January 12, 2009, *www.wired.com/dangerroom/2009/01/help-wanted-hum* (accessed August 8, 2013).

Hodgson, Dorothy, and Richard Schroeder, "Dilemmas of Counter-Mapping Community Resources in Tanzania," *Development and Change* 33(1), 2002, pp. 79–100.

Hoffman, Jon T., *Once a Legend: "Red Mike" Edson of the Marine Raiders*, Presidio Press, Novato, CA, 1994.

Hoffman, Jon T., "'Small Wars' and 'Small Wars Manual,'" in Benjamin R. Beede,

ed., *The War of 1898 and U.S. Interventions, 1898–1934: An Encyclopedia*, Garland Publishing, New York, 1994, pp. 511–516.
Honey, Martha, *Hostile Acts: U.S. Policy in Costa Rica in the 1980s*, University Press of Florida, Gainesville, FL, 1994.
Howe, James, *A People Who Would Not Kneel: Panama, the United States, and the San Blas Kuna*, Smithsonian Institution Press, Washington, DC, 1998.
Hurley, Mary, "The Nisga'a Final Agreement, Revised," Law and Government Division, Parliament of Canada, 2001, www.parl.gc.ca/Content/LOP/ResearchPublications/prb992-e.htm#(9)txt (accessed June 26, 2012).
Iannotta, Ben, "U.S. Seeks to Calm Terrain Team Controversy," *Defense News*, May 3, 2012, www.defensenews.com/article/20120503/C4ISR01/305030005/U-S-Seeks-Calm-Terrain-Team-Controversy (accessed August 8, 2013).
Illueca, Jorge, "The Paseo Pantera Agenda for Regional Conservation," in Anthony Coates, ed., *Central America: A Natural and Cultural History*, Yale University Press, New Haven, CT, 1999, pp. 247–257.
Inter-American Court of Human Rights, "Judgement of August 31, 2001 in the Case of the Mayangna (Sumo) Awas Tingni Community v. Nicaragua," reprinted in *Arizona Journal of International Law* 19(1), 2002, p. 395–442.
de Ita, Ana, "Land Concentration in México After PROCEDE," in Peter Rosset, Raj Patel, and Michael Courville, eds., *Promised Land: Competing Visions of Agrarian Reform*, Food First Books, Oakland (CA), 2006, pp. 148–164.
Jackson, Jean E., "Contested Discourses of Authority in Colombian National Indigenous Politics: The 1996 Summer Takeovers," in Kay B. Warren and Jean E. Jackson, eds., *Indigenous Movements, Self-Representation, and the State in Latin America*, University of Texas Press, Austin, TX, 2002, pp. 81–122.
Jaffee, Dan, *Brewing Justice: Fair Trade Coffee, Sustainability, and Survival*, University of California Press, Berkeley, CA, 2007.
Janelle, Donald, "In Memoriam: William Warntz, 1922–1988," *Annals of the Association of American Geographers* 87(4), 1997, pp. 723–731.
Johnson, Jay, Renee Pualani Louis, and Albertus Hadi Pramono, "Facing the Future: Encouraging Critical Cartographic Literacies in Indigenous Communities," *ACME: An International E-Journal for Critical Geographies*, 4(1), 2006, pp. 80–98.
Johnson, Tim, "Hitting Rock Bottom: Addiction Grips Caribbean Coast," *Miami Herald*, April 2, 1995, A1.
Kaimowitz, David, Angelica Faune, and Rene Mendoza, "Your Biosphere Is My Backyard: The Story of Bosawas in Nicaragua," CIFOR Working Paper No. 23, Bogor, Indonesia, April 2003.
Kelly, John H., Peter H. Herlihy, Derek A. Smith, Aida Ramos Viera, Andrew M. Hilburn, and Gerardo A. Hernández Cendejas, "Indigenous Territoriality at the End of the Social Property Era in Mexico," *Journal of Latin American Geography*, 9(3), 2010, pp. 161–181.
Kerry, John, "Drugs, Law Enforcement and Foreign Policy: A Report Prepared by the Subcommittee on Terrorism, Narcotics and International Operations of the Committee on Foreign Relations, United States Senate," U.S. Government Printing Office, Washington, DC, 1989.
Kipp, Jacob, Lester Grau, Karl Prinslow, and Don Smith, "The Human Terrain

System: A CORDS for the 21st Century," *Military Review*, September–October, 2006, pp. 8–15.

Kopets, Keith, "Why Small Wars Theory Still Matters: The Extension of the Principles on Irregular Warfare and Non-Traditional Missions of the *Small Wars Manual* to the Contemporary Battlespace," *Small Wars Journal* 2(3), 2006, pp. 10–11, *http://smallwarsjournal.com/content/journal-way-back-issues#v6* (accessed June 2, 2012).

"KU, Fort Leavenworth Establish Faculty, Student Exchange Program," *The University of Kansas Oread* 29(1), August 20, 2004, *www.oread.ku.edu/Oread04/Aug20/leavenworth.html* (accessed June 7, 2013).

LaDuke, Winona, *All Our Relations: Native Struggles for Land and Life*, South End Press, Cambridge, MA, 1999.

LaDuke, Winona (with Sean Aaron Cruz), *The Militarization of Indian Country*, Michigan State University Press, East Lansing, MI, 2013.

Langley, Lester, *The Banana Wars: United States Intervention in the Caribbean, 1898–1934*, rev. ed., Scholarly Resources, Wilmington, DE, 2002.

Laqueur, Walter, "The Origins of Guerrilla Doctrine," *Journal of Contemporary History* 10(3), 1975, pp. 341–382.

Latour, Bruno, *Pandora's Hope: Essays on the Reality of Science*, Harvard University Press, Cambridge, MA, 1999.

Lazarus, Arthur Jr., and W. Richard West, Jr., "The Alaska Native Claims Settlement Act: A Flawed Victory," *Law and Contemporary Problems* 40(1), 1976, pp. 132–165.

Légaré, André, "The Process Leading to a Land Claims Agreement and Its Implementation: The Case of the Nunavut Land Claims Settlement," *Canadian Journal of Native Studies* 16(1), 1996, pp. 139–163.

Linebaugh, Peter, and Marcus Rediker, *The Many-Headed Hydra: Sailors, Slaves, Commoners, and the Hidden History of the Revolutionary Atlantic*, Beacon Press, New York, 2001.

Livingstone, David, *The Geographical Tradition: Episodes in the History of a Contested Enterprise*, Blackwell, UK, Oxford, 1992.

Luna, Jaime Martínez, "The Fourth Principle: *Comunalidad*," in Lois Meyer and Benjamín Maldonado, eds., *New World of Indigenous Resistance: Noam Chomsky and Voices from North, South, and Central America*, City Lights, San Francisco, 2010, pp. 85–100.

Lutz, Catherine, and Jane Collins, *Reading National Geographic*, University of Chicago Press, Chicago, 1993.

Lynch, Kevin, *Image of the City*, MIT Press, Cambridge, MA, 1960.

Macdonald, Theodore, "Nicaragua: National Development and Atlantic Coast Indians," *Cultural Survival Quarterly* 5(3), 1981, pp. 9–11, *www.culturalsurvival.org/publications/cultural-survival-quarterly/nicaragua/nicaragua-national-development-and-atlantic-coast* (accessed January 27, 2011).

MacPherson, Alan, "Anti-Americanism in Latin America and the Caribbean: 'False Promise' or Coming Full Circle," in Alan MacPherson and Ivan Krastev, eds., *The Anti-American Century*, Central European University Press, New York, 2007, pp. 49–76.

Maddaloni, Jon-Paul N., *An Analysis of the FARC in Colombia: Breaking the*

Frame of FM 3-24, School of Advanced Military Studies, U.S. Army Command and General Staff College, Fort Leavenworth, KS, 2009.

Main, Alexander, and Annie Bird, "Still Waiting for Justice: An Assessment of the Honduran Public Ministry's Investigation of the May 11, 2012 Killings in Ahuas, Honduras," Center for Economic and Policy Research and Rights Action, Washington, DC, April 2013.

Mamani, Pablo, *Geopolíticas Indígenas,* CADES: Centro Andino de Estudios Estratégicos, El Alto, Bolivia, 2005.

Mander, Jerry, and Victoria Tauli-Corpuz, eds., *Paradigm Wars: Indigenous Peoples' Resistance to Globalization*, Sierra Club Books, San Francisco, 2006.

Manuel, George, *The Fourth World: An Indian Reality*, The Free Press, New York, 1974.

Marcus, Joyce, "Aztec Military Campaigns against the Zapotecs: The Documentary Evidence," in Kent V. Flannery and Joyce Marcus, eds., *The Cloud People: Divergent Evolution of the Zapotec and Mixtec Civilizations*, Academic Press, New York, 1983.

Marcos, Subcomandante, "7 Piezas sueltas del rompecabezas mundial (El neoliberalismo como rompecabezas: la inútil mundial que fragmenta y destruye naciones)," June 1997, *http://palabra.ezln.org.mx/comunicados/1997/1997_06_b.htm* (accessed July 26, 2013).

Matthews, Robert, "Sowing Dragon's Teeth," *NACLA Report on the Americas*, 20(4), July/August 1986, pp. 14–40.

McAfee, Kathleen, "Corn Culture and Dangerous DNA: Real and Imagined Consequences of Maize Transgene Flow in Oaxaca," *Journal of Latin American Geography* 2(1), 2003, pp. 18–42.

McCoy, Alfred, *The Politics of Heroin: CIA Complicity in the Global Drug Trade*, 2nd ed., Lawrence Hill Books, Chicago, 2003.

McCoy, Alfred, *Policing America's Empire: The United States, the Philippines, and the Rise of the Surveillance State*, University of Wisconsin Press, Madison, WI, 2009.

McFate, Montgomery, and Andrea Jackson, "An Organizational Solution for DOD's Cultural Knowledge Needs," *Military Review*, July–August, 2005, pp. 18–21.

McMonagle, Richard C., "The Small Wars Manual and Military Operations Other Than War," master's thesis, U.S. Army and General Staff College, Ft. Leavenworth, KS, 1996.

McSweeney, Kendra, and Zoe Pearson, "Prying Native People from Native Lands: Narco Business in Honduras," *NACLA Report on the Americas*, 46(4), 2013, pp. 7–12.

Meringer, Eric R., "The Local Politics of Indigenous Self-Representation: Intraethnic Political Division among Nicaragua's Miskito People during the Sandinista Era," *Oral History Review* 37(1), 2010, pp. 1–17.

Milk, Theresa, *Haskell Institute: 19th Century Stories of Sacrifice and Survival, with Haskell Cemetery Walking Tour*, Mammoth Publications, Lawrence, KS, 2007.

Miller, Yvonne E., "Haskell Cemetery once was Site of Impressive Services" *Lawrence Journal–World*, May 28, 1973, p. 1.

Miller, Yvonne E., "Tribes That Slumber," *True West* 27(5), 1980, pp. 38–41.

MISURASATA, "Plan of action 1981," in Klaudine Ohland and Robin Schneider, eds., *National Revolution and Indigenous Identity: The Conflict Between the Sandinistas and Miskito Indians on Nicaragua's Atlantic Coast*, Document 47, International Work Group for Indigenous Affairs, Copenhagen, Denmark, November, 1983.

Mitchell, Timothy, *Rule of Experts*, University of California Press, Berkeley, CA, 2002.

Mitchell, Timothy, "The Work of Economics: How a Discipline Makes Its World," *European Journal of Sociology* 46(2), 2006, pp. 297–320.

Molieri, Jorge Jenkins, *El Desafío Indígena en Nicaragua: El Caso de los Mískitos*, Editorial Vanguardia, Managua, Nicaragua, 1986.

Mollet, Sharlene, "Racial Narratives: Miskito and Colono Land Struggles in the Honduran Mosquitia," *Cultural Geographies* 18(1), 2011, pp. 43–62.

Moos, Felix, "Some Thoughts on Anthropological Ethics and Today's Conflicts," *Anthropology News* 46(6), 2005, pp. 40–42.

Morin, Karen, "Charles P. Daly's Gendered Geography, 1860–1890," *Annals of the Association of American Geographers* 98(4), 2008, pp. 897–919.

Morin, Karen, *Civic Discipline: Geography in America, 1860–1890*, Ashgate, Farnham, Surrey, UK, 2011.

Moss, Ambler, "Preface," in Geoffrey Demarest's *Mapping Colombia: The Correlation between Land Data and Strategy*, Strategic Studies Institute, U.S. Army War College, Carlisle Barracks, PA, 2003.

Mutersbaugh, Tad, "Migration, Common Property, and Communal Labor: Cultural Politics and Agency in a Mexican Village," *Political Geography* 21(4), 2002, pp. 473–494.

Nahanni, Phoebe, "The Mapping Project," in Mel Watkins, ed., *Dene Nation: The Colony Within*, University of Toronto Press, Toronto, ON, Canada, 1977, pp. 21–27.

Ng'weno, Bettina, *Turf Wars: Territory and Citizenship in the Contemporary State*, Stanford University Press, Stanford, CA, 2007.

Nietschmann, Bernard Q., *Between Land and Water: The Subsistence Ecology of the Miskito Indians, Eastern Nicaragua*, Seminar Press, New York, 1973.

Nietschmann, Bernard Q., "When the Turtle Collapses, the World Ends," *Natural History* 83(6), 1974, pp. 34–43.

Nietschmann, Bernard Q., *Caribbean Edge: The Coming of Modern Times to Isolated People and Wildlife*, Bobbs–Merrill, New York, 1979.

Nietschmann, Bernard Q., "Turtles and the Revolution," *Natural History* 89(2), 1980, pp. 8, 12.

Nietschmann, Bernard Q., "The Unreported War Against the Sandinistas," *Policy Review* 29, 1984, pp. 32-39.

Nietschmann, Bernard Q., "The Third World War," *Cultural Survival Quarterly* 11(3), 1987, pp. 1–16.

Nietschmann, Bernard Q., *The Unknown War: The Miskito Nation, Nicaragua and the United States*, Freedom House, New York, 1989.

Nietschmann, Bernard Q., "Bruno Gabriel: A Miskito Nationalist and Revolutionary," *Fourth World Journal* 2(3), 1990, p. 164–184.

Nietschmann, Bernard Q., "Conservation by Self-Determination," *Research and Exploration* 7(3), 1991, pp. 372–373.

Nietschmann, Bernard Q., "Field Notes: Miskito Coast Protected Area," *Research and Exploration* 7(2), 1991, pp. 232–237.

Nietschmann, Bernard Q., "Geographical Security: The Co-Existence of Biological and Cultural Diversity," in the *Briefing Book on International Security: The Environmental Dimension*, Tufts University, Boston, 1992, pp. 96–98.

Nietschmann, Bernard Q., "The Fourth World: Nations versus States," in George Demko and William B. Wood, eds., *Reordering the World: Geopolitical Perspectives on the Twenty-first Century*, Westview, Boulder, CO, 1994, pp. 225–236.

Nietschmann, Bernard Q., "Defending the Miskito Reefs with Maps and GPS: Mapping with Sail, Scuba and Satellite," *Cultural Survival Quarterly* 18(4), 1995, p. 34–37.

Nietschmann, Bernard Q., "Protecting Indigenous Coral Reefs and Sea Territories, Miskito Coast, RAAN, Nicaragua," in Stan Stevens, ed., *Conservation through Cultural Survival*, Island Press, Washington, DC, 1997, pp. 193–224.

Noyes, C. Reinhold, *The Institution of Property: A Study of the Development, Substance and Arrangement of the System of Property in Modern Anglo-American Law*, Longmans, Green, New York, 1936.

Offen, Karl, "The Sambo and Tawira Miskitu: The Colonial Origins and Geography of Intra-Miskitu Differentiation in Eastern Nicaragua and Honduras," *Ethnohistory* 49(2), 2002, pp. 319–372.

Offen, Karl, "The Territorial Turn: Making Black Territories in Pacific Colombia," *Journal of Latin American Geography* 2(1), 2003, pp. 43–73.

Offen, Karl, "The Geographical Imagination, Resource Economies and Nicaraguan Incorporation of the Mosquitia, 1838–1909," in Christian Brannstrom, ed., *Territories, Commodities and Knowledges: Latin American Environmental History in the Nineteenth and Twentieth Centuries*, Institute for the Study of the Americas, London, 2004, pp. 50–89.

Olund, Eric, "From Savage Space to Governable Space: The Extension of United States Judicial Sovereignty over Indian Country in the Nineteenth Century," *Cultural Geographies* 9(2), 2002, pp. 129–157.

O'Malley, Martin, *Past and Future Land: An Account of the Berger Inquiry into the Mackenzie Valley Pipeline*, P. Martin Associates, Toronto, ON, Canada, 1976.

"The Online Guide to Thomas R. Berger," http://web.uvic.ca/~mharbell/a1/workshop2/mvp_inquiry.html (accessed 14 August 2012).

Owen, Robert, to Oliver North, "Costa Rica Trip," May 20, 1985, in National Security Archives, http://gateway.proquest.com/openurl?url_ver=Z39.88-2004&res_dat=xri:dnsa&rft_dat=xri:dnsa:article:CIC01153 (accessed August 9, 2013).

Owolabi, Kunle, "¿La Legalización de los 'Usos y Costumbres' ha Contribuido a la Permanencia del Gobierno Priísta en Oaxaca?: Análisis de las Elecciones para Diputados y Gobernadores, de 1992 a 2001," *Foro Internacional 177*, 2004, pp. 474–508.

Paddock, John, "Oaxaca in Ancient Mesoamerica," in John Paddock, ed., *Ancient Oaxaca: Discoveries in Mexican Archeology and History*, Stanford University Press, Stanford, CA, 1966, pp. 83–242.

Parker, Geoffrey, "Maps and Ministers: The Spanish Habsburgs," in David Buisseret, ed., *Monarchs, Ministers, and Maps: The Emergence of*

Cartography as a Tool of Government in Early Modern Europe, University of Chicago Press, Chicago, 1992, pp. 124–152.

Peluso, Nancy, "Whose Woods Are These?: Counter-Mapping Forest Territories in Kalimantan, Indonesia," *Antipode 27*(4), 1995, pp. 383–406.

Penikett, Tony, *Reconciliation: First Nations Treaty Making in British Columbia*, Douglas & McIntyre, Vancouver, BC, Canada, 2006.

Pérez Chiriboga, Isabel, *Espíritus de Vida y Muerte: Los Miskitu Hondureños en Época de Guerra*, Editorial Guaymuras, Tegucigalpa, Honduras, 2002.

Perramond, Eric, "The Rise, Fall, and Reconfiguration of the Mexican Ejido," *Geographical Review 98*(3), 2008, pp. 356–371.

Petraeus, Lt. Gen. David H., "Learning Counterinsurgency: Observations from Soldiering in Iraq," *Military Review*, January–February 2006, pp. 45–55.

Poole, Peter, *Developing a Partnership of Indigenous Peoples, Conservationists, and Land Use Planners in Latin America*, World Bank Publications, Washington, DC, 1989.

Poole, Peter, ed., "Geomatics: Who Needs It? (Special Issue)," *Cultural Survival Quarterly 18*(4), 1994.

Poole, Peter, "Indigenous Peoples, Mapping and Biodiversity Conservation: An Analysis of Current Activities and Opportunities for Applying Geomatics Technologies," Biodiversity Support Program, 1995, *http://rmportal.net/library/content/frame/indigenous-people.pdf/view* (accessed October 24, 2012).

Powelson, John P., *The Story of Land: A World History of Land Tenure and Agrarian Reform*, Lincoln Institute of Land Policy, Cambridge, MA, 1988.

Powelson, John P., and Richard Stock, *The Peasant Betrayed: Agriculture and Land Reform in the Third World*, 2nd ed., Cato Institute, Washington, DC, 1990.

Pratt, Richard H., "The Advantages of Mingling Indians with Whites," in Francis P. Prucha, ed., *Americanizing the American Indian: Writings by "Friends of the Indian," 1890–1900*, Harvard University Press, Cambridge, MA, 1973, pp. 260–271.

Pratt, Richard H., *Battlefield and Classroom: Four Decades with the American Indian, 1867–1904*, University of Oklahoma Press, Norman, OK, 2004/1964.

Price, David, *Weaponizing Anthropology: Social Science in Service of the Militarized State*, CounterPunch and AK Press, Petrolia, CA, 2011.

Proske, Beatrice Gilman, *Archer Milton Huntington*, Hispanic Society of America, New York, 1963.

"Public Diplomacy Action Plan: Support for the White House Educational Campaign," Confidential project proposal, March 12, 1985, *http://gateway.proquest.com/openurl?url_ver=Z39.88-2004&res_dat=xri:dnsa&rft_dat=xri:dnsa:article:CNI02394* (accessed August 9, 2013).

Puxley, Peter, "The Colonial Experience," in Mel Watkins, ed., *Dene Nation: The Colony Within*, University of Toronto Press, Toronto, ON, Canada, 1977, pp. 103–119.

Quist, David, and Ignacio H. Chapela, "Transgenic DNA introgressed into traditional maize landraces in Oaxaca, Mexico," *Nature 414*(29), pp. 541–543.

Rappaport, Joanne, *Intercultural Utopias: Public Intellectuals, Cultural*

Experimentation, and Ethnic Pluralism in Colombia, Duke University Press, Durham, NC, 2005.

Rauber, Paul, "The Nietschmann File," *The Express* (Berkeley, CA), 6(46), August 31, 1984, pp. 1, 16–23.

Reyes, Alvaro, and Mara Kaufman, "Sovereignty, Indigeneity, Territory: Zapatista Autonomy and the New Practices of Decolonization," *South Atlantic Quarterly 110*(2), 2011, pp. 505–525.

Reyes, Reynaldo, *Rafaga: The Life Story of a Nicaraguan Miskito Comandante*, University of Oklahoma Press, Norman, OK, 1992.

Rheinberger, Hans-Jörg, *Toward a History of Epistemic Things: Synthesizing Protein in the Test Tube*, Stanford University Press, Stanford, CA, 1997.

Richardson, Boyce, *Strangers Devour the Land: A Chronicle of the Assault upon the Last Coherent Hunting Culture in North America, the Cree Indians of Northern Québec, and Their Vast Primeval Homelands*, Alfred A. Knopf, New York, 1976.

Richter, Paul, and Greg Miller, "Colombia Army Chief Linked to Outlaw Militias," *Los Angeles Times*, March 25, 2007, Home Edition, A1.

Riewe, Rick, ed., *Nunavut Atlas*, Canadian Circumpolar Institute and the Tungavik Federation of Nunavut, Edmonton, AB, Canada, 1992.

Rivera Cusicanqui, Silvia, *Ch'ixinakax Utxiwa: Una Reflexión Sobre Prácticas y Discursos Descolonizadores*, Tinta Limón, Buenos Aires, Argentina, 2010.

Robinson, William, *Transnational Conflicts: Central America, Social Change and Globalization*, Verso, London, 2003.

Robson, Seth, "Military to Unveil New Counterinsurgency Field Manual," *Stars and Stripes*, January 28, 2013, www.stripes.com/news/military-to-unveil-new-counterinsurgency-field-manual-1.205579 (accessed February 2, 2013).

Rodríguez-Piñero, Luis, *Indigenous Peoples, Postcolonialism, and International Law: The ILO Regime, 1919–1989*, Oxford University Press, Oxford, UK, 2005.

Roldán Ortega, Roque, *Legalidad y Derechos Étnicos en la Costa Atlántica de Nicaragua*, Programa de Apoyo Institucional a los Consejos Regionales y las Administraciones Regionales de la Costa Atlántica RAAN-ASDI-RAAS, I/M Editores, Bogotá, Colombia, 2000.

"Role of the Canadian Courts in Aboriginal Rights," Grand Council of the Crees, www.gcc.ca/archive/article.php?id=103 (accessed August 13, 2012).

Rose, Alex, *Spirit Dance at Meziadin: Chief Joseph Gosnell and the Nisga'a Treaty*, Harbor Publishing, Madeira Park, BC, Canada, 2000.

Rose, Nikolas, *Powers of Freedom; Reframing Political Thought*, Cambridge University Press, Cambridge, UK, 1999.

Rothenberg, Tamar Y., "Voyeurs of Imperialism: *The National Geographic Magazine* before World War II," in Anne Godlewska and Neil Smith, eds., *Geography and Empire*, Blackwell, Oxford, UK, 1994, pp. 155–172.

Ruiz y Ruiz, Frutos, *Informe, del Doctor, Comisionado del Poder Ejecutivo en la Costa Atlántica de Nicaragua*, C. Heuberger, Managua, 1927.

Rushforth, Scott, "Country Food," in Mel Watkins, ed., *Dene Nation: The Colony Within*, University of Toronto Press, Toronto, ON, Canada, 1977, pp. 32–45.

Rÿser, Rudolph C., "Peaceful Warriors Passing through," *Fourth World Eye*,

December 2000, http://cwis.org/publications/FWE/archive-2000-2003/peaceful-warriors-passing-through (accessed August 9, 2013).

Sahlins, Marshall, "Preface," in Network of Concerned Anthropologists Steering Committee, *The Counter-Counterinsurgency Manual: Or, Notes on Demilitarizing American Society*, Prickly Paradigm Press, Chicago, 2009, pp. ii–iii.

Salisbury, Richard, *Not by Bread Alone*, Indians of Québec Association, Montréal, QC, Canada, 1972.

Scales, Robert, "Statement of Major General Robert Scales, USA (ret.), Testifying before the House Armed Services Committee on July 15, 2004," www.au.af.mil/au/awc/awcgate/congress/04-07-15scales.pdf (accessed August 16, 2013).

Schaffer, Ronald, "The 1940 Small Wars Manual and the 'Lessons of History,'" *Military Affairs 36*, 1972, 46–51.

Scharfe, Wolfgang, "Max Eckert's 'Kartenwissenschaft': The Turning Point in German Cartography," *Imago Mundi 38*, 1986, pp. 61–66.

Schlosser, Nicholas, "'The Marine Corps' *Small Wars Manual*: An Old Solution to a New Challenge," *Fortitudine 35*(1), 2010, pp. 4–9.

Schroeder, Michael, *"To Defend Our Nation's Honor": Toward a Social and Cultural History of the Sandino Rebellion in Nicaragua, 1927–1934*, PhD dissertation, University of Michigan, Ann Arbor, MI, 1993.

Schroeder, Michael, "Horse Thieves to Rebels to Dogs: Political Gang Violence and the State in the Western Segovias, Nicaragua, in the Time of Sandino, 1926–1934," *Journal of Latin American Studies 28*(2), 1996, pp. 383–434.

Schulten, Susan, *The Geographical Imagination in America, 1880–1950*, University of Chicago Press, Chicago, 2001.

Schwartz, Mattathias, "A Mission Gone Wrong: Why Are We Still Fighting the Drug War?," *The New Yorker*, January 6, 2014, pp. 44–55.

Scott, Patrick, *Stories Told: Stories and Images of the Berger Inquiry*, Edzo Institute, Yellowknife, NT, Canada, 2007.

Sedillo, Simon, director, *The Demarest Factor*, 2010, http://elenemigocomun.net/2011/09/demarest-factor-military-mapping-indigenous-communities (accessed March 8, 2012).

Seoane, María, "Los Secretos de la Guerra Sucia Continental de la Dictadura," *Clarín Especiales: A 30 Años de la Noche Más Larga*, March 24, 2006, www.espacioalternativo.org/node/1270 (accessed May 13, 2014).

Shanker, Thom, "Lessons of Iraq Help U.S. Fight a Drug War in Honduras," *New York Times*, May 5, 2012, p. 1, www.nytimes.com/2012/05/06/world/americas/us-turns-its-focus-on-drug-smuggling-in-honduras.html?_r=2&pagewanted=all& (accessed July 24, 2013).

Slater, Candace, ed., *In Search of the Rain Forest*, Duke University Press, Durham, NC, 2004.

Smith, Derek, Peter Herlihy, John Kelly, and Aida Ramos Viera, "The Certification and Privatization of Indigenous Lands in México," *Journal of Latin American Geography 8*(2), 2009, pp. 175–207.

Smith, Neil, *American Empire: Roosevelt's Geographer and the Prelude to Globalization*, University of California Press, Berkeley, CA, 2004.

Sonnenfeld, Joseph, *Changes in Subsistence among Barrow Eskimo*, Project No. ONR–140, Arctic Institute of North America, 1956.

de Soto, Hernando, *The Other Path: The Invisible Revolution in the Third World*, Harper & Row, New York, 1989.
de Soto, Hernando, *The Mystery of Capital: Why Capitalism Triumphs in the West and Fails Everywhere Else*, Basic Books, New York, 2000.
Sparke, Matthew, "A Map That Roared and an Original Atlas: Canada, Cartography, and the Narration of Nation," *Annals of the Association of American Geographers 88*(3), 1998, pp. 463–495.
Squier, Ephraim George, *Waikna: Adventures on the Mosquito Shore*, University of Florida Press, Gaineville, FL, 1965/1855. (Published under the pseudonym of Samuel A. Bard.)
Stanton, John, *General David Petraeus' Favorite Mushroom: Inside the U.S. Army's Human Terrain System*, Wiseman Publishing, 2009.
Starr, Frederick, *Indians of Southern Mexico: An Ethnographic Volume*, Lakeside Press, Chicago, 1899.
Starr, Frederick, *In Indian Mexico: A Narrative of Travel and Labor*, Forbes, Chicago, 1908.
Stephen, Lynn, *Zapata Lives!: Histories and Cultural Politics in Southern Mexico*, University of California Press, Berkeley, CA, 2001.
Sterritt, Neil, "The Nisga'a Treaty: Competing Claims Ignored!", *BC Studies 120*, 1998–1999, pp. 73–94.
Sterritt, Neil, Susan Marsden, Robert Galois, Peter Grant, and Richard Overstall, *Tribal Boundaries in the Nass Watershed*, University of British Columbia Press, Vancouver, BC, Canada, 1998.
Stocks, Anthony, "Mapping Dreams in Nicaragua's Bosawas Reserve," *Human Organization 62*(4), 2003, pp. 344–356.
Striffler, Steve, and Mark Moberg, *Banana Wars: Power, Production, and History in the Americas*, Duke University Press, Durham, NC, 2003.
Submission to the Royal Commission on Aboriginal Peoples of Makavik and Inuit Tapirisat of Canada on Behalf of the Inuit Relocated to the High Arctic (Griese Fiord and Resolute Bay) in the 1950s by the Federal Government, June 28, 1993, *http://pubs.aina.ucalgary.ca/makivik/CI034.pdf* (accessed June 11, 2012).
Sundberg, Juanita, "NGO Landscapes in the Maya Biosphere Reserve, Guatemala," *Geographical Review 88*(3), 1998, pp. 388–412.
Sutton, Imre, ed., *Irredeemable America: The Indians' Estate and Land Claims*, Olympic Marketing, Minnetonka, MN, 1986.
Sutton, Imre, "Preface to Indian Country: Geography and Law," *American Indian Culture and Research Journal 15*(2), 1991, pp. 3–36.
Tennant, Paul, *Aboriginal Peoples and Politics: The Indian Land Question in British Columbia, 1849–1989*, University of British Columbia Press, Vancouver, BC, Canada, 1990.
Tester, Frank, and Peter Kulchyski, *Tammamiit (Mistakes): Inuit Relocation in the Eastern Arctic, 1939–1963*, University of British Columbia Press, Vancouver, BC, Canada, 1994.
Tillman, Benjamin F., *Imprints on Native Lands: The Miskito–Moravian Settlement Landscape in Honduras*, University of Arizona Press, Tucson, AZ, 2011.
Tobias, Terry, *Chief Kerry's Moose: A Guidebook to Land Use and Occupancy*

Mapping, Research Design and Data Collection, Union of British Columbia Indian Chiefs and Ecotrust Canada, Vancouver, BC, Canada, 2000.

Tobias, Terry, *Living Proof: The Essential Data-Collection Guide for Indigenous Use-and-Occupancy Map Surveys*, Ecotrust Canada and the Union of British Columbia Indian Chiefs, Vancouver, BC, Canada, 2009.

T'Seleie, Frank, "Statement to the Mackenzie Valley Pipeline Inquiry," in Mel Watkins, ed., *Dene Nation: The Colony Within*, University of Toronto Press, Toronto, ON, Canada, 1977, pp. 12–17.

Turnbull, David, "Maps, Narratives, and Trails: Performativity, Hodology, and Distributed Knowledges in Complex Adaptive Systems—an Approach to Emergent Mapping," *Geographical Research*, 45(2), 2007, pp. 140–149.

Turner, Dale, *This Is Not a Peace Pipe: Towards a Critical Indigenous Philosophy*, University of Toronto Press, Toronto, ON, Canada, 2006.

Tychon, G. G., "The Dene Mapping Project: Past and Present," presentation at the 7th annual symposium on Geographic Information Systems in Forestry, Environment and Natural Resources Management, Vancouver, BC, Canada, 1993.

Ulloa, Astrid, *The Ecological Native: Indigenous Peoples' Movements and Eco-Governmentality in Colombia*, Routledge, New York, 2005.

Urcid, Javier, *Zapotec Hieroglyphic Writing, Studies in Pre-Columbian Art and Archaeology 34*, Dumbarton Oaks, Washington, DC, 2001.

Urcid, Javier, *Zapotec Writing: Knowledge, Power, and Memory in Ancient Oaxaca*, Foundation for the Advancement of Mesoamerican Studies, 2005, www.famsi.org/zapotecwriting/index.html (accessed March 6, 2012).

Usher, Peter, *The Bankslanders: Economy and Ecology of a Frontier Trapping Community*, Indian Affairs and Northern Development, Ottawa, ON, Canada, 1971.

Usher, Peter, "Environment, Race, and Nation Reconsidered: Reflections on Aboriginal Land Claims in Canada," *Canadian Geographer* 47(4), 2003, pp. 365–382.

Usher, Peter, Frank Tough, and Robert Galois, "Reclaiming the Land: Aboriginal Title, Treaty Rights and Land Claims in Canada," *Applied Geography* 12(2), 1992, pp. 109–132.

U.S. Agency for International Development, *Country Development Strategy Statement: USAID/Nicaragua 1991–1996*, Washington, DC, June 14, 1991.

U.S. Army Command and General Staff College, *Field Circular 100-20: Low Intensity Conflict*, Fort Leavenworth, KS, July 16, 1986, www.cgsc.edu/carl/docrepository/FC_100_20_1986.pdf

U.S. Army, *Counterinsurgency/FM 3-24/MCWP 3.33.5*, Department of the Army, U.S. Government Printing Office, Washington, DC, 2006.

U.S. Army, *The U.S. Army/Marine Corps Counterinsurgency Field Manual*, Foreword, John Nagl, and Introduction, Sarah Sewell, University of Chicago Press, Chicago, 2007.

U.S. Department of State, *Human Rights in Nicaragua under the Sandinistas: From Revolution to Repression*, U.S. Department of State, Washington, DC, 1986.

U.S. Marine Corps, *Small Wars Manual*, NAVMC 2890, U.S. Government Printing

Office, Washington, DC, 1940, *www.marines.mil/Portals/59/Publications/ FMFRP%2012-15%20%20Small%20Wars%20Manual.pdf*.
Utley, Harold, "An Introduction to the Tactics and Technique of Small Wars," *Marine Corps Gazette*, May 1931, pp. 50–53.
Utley, Harold, "The Tactics and Technique of Small Wars: Part II–Intelligence," *Marine Corps Gazette*, August 1933, pp. 44–48.
Utley, Harold, "The Tactics and Technique of Small Wars: Part III–Functions of the Personnel (First) Section of the Staff," *Marine Corps Gazette*, November 1933, pp. 43–46.
van Creveld, Martin, *Supplying War: Logistics from Wallenstein to Patton*, Cambridge University Press, Cambridge, UK, 1977.
Varese, Stefano, *Witness to Sovereignty: Essays on the Indian Movement in Latin America*, International Work Group for Indigenous Affairs, Copenhagen, Denmark, 2006.
Vincent, Theodore G., *Black Power and the Garvey Movement*, Black Classic Press, Baltimore, MD, 2006.
Vogt, Evon, *Aerial Photography in Anthropological Field Research*, Harvard University Press, Cambridge, MA, 1974.
Wainwright, Joel, *Decolonizing Development: Colonial Power and the Maya*, Blackwell, Malden, MA, 2008.
Wainwright, Joel, *Geopiracy: Oaxaca, Militant Empiricism, and Geographical Thought*, Palgrave, New York, 2012.
Wainwright, Joel, and Joe Bryan, "Cartography, Territory, Property: Postcolonial Reflections on Indigenous Counter-Mapping in Nicaragua and Belize," *Cultural Geographies* 16(2), 2009, pp. 153–178.
Wali, Alaka, and Shelton H. Davis, *Protecting Amerindian Lands: A Review of World Bank Experience with Indigenous Land Regularization Programs in Lowland South America*, Latin America and the Caribbean Regional Office, Technical Department, Regional Studies Program, the World Bank, Washington, DC, 1992.
Watkins, Mel, ed. *Dene Nation: The Colony Within*, University of Toronto Press, Toronto, ON, Canada, 1977.
Webb, Gary, *Dark Alliance: The CIA, the Contras, and the Crack Cocaine Explosion*, Seven Stories Press, New York, 1999.
Weinberg, Bill, "The Battle for the Miskito Coast: Piracy and Ecology on Nicaragua's Wild Frontier," *Native Americas: Akwe:kon's Journal of Indigenous Issues*, Fall 1995, pp. 22–33.
Weinstein, Martin, *What the Land Provides: An Examination of the Fort George Subsistence Economy and the Possible Consequences on It of the James Bay Hydroelectric Project*, Grand Council of the Crees of Québec, Montréal, QC, Canada, 1976.
Whisnant, David, *Rascally Signs and Sacred Places: The Politics of Culture in Nicaragua*, University of North Carolina Press, Chapel Hill, NC, 1995.
White, Richard, *Railroaded: The Transcontinentals and the Making of Modern America*, Norton, New York, 2011.
Williams, Robert A., *Linking Arms Together: American Indian Treaty Visions of Law and Peace, 1600–1800*, Oxford University Press, New York, 1997.
Wilshusen, Peter R., "The Receiving End of Reform: Everyday Responses to

Neoliberalisation in Southeastern Mexico," *Antipode* 42(3), 2010, pp. 767–799.

Winks, Robin, *Cloak and Gown Scholars in the Secret War, 1939–1961*, 2nd ed., William Morrow, New York, 1996.

Woldenberg, Michael, "Energy Flow and Spatial Order: Mixed Hexagonal Hierarchies of Central Places," *Geographical Review* 58(4), 1968, pp. 552–574.

Wood, Denis, *Fleeting Glimpses*, Clark University Cartographic Laboratory, Worcester, MA, 1971.

Wood, Denis, *I Don't Want to, But I Will*, Clark University Cartographic Laboratory, Worcester, MA, 1973.

Wood, Denis, *The Power of Maps*, Guilford Press, New York, 1992.

Wood, Denis, *Rethinking the Power of Maps*, Guilford Press, New York, 2010.

Wood, Denis, and John Fels, *The Natures of Maps*, University of Chicago Press, Chicago, 2008.

Wright, John Kirtland, *Geography in the Making: The American Geographical Society, 1851–1951*, American Geographical Society, New York, 1952.

Wright, John Kirtland, "Geography, Experience, and Imagination: Towards a Geographical Epistemology," *Annals of the Association of American Geographers* 51(3), 1961, pp. 241–260.

Wright, John Kirtland, *Human Nature in Geography: Fourteen Papers, 1925–1965*, Harvard University Press, Cambridge, MA, 1966.

Yannakakis, Yanna, *The Art of Being in-between: Native Intermediaries, Indian Identity, and Local Rule in Colonial Oaxaca*, Duke University Press, Durham, NC, 2008.

YATAMA, La Nueva Alternativa—YATAMA, "A Proposal to the National Endowment for Democracy," 1988, *http://cwis.org/GML/?post=494* (accessed August 9, 2013).

Ybarra, Megan, "Taming the Jungle, Saving the Maya Forest: Sedimented Counterinsurgency Practices in Contemporary Guatemalan Conservation," *Journal of Peasant Studies* 39(2), 2012, pp. 479–502.

Index

An *f* following a page number indicates a figure; an *n* following a page number indicates a note.

Afghanistan war, 52, 133–138, 168–169
African Command (AFRICOM), 157–158, 229*n*
African Genesis (Ardrey, 1966), 128
Afro-Colombian communities, 113–114, 146
Agenda 21, 107
Agrarian reforms, 152–153
AGS conference at Haskell, 170–172
Aguilar Robledo, Miguel, 145, 147
Alaska Native Claims Settlement Act, 59–60, 200*n*–201*n*
"All Our Eggs in a Broken Basket" (Constable, 2009), 138
ALPROMISU political organization, 83–84, 209*n*–210*n*
American Anthropological Association, 137–138
American Civil War, 25
American Counterinsurgency (González, 2009), 137–138
American Geographers Union, 20
American Geographical Society (AGS)
 AGS conference at Haskell, 170–172
 exploration and, 18–20
 Foreign Military Studies Office of the U.S. Army and, 145–146
 gendered history of, 186*n*
 geography as a science and AGS's activities, 24–27
 history of, 17–33
 honorary appointments and medal awards and, 22–23, 29
 mapping of Tiltepec and, 14–15
 México Indígena Project and, 142, 143–144, 147
 overview, 17–18, 18*f*, 133, 185*n*–186*n*
 publications, 21, 28, 30–32
 Transcontinental Excursion of 1912 and, 23–24
 U.S. Army and, 32–33
 war and, 25–26
 World Conference on Human Geography and, 165
American Numismatic Society, 26–27
American Spanish War, 25
Anaya, S. James, 120–124
"An Organizational Solution for DOD's Cultural Knowledge Needs" (McFate & Jackson, 2005), 136–137
Ardrey, Robert, 128
Army, U.S.
 American Geographical Society (AGS) and, 32–33
 Bowman Expeditions and, 234*n*

funding of México Indígena Project and, 14–15
Human Terrain System and, 138–140
mapping of Tiltepec and, 6–9
mapping technology and, 97–98
Assimilation, 163–166
Association of American Geographers, 20, 133, 161
Awas Tingni, 118, 120–126, 122*f*, 125*f*
Aztecs, 4

B

BAE Systems, 137–138
Baffin, William, 16–17
Balboa, 16–17
Barbados Declaration, 167
Barrios, Luis de, 4
Barton, Charles, 55
Basso, Keith, 102
Belize, 100–108, 101*f*, 103*f*
Bell, Alexander Graham, 19
Berger, Thomas, 56–57, 67–68, 71–72
Berger Inquiry, 71–72
Biopiracy, 159
Black communities
  Colombia, 113–114, 146
  Nicaragua, 39, 84, 86, 117–121, 220*n*
Boas, Franz, 60
Borowiecki, Barbara, 30
Borrow, George, 26–27
Bosawas Protected Area, 109
Boundaries, 25, 26
Bourgois, Philippe, 84, 85*f*
Bowman, Isaiah, 9, 21, 23–24, 25–26, 98, 183*n*
Bowman Expeditions
  AGS conference at Haskell, 170–171
  expansion of the México Indígena Project and, 158–159, 160–161
  Human Terrain System and, 138–141
  Indian wars and, 165–166
  México Indígena Project and, 143–144
  overview, 9, 167–168, 173–174, 183*n*, 230*n*–231*n*, 233*n*–234*n*
  Small Wars Manual and, 53
  U.S. Army and, 32–33
Bowman Expeditions/México Indígena (BEMI), 139–140
Breckinridge, James C., 48–49
Brigham, Albert Perry, 20
British rule, 39–40
Brody, Hugh, 60–61, 63, 203*n*–204*n*
Brooks, David, 43–44
"Bush Brigades" (Ellis), 48
Bush Warfare (Heneker, 1907), 48, 195*n*

C

Cabot, 16–17
Calder, Arthur, 56
Calder, Frank Arthur, 54–55, 56
Calder, Job, 56
Calder v. the Attorney-General of British Columbia (1973), 56–59, 72–73
Callwell, C. E., 48, 49, 134–135
Canada. *See also* Indigenous mapping in Canada
  Calder v. the Attorney-General of British Columbia (1973) and, 56–59
  indigenous mapping and, 72–73, 169
  Inuit Land Use and Occupancy Project, 58–64, 62*f*
  overview, 55
Canadian Arctic Archipelago, 58
Canal routes, 17
Caribbean Conservation Corporation, 107–108, 215*n*
Caribbean Central America Research Council, 119, 172
Caribbean Edge: The Coming of Modern Times to Isolated People and Wildlife (Nietschmann, 1979), 88–89
Cartier, Jacques 16–17
Center for Native Lands, 111
Central America
  counterinsurgency and, 98–100
  property and peace and, 108–118, 119*f*
  Research and Exploration journal (National Geographic Society), 100–108, 101*f*, 103*f*
Central Intelligence Agency (CIA), 52
Chamorro, Emiliano, 37
Champlain, Samuel de, 16–17
Chapin, Mac, 102–104, 111, 220*n*
Chrétien, Jean, 63–64
Civilization, 164
Coco. *See* Río Coco
The Coexistence of Indigenous Peoples and Natural Ecosystems in Central America, 100
Cold War, 82, 97–98
Colombia, 112–118, 119*f*
Colombian Constitution, 113–114, 114
Colombian Geographic Society (CGS), 133

Colonization, 81–82
Columbus, 16–17
Committee on Original Peoples Entitlement, 57–58, 200*n*
Communal property, 148–149, 153. *See also* Property rights
Community mapping, 114–115
Complementary Agreement No. 13, 67
Computer hardware, 9–14, 10*f*
Comunalidad, 153–154
Conference of Latin Americanist Geographers in Morelia (2005), 151
Conklin, Harold, 60
Conquest, 164
Conservation
 cultural survival and, 107
 Nietschmann's focus on, 80–82
 property and peace and, 115–116
 Research and Exploration journal (National Geographic Society), 106–107
 turtling and, 79–80
Constable, Ben, 138
Contra war
 freedom fighters and, 87–93
 maps and the claiming of indigenous territories and, 86–87
 property rights and, 93–94
Contras, 109
Convention 169, 106, 120, 216*n*–217*n*
Convention on Biological Diversity, 107
CORDS, 223*n*–224*n*
Corn, domestication of, 3
Coronado, 16–17
Cortéz, Hernán, 4
Costa Rica
 Nicaraguan canal routes and, 38
 Research and Exploration journal (National Geographic Society), 100–108, 101*f*, 103*f*
Counterinsurgency Field Manual, 139–140
Counterinsurgency tactics
 mapping technology and, 98
 overview, 98–100, 139–140
 PROCEDE (Program for Certification of Ejidal Rights and Titling of Urban Lots) and, 149
 property and peace and, 115–116
Counter-mappers, 96–98, 124–126, 125*f*
Counterrevolution, 98–100
Craib, Raymond, 148

Cree territories, 75–76
Creole communities (Nicaragua), 39, 84, 86, 117–121, 220*n*
Cultural ecology
 conservation and, 80
 maps and the claiming of indigenous territories and, 83
 overview, 60
Cultural Knowledge Consortium, 165
Cultural Survival organization, 86, 102, 120–121
Cultural Survival Quarterly journal, 102, 103–104
Cultural terrain, 158

D

Daly, Charles, 17, 29
Davis, Shelton "Sandy," 112–113, 115
Davis, William Morris, 16–17, 19–20, 23–24, 27
Declaration of Barbados, 81–82
Declaration of Quito (1990), 97–98
Decolonization
 Declaration of Quito (1990) and, 97–98
 property and peace and, 115–116
 Research and Exploration journal (National Geographic Society), 100–108, 101*f*, 103*f*
 rule of law and, 124
Demarest, Geoffrey, 127–133, 146–147, 167–168, 173, 183*n*–184*n*, 221*n*–222*n*
 expansion of the México Indígena Project, 149–161, 152*f*, 154*f*
 Human Terrain System and, 139
 México Indígena Project and, 142–144
 PROCEDE (Program for Certification of Ejidal Rights and Titling of Urban Lots) and, 149
Dene Nation: The Colony Within (Watkins), 68
Dene Studies, 64–72, 65*f*, 70*f*, 169, 179. *See also* James Bay Cree
Dene territories, 75–76, 205*n*–206*n*
Dias, Bartolomeu, 16–17
Díaz, Porfirio, 3
Distant Early Warning (DEW) Line, 58
Dobson, Jerome ("Jerry"), 24, 52–53, 145–147, 169–170, 228*n*, 235*n*
 expansion of the México Indígena Project, 149–161, 152*f*, 154*f*

Human Terrain System and, 139–140
Indian wars and, 165–166
México Indígena Project and, 143
Drug Enforcement Administration (DEA), 172–173
Drug trafficking, 109–110, 170–173
Drug war, 142

E

Earth Summit in Rio de Janeiro (1992), 107
Economic restructuring, 97–100, 115–116
Edson, Merritt A. "Red Mike," 33, 34–35, 168–169, 194n–195n
  bandits, 35–38
  Bosawas Protected Area and, 109
  Eastern Nicaragua, 38–40
  Indians, 40–47, 42f, 46f
  Iraq and Afghanistan wars and, 134–135
  maps and the claiming of indigenous territories and, 77
  Small Wars Manual, 47–53
El Salvador, 99, 100–108, 101f, 103f
Environmental conservation. *See* Conservation
Espino Negro Accords, 37
The Ethnographic Atlas of Ifugao (Conklin, 1980), 60, 88, 203n
Expansion goals, 17
Exploration, 16–18
Explorer societies, 18–20. *See also* Geographical societies
Explorers, 16–17

F

Fagoth, Steadman, 86–87
Field Manual 31-5 (Landing Operation on Hostile Shores), 52
Finley, John Huston, 31–32
Flad, Harvey, 29
Fleet Training Publication 167 (Landing Operations Doctrine), 52
"Fliers' and Explorers' Globe," 31–32
Ford, James B., 27
Ford Foundation
  community mapping and, 115
  overview, 97, 98
  property and peace and, 115–116
Foreign Military Studies Office of the U.S. Army
  American Geographical Society (AGS) and, 145–146
  Bowman Expeditions and, 174
  expansion of the México Indígena Project and, 160, 161
  funding of México Indígena Project and, 14–15
  Human Terrain System and, 136–137
  indigenous mapping and, 178
  mapping of Tiltepec and, 10–14
  México Indígena Project and, 143, 144, 147
Foreign-Deployed Advisory Support Team (FAST), 172
Fort George Resource Use and Subsistence Economy Study, 64–66, 65f
Fort Leavenworth
  expansion of the México Indígena Project and, 161
  Human Terrain System and, 139
  Indian wars and, 165–166
  México Indígena Project and, 147
  overview, 136, 146–147
Fougeu, Jacques, 44–45
Fourth World
  decolonization and, 97–98
  overview, 71, 92f, 93–95, 103–104, 214n
  property and peace and, 110–111
  Research and Exploration journal (National Geographic Society), 103–108
The Fourth World: An Indian Reality (Manuel, 1974), 71
Fox, Vicente, 155–156
Freedom fighters, 87–93

G

Garvey, Marcus, 40
Gendered geography, 186n, 187n–188n
General David Petraeus' Favorite Mushroom (Stanton, 2009), 138–139
Geographic information systems (GIS) software
  American Geographical Society (AGS) and, 145–146
  Bowman Expeditions and, 141
  expansion of the México Indígena Project and, 159–160
  mapping of Tiltepec and, 9–14, 10f
  rule of law and, 120–121
Geographical Review, 21, 28, 189n

Geographical societies, 19–20. *See also* Explorer societies
Geography of the Central Andes (Ogilvie, 1992), 28
Geography overview, 24–27
Geological Society of America, 20
Geopiracy, 159
Geoproperty: Foreign Affairs, National Security, and Property Rights (Demarest, 1998), 128–130
Global positioning system (GPS)
   community mapping and, 115
   expansion of the México Indígena Project and, 159–160
   mapping of Tiltepec and, 9–14, 10*f*
Gómez, Ismael, 37
González, Aldo, 9, 10–14, 157–159
González, Roberto, 137–138
Google Maps, 169–170, 174, 235*n*
Grand Cayman, 79
Grünberg, Georg, 81, 84, 85*f*
Guatemala
   boundary disputes and, 26
   counterinsurgency and, 99
   Research and Exploration journal (National Geographic Society), 100–108, 101*f*, 103*f*
Guerilla (small war), 196*n*
Guns and the claiming of territory, 74–75, 95
Gurdián, Galio, 84
Gurney, Arthur, 55

H

Habitants, 81–82
Handbook of South American Indians (Steward), 60
Harrington, Samuel, 48
Harvard Chiapas Project, 60, 213*n*
Haskell Indian Nations University, 163–166, 167, 231*n*–232*n*
Herlihy, Peter, 4–6, 14–15, 108, 144–145, 147, 168, 169–170, 183*n*, 185*n*, 214*n*, 228*n*
   Drug Enforcement Administration (DEA) and, 172–173
   expansion of the México Indígena Project, 149–161, 152*f*, 154*f*
   Human Terrain System and, 139–140
   Indian wars and, 165–166
   indigenous mapping and, 73
   México Indígena Project and, 142–144
   promise of free computer hardware and, 9–14
   property and peace and, 110–111
   Small Wars Manual and, 52–53
Hispanic Society on Audubon Terrace, 26–28
Homestead Act (1862), 130–131
Honduras
   AGS conference at Haskell, 170–172
   boundary disputes and, 26
   counterinsurgency and, 99
   maps and the claiming of indigenous territories and, 77–78, 234*n*–235*n*
   overlaps, 220*n*
   property and peace and, 108–118, 119*f*
   Research and Exploration journal (National Geographic Society), 100–108, 101*f*, 103*f*
   rule of law and, 123
Huasteca Potosina in San Luis Potosí
   expansion of the México Indígena Project and, 156, 160
   Herlihy and, 145, 147
   México Indígena Project and, 142, 144
   PROCEDE (Program for Certification of Ejidal Rights and Titling of Urban Lots) and, 149
Hudson, 16–17
Huichol communities, 151
Human Terrain Mapping (HTM), 139–140
Human Terrain System (HTS)
   Bowman Expeditions and, 138–141
   expansion of the México Indígena Project and, 158–159, 161
   overview, 136–138, 167, 223*n*–224*n*, 230*n*
   Small Wars Manual and, 53
   World Conference on Human Geography and, 165
Humbolt, Alexander von, 16–17
Huntington, Anna Hyatt, 23
Huntington, Arabella Duval, 23
Huntington, Archer Milton, 17, 19, 23–24, 26–27
Huntington, Charles P., 23
Huntington, Collis P., 17

I

Illueca, Jorge, 108
Imperial Geographical Society of St. Petersburg, 18
Indian Act (Canada), 55–56

Indian boarding schools, 163–166, 167, 231*n*
Indian Brotherhood of the Northwest Territories, 68*f*
Indian Claims Commission, 83
Indian Law Resource Center, 120, 124–126
Indian Rights Association, 55–56
Indian wars, 163–166
Indigenous areas, 6–9, 14–15, 33
Indigenous mapping. *See also* Canada
  in the 1990s, 212*n*
  counter-mappers and, 96–98
  counterrevolution and, 98–100
  expansion of the México Indígena Project and, 151–161, 152*f*, 154*f*
  Inuit Land Use and Occupancy Project, 58–64, 62*f*
  James Bay Cree, 64–72, 65*f*, 70*f*
  México Indígena Project and, 142–144
  overview, 72–73, 166–175, 169, 177–179
  property and peace and, 112–113
  Research and Exploration journal (National Geographic Society) and, 100–108, 101*f*, 103*f*
Indigenous mapping in Canada. *See also* Canada
  Inuit Land Use and Occupancy Project, 58–64, 62*f*
  James Bay Cree, 64–72, 65*f*, 70*f*
  overview, 72–73, 169
Indigenous peoples, 181*n*
Indigenous territories. *See also* Territory
  assimilation and, 163–166
  conservation and, 80–82
  counterinsurgency and, 99–100
  Fourth World and, 92*f*, 93–95
  maps and the claiming of, 74–80, 83–87, 85*f*
  property and peace and, 108–118, 119*f*
  Research and Exploration journal (National Geographic Society), 100–108, 101*f*, 103*f*
  rule of law and, 118, 120–126, 122*f*, 125*f*
  weaponizing maps and, 83–87
Inquiry Commission, 24–25
"The Inquiry," 21, 25–26, 98
The Institution of Property (Noyes, 1936), 128–129
Institutionalized racism, 81–82

Instituto Nicaragüense de Reforma Agraria, 85*f*
Inter-American Commission on Human Rights, 120–121, 123–124
Inter-American Court of Human Rights, 118, 120
Inter-American Development Bank, 99
International Labor Organization Convention 169, 106, 120, 216*n*–217*n*
Inuit Land Use and Occupancy Project
  maps and the claiming of indigenous territories and, 83
  overview, 58–64, 62*f*, 73, 167, 203*n*–204*n*
Inuit Tapirisat of Canada, 58–64, 62*f*
Inuit territories, 75–76
Iraq invasions, 52
Iraq war, 133–138, 168–169

J

Jackson, Andrea, 136–137
Jamaica, 79
James Bay hydroelectric project, 57–58, 64–72, 65*f*
James Bay Cree, 64–72, 65*f*, 70*f*, 72–73. *See also* Dene Studies
James Bay Northern Québec Agreement, 67

K

Keeling, David, 141
Kittle, Arthur, 47
Kuna in Panama, 112

L

La Salle, 16–17
Lachixila, 156
Lake Nicaragua, 38–39
Land ownership, 129–130. *See also* Property rights
Land tenure, 112
Land Use and Occupancy, 59
Land Use Atlas, 59, 63
LandScan Database for the Department of Defense, 146–147
Lawrence, T. E., 135
Lean-tos, 41, 43
Lebanon, 52
Livingstone, David, 16–17
Lowenthal, David, 29–30
Luna, Jaime Martínez, 153–154, 179

M

Macdonald, John A., 55
Macdonald, Theodore (Ted), 120–121
Mackenzie Valley Pipeline Inquiry, 73
Magellan, Ferdinand, 16–17
Manuel, George, 74, 82
Map of Hispanic America, 28
Mapping, indigenous. *See also* Canada
  in the 1990s, 212*n*
  counter-mappers and, 96–98
  counterrevolution and, 98–100
  expansion of the México Indígena
    Project and, 151–161, 152*f*, 154*f*
  Inuit Land Use and Occupancy Project,
    58–64, 62*f*
  James Bay Cree, 64–72, 65*f*, 70*f*
  México Indígena Project and, 142–144
  overview, 72–73, 166–175, 169,
    177–179
  Research and Exploration journal
    (National Geographic Society) and,
    100–108, 101*f*, 103*f*
Mapping, military uses of. *See also* Army,
    U.S.
  Bowman Expeditions and, 234*n*
  counterinsurgency and, 98–100
  expansion of the México Indígena
    Project and, 151–161, 152*f*, 154*f*
  mapping technology and, 97–98
  overview, 168
  property rights and, 127, 130
*Mapping Colombia: The Correlation
    between Land Data and Strategy*
    (Demarest, 2003), 130, 132–133,
    141, 183*n*–184*n*
Mapping of indigenous territories. *See*
    Indigenous mapping
Mapping technologies. *See also*
    Geographic information systems
    (GIS) software; Global positioning
    system (GPS)
  American Geographical Society (AGS)
    and, 145–146
  community mapping and, 115
  expansion of the México Indígena
    Project and, 159–160
  overview, 97–98
  property rights and, 130
Maps and the claiming of territory
  conservation and, 80–82
  Declaration of Barbados and, 81–82
  overview, 74–80, 83–87, 166–175
  weaponizing maps, 83–87

Marco Polo, 16–17
Marines
  bandits and, 35–38
  Indians, 40–47, 42*f*, 46*f*
  overview, 195*n*
  Small Wars Manual, 47–53
Mattis, James, 134
Maury, Matthew, 22
The Maya Atlas, 124–126, 125*f*
Maya communities, 123–124
Mayangna communities, 93–94, 109,
    116–117
McClendon, Brian, 170
McFate, Montgomery, 136–137
Medellín government, 140–141
Mercator, 16–17
Merchants, 17–18
Mérida Initiative, 173
Meso-American Biological Corridor,
    116–117. *See also* Paseo Pantera
    project
Mexican Revolution, 142, 148, 152–153
México Indígena Project
  expansion of, 149–161, 152*f*, 154*f*
  Human Terrain System and, 138–141
  mapping of Tiltepec and, 4–9
  overview, 14–15, 142–144, 147,
    228*n*–229*n*
  PROCEDE (Program for Certification
    of Ejidal Rights and Titling of Urban
    Lots) and, 148–149, 150*f*
  promise of free computer hardware
    and, 10–14, 10*f*
  Small Wars Manual and, 52–53
  World Conference on Human
    Geography and, 165
Militarizing Culture (González, 2010), 138
"Military Operations Other Than War,"
    52
Military uses of mapping. *See also* Army,
    U.S.
  Bowman Expeditions and, 234*n*
  counterinsurgency and, 98–100
  expansion of the México Indígena
    Project and, 151–161, 152*f*, 154*f*
  mapping technology and, 97–98
  overview, 168
  property rights and, 127, 130
Millionth Map, 26–28, 33
Minerva project, 174
Miskito
  conservation and, 80–82
  Cultural Survival organization and, 102

Fourth World and, 92f, 93–95
maps and the claiming of indigenous territories and, 75, 77, 83–84, 85f, 86–87
overview, 39
property and peace and, 109, 116–117
property rights and, 93–94
rule of law and, 123
turtling and, 79–80
Wanks Reconnaissance and, 43–44
war and, 87–93
Miskito Coast Protected Area, 106–107, 109–110, 215n
Missionaries, 17–18, 39
MISURASATA, 84, 86–87, 89, 210n
Moravian Church (Nicaragua), 39f, 83f, 192n, 194n
Moos, Felix, 146–147
Mosquito Reserve, 39–40
Mosquito Shore, 39
The Mystery of Capital (de Soto), 130–131

N

Nahanni, Phoebe, 69–71, 70f, 74, 82, 169
Napoleon, 135
National frontiers, 99–100
National Geographic Magazine, 19, 186n, 220n
National Geographic Society
overview, 186n–187n, 220n
Research and Exploration journal (National Geographic Society), 100–108, 101f, 103f
National Geographic Society of Washington, 18, 19
National Indian Brotherhood (Canada), 71, 82f, 205n
National lands, 99–100
Native Americans, 163–166
Nature Conservancy, 110
Navidad Roja (Red Christmas) campaign, 87
Nicaragua
bandits and, 35–38
Bosawas Protected Area and, 109
conservation and, 80–82
counterinsurgency and, 99
Fourth World and, 92f, 93–95
indigenous mapping and, 73, 208n
mapping technology and, 98
maps and the claiming of indigenous territories and, 83
property and peace and, 108–118, 119f

Research and Exploration journal (National Geographic Society), 100–108, 101f, 103f
rule of law and, 118, 120–126, 122f, 125f
turtling and, 79
Nicaragua Canal Company, 17
Nicaraguan canal routes
bandits and, 37–38
Eastern Nicaragua, 38–40
exploration and, 17
Small Wars Manual, 47–53
Nicaraguan Long Leaf Pine Company (NIPCO), 77–78
Nicaraguan Ministry of the Environment, 110
Nicaraguan National Assembly, 121
Nietschmann, Bernard, 73, 96–97, 102–107, 168–169, 208n, 208n–209n
conservation and, 80–82
Fourth World and, 92f, 93–95
maps and the claiming of indigenous territories and, 74–80, 83–87, 85f
Maya communities and, 125f
Miskito Coast Protected Area and, 106–107, 109–110
property and peace and, 110–111
Sandinistan revolution and, 87–93
Nietschmann, Judith, 76
Nisga'a, 54–56, 198n
Nisga'a Land Committee, 55–56
Nisga'a Nation Tribal Council, 54–56
North American Free Trade Agreement (NAFTA), 3–4, 108, 149, 150f
Northern Frontier, Northern Homeland: The Report of the Mackenzie Valley Pipeline Inquiry, 68
Northern Québec Agreement, 67, 73
Noyes, C. Reinhold, 128–129
Nunavut Atlas, 64
Nunavut Atlas Project, 64
Nunavut Settlement Boundaries, 64

O

Oak Ridge National Laboratory, 146–147
Oaxaca. *See* Sierra Juárez of Oaxaca
Occupancy practices, 112
Office of Strategic Services (OSS), 52
Operation Orion, 141
Operational Environment, 160
The Other Path: The Invisible Revolution in the Third World (de Soto, 1989), 130

P

Panama, 100–108, 101f, 103f
Panamanian canal route, 17, 38
Paris Peace Conference, 26
Participatory research mapping (PRM) methodology, 144–145
Paseo Pantera project, 108, 116–117, 120
Pat Roberts Intelligence Scholars Program, 146–147, 226n–227n
Patsah, Joseph, 60–61
Pearl Lagoon, 75–76
Peary, Robert, 19
The Peasant Betrayed (Powelson, 1990), 129
Peoples' Popular Assembly of Oaxaco (APPO), 155–157
Petraeus, David, 134, 136, 137, 139, 158
Puerto Cabezas, 40, 77, 86f, 193n
Philanthropy, 18–19
Philippine-American War of 1899–1904, 168
Plan Colombia, 141
Polanco, Mauricio, 86, 117
Political restructuring, 97–100
Poole, Peter, 112, 115
Port Alberni conference, 71, 82f, 106
Pratt, Richard H. 163–164, 231n
Prinslow, Karl, 160
Private property, 149, 153. *See also* Property rights
PROCEDE (Program for Certification of Ejidal Rights and Titling of Urban Lots), 148–161, 152f, 154f, 227n–228n
Program of Certification of Ejido Rights and Urban Lots (PROCEDE) program, 15
Property and Peace (Demarest, 2007), 131–132
Property rights. *See also* Territory
 expansion of the México Indígena Project, 149–161, 152f, 154f
 Fourth World and, 92f, 93–95
 indigenous mapping and, 80–82, 83–87, 85f
 overview, 108–118, 119f, 127–133, 167–168
 PROCEDE (Program for Certification of Ejidal Rights and Titling of Urban Lots) and, 148–149
 property and peace and, 116
 regularization of, 113–121, 148–149
 rule of law and, 118, 120–126, 122f, 125f
Ptolemy, Claudius,16–17

Q

Quintín Lame Armed Movement, 114, 217n

R

Rama, 39, 85f
Ramírez, Gustavo, 9, 10–14, 151–152, 156–157, 158, 183n
Rangel, Rodrigo, 4
Raramuri communities, 151, 156
Reagan, Ronald, 87, 91
Refugee camps, 109
Research and Exploration journal (National Geographic Society), 100–108, 101f, 103f
Research and Exploration map, 126
Resguardo system in Colombia, 112
Resource conservation. *See* Conservation
Resource management
 indigenous mapping and, 96–98
 property and peace and, 108–118, 119f
 rule of law and, 118, 120–126, 122f, 125f
Rincón communities
 expansion of the México Indígena Project and, 155, 156, 159
 indigenous mapping and, 179
 mapping of Tiltepec and, 14–15
 México Indígena Project and, 142–143
 overview, 2f, 3–4
 promise of free computer hardware and, 13–14
 struggle and, 4–9
Rincón de Ixtlán in Oaxaca, 142
Río Coco
 Bosawas Protected Area and, 109
 Indians, 40–47, 42f, 46f
 maps and the claiming of indigenous territories and, 75–76, 77–78, 86–87
 property and peace and, 110, 116–117
 Small Wars Manual and, 49–53
Rio Earth Summit (1992), 107
Río San Juan, 35–36, 38
Ritter, Karl, 16–17
Robertson, Pat, 91
Roselle, Bill, 30

Royal Geographical Society of London, 18, 186*n*
Royal Proclamation of 1763, 56, 199*n*
"Rule of Indigenous Environments" (Nietschmann), 104

S

Sahlins, Marshall, 135–136
San Juan. *See* Río San Juan
San Luis Potosí, Huasteca Potosina in
    expansion of the México Indígena Project and, 156, 160
    Herlihy and, 145, 147
    México Indígena Project and, 142, 144
    PROCEDE (Program for Certification of Ejidal Rights and Titling of Urban Lots) and, 149
San Miguel Tiltepec. *See* Tiltepec
Sandinista government
    maps and the claiming of indigenous territories and, 75, 83, 87
    rule of law and, 121
    war and, 87–93
Sandinista revolution, 86–93
Sandinistas
    defeat of, 93
    Fourth World and, 92*f*, 93–95
    property rights and, 93–94
Sandino, Augusto, 33, 45, 47, 191*n*
    bandits and, 36–38
    maps and the claiming of indigenous territories and, 78
    Small Wars Manual, 47–53
Sandoval, Gonzalo de, 4
Sandy Bay, 109–110
Santiago Teotlaxco, 156–157
Schwartz, Lee, 165
Scientific data collection goals, 17–18
Scott, Patrick, 68
Scott, Tom, 17
Second Gulf War, 52
Self-government, 81–82
Serrano communities, 153
Sevilla, Julián, 37
Sierra Juárez of Oaxaca
    AGS conference at Haskell, 170
    expansion of the México Indígena Project, 149–161, 152*f*, 154*f*
    Human Terrain System and, 138–139
    indigenous mapping and, 73
    mapping of Tiltepec and, 14–15
    México Indígena Project and, 143, 144
    overview, 1–4, 2*f*, 183*n*, 235*n*
    struggle and, 4–9
Sierra Madre, 151
Sierra Tarahumara, 156
Small Wars Manual, 49–53, 134–138, 196*n*–197*n*
Small Wars Operations, 49
Small Wars Operations Research Directorate (SWORD), 95
Small Wars: Their Principles and Practice (Callwell, 1896), 48
Smithe, William, 55
Société de Géographie of Paris, 18
Somalia, 168
Somozas, 78–80, 83
Somoza García, Anastasio, 78
Sonnenfeld, Joseph, 60
de Soto, Hernando, 16–17, 130–131, 148–149, 167–168, 222*n*
Soviet Union, 97–98, 99
Standard Fruit Company, 36, 40, 77
Stanton, John, 138–139
Starr, Frederick, 4–6, 12, 182*n*–183*n*
Statement of the Government of Canada on Indian Policy, 57
Steward, Julian, 60, 83
Stocks, Anthony, 110–111
The Story of Land (Powelson, 1988), 128–129
Stout, Francis, 29
"Strategy and Tactics" (Harrington, 1921/1922), 48
Sumu (Sumo), 39, 85*f*. *See also* Mayanagna communities
Supporting Studies, 63

T

"The Tactics and Techniques of Small Wars" (Utley 1931, 1933), 48
Tasbapauni, 75–77, 78–79, 87–93
Tawahka, 110–111, 116–117
Technologies, mapping. *See also* Geographic information systems (GIS) software; Global positioning system (GPS)
    American Geographical Society (AGS) and, 145–146
    community mapping and, 115
    expansion of the México Indígena Project and, 159–160
    overview, 97–98
    property rights and, 130

Tentative Manual for Landing Operations, 1934, 48–49, 52
Tenure rights, 118, 119*f*, 120–126, 122*f*, 125*f*
Teotlaxco, 156, 156–157
The Territorial Imperative (Ardrey,1966), 128–129
Territory. *See also* Indigenous territories; Property rights
  conservation and, 80–82
  Declaration of Barbados and, 81–82
  decolonization and, 97–98
  expansion of the México Indígena Project, 151–161, 152*f*, 154*f*
  overview, 74–80, 83–87, 166–175
  property and peace and, 108–118, 119*f*
  Research and Exploration journal (National Geographic Society), 107
  weaponizing maps and, 83–87
Tiltepec
  computer hardware and, 9–14
  México Indígena Project and, 143–144, 159–160
  overview, 1–4, 2*f*, 185*n*
  struggle and, 4–9
Trade goals, 17
TRADOC (Training and Doctrine Command), 138–140
Trafficking, 109–110, 170–173
Trans-Alaska Pipeline, 116
Transcontinental Excursion of 1912, 21, 23–24
Trans-isthmian canal, 37–38
Tribal councils, 116
Trudeau, Pierre Elliot, 54–55, 56–57
Turtling, 79, 80–82, 87–88

U

Union of Organizations of the Sierra Juarez of Oaxaca (UNOSJO)
  expansion of the México Indígena Project, 151–161, 152*f*, 154*f*
  Human Terrain System and, 138–140
  mapping of Tiltepec and, 9
  promise of free computer hardware and, 12–13
United Fruit Company, 40, 77
United Nations Declaration on the Rights of Indigenous Peoples, 14, 82, 96, 106
United Negro Improvement Association (UNIA), 40
United States, 90–91, 99, 133–138
United States Indian Industrial Training School, 163. *See also* Haskell Indian Nations University
University of Kansas, 146–147
The Unknown War: The Miskito Nation, Nicaragua, and the United States (Nietschmann, 1989), 91–93
"Urban Land Use by Illegal Armed Groups in Medellin" (Demarest, 2011), 140
Urbano, Gregorio, 11–12, 69, 71, 155, 175
U.S. Agency for International Development (USAID)
  community mapping and, 115
  counterinsurgency and, 99
  overview, 97, 98
  property and peace and, 110
  Research and Exploration journal (National Geographic Society) and, 107–108
U.S. Army TRADOC, 138–140
Usher, Peter, 59–61
Utley, Harold, 41, 48

V

Vespucius, 16–17
Vietnam, 52
Violence, 167–168
Vogt, Evon, 60

W

Wanks Reconnaissance, 41–47, 42*f*, 46*f*
War. *See also* Weaponizing maps
  American Geographical Society (AGS) and, 25–26
  counterinsurgency and, 98–100
  freedom fighters, 87–93
  indigenous territories and, 100
  maps and the claiming of indigenous territories and, 86–87
  overview, 195*n*–196*n*
  property and, 108–118, 119*f*
  small wars, 103–104
War on Terror, 161
Warntz, William, 29–30
Weaponizing Anthropology (Price, 2011), 138, 140
Weaponizing maps, 83–93, 142–144. *See also* War

Wesley, John, 55
What the Land Provides, 66–67
Wildlife Conservation Society, 107–108
Wilson, President Woodrow, 21
Winks, Robin, 140
Winning Insurgent War (Demarest, 2011), 132–133
Woldenberg, Michael, 30
World Bank
  counterinsurgency and, 99
  overview, 97, 98, 169
  property and peace and, 111–112, 112–118, 119f
  property rights and, 131
  rule of law and, 118, 120–126, 122f, 125f
World Conference on Human Geography, 165
World Council of Churches, 81–83
World Council of Indigenous Peoples, 71
World War I, 25, 82, 98
World War II, 25, 82
World Wildlife Fund, 118, 120, 219n
World Wildlife Fund's Biodiversity Support Program, 115
Wright, J. K., 25, 30
Wrigley's Chewing Gum Company, 77–78

Y

Yagavila
  computer hardware and, 9–14, 10f
  expansion of the México Indígena Project and, 156, 157–158
  overview, 5–6
Yagila
  computer hardware and, 9–14
  expansion of the México Indígena Project and, 155, 156, 158, 161
  overview, 5–6
Yapti Tasba
  Fourth World and, 92f, 93–95
  property rights and, 93–94
  war and, 91–93
YATAMA, 90, 92f, 93–95, 211n
Yemen, 168

Z

Zapatista uprising, 142
Zapotec
  expansion of the México Indígena Project and, 159–160
  overview, 2–4, 183n
  promise of free computer hardware and, 13–14
  struggle and, 4–9
Zoogochi, 9, 155, 156

About the Authors

Joe Bryan, PhD, is Assistant Professor of Geography at the University of Colorado Boulder. He is the author of numerous articles, book chapters, and papers on participatory mapping and indigenous rights that draw from his research with indigenous communities in the United States, Honduras, Nicaragua, and Mexico. He has also participated in mapping projects with indigenous communities in the United States and Central America as an independent consultant.

Denis Wood, PhD, is an independent scholar living in Raleigh, North Carolina. He lectures widely and is the author of a dozen books and over 150 papers. From 1974 to 1996, he taught in the School of Design at North Carolina State University. In 1992, he curated The Power of Maps exhibition for the Cooper-Hewitt National Museum of Design (remounted at the Smithsonian in Washington, DC, in 1994), for which he wrote the bestselling *The Power of Maps*. His other books include *Rethinking the Power of Maps* and *Making Maps, Second Edition* (coauthored with John Krygier).